THE METALS BLUE BOOK™

Welding Filler Metals

Volume 3
The Metals Data Book Series™

Co-Published By:

CASTI **Publishing Inc.**
14820 - 29 Street
Edmonton, Alberta, T5Y 2B1, Canada
Tel: (403) 478-1208 Fax: (403) 473-3359

and

American Welding Society
550 N.W. LeJeune Rd.
Miami, Florida 33126, USA
Tel: (305) 443-9353 Fax: (305) 443-7559

ISBN 0-9696428-2-2
Printed in Canada

Important Notice

The material presented herein has been prepared for the general information of the reader and should not be used or relied upon for specific applications without first securing competent technical advice. Nor should it be used as a replacement for current complete engineering standards. In fact, it is highly recommended that current engineering standards be reviewed in detail prior to any decision-making. See the list of technical societies and associations in Appendix 4, many of which prepare engineering standards, to acquire the appropriate metal standards or specifications.

While the material in this book was compiled with great effort and is believed to be technically correct, *CASTI* Publishing Inc. and the American Welding Society and their staffs do not represent or warrant its suitability for any general or specific use and assume no liability or responsibility of any kind in connection with the information herein.

Nothing in this book shall be construed as a defense against any alleged infringement of letters of patents, copyright, or trademark, or as defense against liability for such infringement.

First printing, March 1995
ISBN 0-9696428-2-2   Copyright © 1995

Authors

The welding metallurgy section was written by Dr. Barry M. Patchett, P.Eng., FAWS, NOVA Professor of Welding Engineering, University of Alberta, Edmonton, Alberta, Canada.

The welding filler metal data section was researched, compiled and edited by John E. Bringas, P.Eng., Publisher and Executive Editor, *CASTI* Publishing Inc.

Acknowledgments

CASTI Publishing Inc. and the American Welding Society have been greatly assisted by Richard A. LaFave, P.E. and Richard A. Huber for their technical review of *The Metals Blue Book™* - Welding Filler Metals. Grammatical editing was performed by Jade DeLang Hart and Carol Issacson. These acknowledgments cannot, however, adequately express the publishers' appreciation and gratitude for their valued assistance, patience, and advice.

CASTI Publishing Inc. also acknowledges the invaluable assistance of Robert L. O'Brien in co-publishing this book with the American Welding Society.

A special thank you is extended to Christine Doyle, who entered all the data in the book with care and diligence.

Our Mission

Our mission at *CASTI* Publishing Inc. is to provide industry and educational institutions with practical technical books at low cost. To do so, the book must have a valuable topic and be current with today's technology. *The Metals Blue Book*™ - Welding Filler Metals is the 3rd volume in *The Metals Data Book Series*™, containing over 400 pages with more than 120,000 pieces of practical metals data. Since accurate data entry of more than 120,000 numbers is contingent on normal human error, we extend our apologies for any errors that may have occurred. However, should you find errors, we encourage you to inform us so that we may keep our commitment to the continuing quality of *The Metals Data Book Series*™.

If you have any comments or suggestions we would like to hear from you:

CASTI Publishing Inc., 14820 - 29 Street,
Edmonton, Alberta, T5Y 2B1, Canada,
tel: (403) 478-1208, fax: (403) 473-3359.

Contents

SECTION I WELDING METALLURGY

SECTION II WELDING DATA

SECTION II WELDING DATA (Continued)

SECTION II WELDING DATA (Continued)

SECTION II WELDING DATA (Continued)

SECTION III WELDING TERMS

APPENDICES & INDEX

Chapter

1

INTRODUCTION TO WELDING

Before the scientific era, the joining of metals was usually accomplished without fusion of the parent metal. Metals were fabricated in ancient times by riveting, brazing, soldering, and forge welding. None of these techniques involved melting of the metals that were joined.

Welding on an atomic scale, in the absence of melting, is prevented by the surface oxide layers and adsorbed gases present on virtually all metals. In the absence of such films, or via disruption of the film, intimate contact of the surfaces of two pieces of metal will cause welding of the two pieces into one. Very little pressure is required for truly clean metals. Welding can be accomplished by cleaning the surfaces to be joined in a hard vacuum (to prevent reoxidation). An alternative is to deform the metal mechanically while the surfaces to be joined are in contact, which causes the brittle oxide layer to break. Clean metal is exposed which will bond on contact. This is how forge welding is accomplished. Fluxing may assist in disrupting and dispersing the surface contaminants. Sand fluxes surface oxide in the forge welding of iron.

Welding involving melting of the parent metal requires the attainment of quite elevated temperatures in a concentrated area. This requirement is the primary reason why it took until nearly the dawn of the 20th century for fusion welding to appear. The precursors of fusion welding, namely brazing and soldering, involve the fusion of a filler metal which melts at a temperature *below* the bulk solidus of the parent metal and flows via capillary action into a narrow gap between the parent metal sections. It then solidifies to complete the joint. Soldering, initially using tin and tin-lead alloys, takes place at lower temperatures than does brazing; the American Welding Society arbitrarily differentiates between them at 450°C (840°F), leaving brazing as occurring from 450°C (840°F) up to near the melting temperature of the parent metal.

Processes using tin-based alloys in the lower temperature range are still referred to as soft soldering, probably because the filler metals are quite soft. Silver was one of the first brazing filler metals. Silver brazing takes place at temperatures in excess of 700°C (1290°F). The process is often referred to as silver soldering, or hard soldering, rather than brazing. This type of confusing labelling is not uncommon in the joining field, and it is necessary to keep a wary eye open. Part of the reason for such confusion may be that science has only recently discovered joining technology, and this has resulted in the retention of some inaccurate terminology from earlier times.

The history of the joining of materials is a long one, but can be conveniently divided into three eras. Prior to 1880, only forge welding took place, along with soldering and brazing - a *blacksmith* era. From 1880 to 1940, many advances were made and true fusion welding was possible, but most of the advances were accomplished by invention and inspired empirical observation. From 1940 to the present, scientific principles have played a significant role in the advancement of welding technology. The present emphasis on quality assurance and automated welding, including the use of robots, depends on the use of many scientific and engineering disciplines.

Welding metal together, without the use of a low melting filler metal, requires clean surfaces to allow atomic bonding. Mechanical deformation of two surfaces just prior to forcing the surfaces together disrupts the brittle oxide layer, exposing virgin metal and allowing pressure to weld the metal together. The first known joining of this type involved hammer welding of gold, which does not oxidize significantly, and therefore will readily weld to itself in the solid state with a little mechanical encouragement at ambient temperature. Gold also has a very low yield stress and is very ductile - it was and is regularly beaten into foil only 0.1 mm (0.004 in.) thick. Gold is now routinely welded by heating in a neutral (non-oxidizing) flame at relatively low temperature in order to remove adsorbed surface gases before pressure welding. Other precious metals are not as easy to pressure weld, because the surface oxides interfere with bonding of the metal atoms and higher yield stresses make large one-step deformation difficult.

Thus ferrous metals were not readily welded because of the tendency of iron to oxidize rapidly and also because of its high melting temperature and relatively high yield strength. The early welding of iron was probably accomplished in the solid state via hammering at high temperature, which in time has led to forge welding. This is still used for decorative iron work today. In forge welding, silica sand is used as a flux to remove oxide from the interface. Iron is one of the few metals whose oxide melts at a lower temperature than the pure metal, 1378°C (2500°F) and 1535°C (2800°F) respectively, providing the basis for at least some self-fluxing during heating. Adding silica sand forms iron silicates, which melt at even lower temperatures than the iron oxide. The main problem in the early days was achieving a temperature high enough to allow simultaneous flux/oxide melting and sufficient and easily accomplished deformation.

Early iron was smelted at about 1000°C (, and hammered at that temperature to consolidate the bloom. The yield stress of wrought iron is about 200 MPa (30 ksi) at ambient temperature, but drops to about 20 MPa (3 ksi) above 1000°C (1830°F). The appropriate temperature for welding is in the range of 1000-1200°C (1830-2190°F), which is possible in a blown charcoal fire. The melting point of the slag inclusions in iron is probably above 1000°C (1830°F), since it is largely fayalite or Fe_2SiO_4, which melts at 1146°C (2050°F). The joining of iron by the forge or hammer welding process developed empirically until eventually quite large fabrications could be made by skilled artisans. Examples are stern plates of over 25 tonnes (28 tons) for ships. Total deformations of up to 30-35% were used. An identical process produced a propeller shaft which weighed more than 30 tonnes (33 tons) for Isambard Kingdom Brunel's "Great Eastern".

The appearance of steel in large quantities during the 19th century made life somewhat more complicated, since very little slag is present (in comparison with wrought iron), and only very skilled forge welders could join it without significant numbers of flaws appearing in the completed joint. Sand fluxing minimized the problem on small fabrications, but not in large ones, where total deformations were smaller and the risk of slag inclusions was greater. Steel, or carburized wrought iron, had traditionally been difficult to make with sufficient hardness and toughness together.

The history of scientific contributions to welding is quite recent and fairly short. The scientific development of welding processes was not common until the middle of the 20th century, but scientific investigations started much earlier. The use of electricity to weld metals together began with two of the foremost practical geniuses in scientific history, Sir Humphrey Davy and his student, Michael Faraday. In the first decade of the 19th century, Davy investigated the nature of electricity. At one point, he touched two carbon electrodes together and passed a current through them from a large battery. When the electrodes were drawn apart, the current jumped the gap, forming an electrical discharge. Its path from one electrode to the other was curved, not straight, hence the name *arc*. Faraday was involved in the derivation in 1831 of the principles required to make electric power sources. The introduction of practical generators and motors had to wait about 50 years between scientific principle and technological application. In 1856, Joule suggested the possible use of electricity to join metal using contact resistance heating. This was ignored by the technological community of the day, although electric resistance welding is now a major fabrication process.

During a lecture at the Franklin Institute in Philadelphia in 1887, Elihu Thomson accidentally produced a resistance weld and developed a welding process based on this lucky event. He circumvented the problem of acceptance of his ideas by forming his own company to exploit the technique. He was remarkably successful and dominated the industry for many years, despite an obsession with wire welding rather than resistance welding in general. After an early but unexploited patent for electric arc welding, de Meritens used an electric arc generated with batteries and a carbon anode to weld lead battery plates. He was soon followed by the Russian inventor Benardos (working in France) who used the process to weld steel. He avoided problems of carbon oxidation by using electrode positive polarity. Although the carbon anode was slowly evaporated and would form CO_2 to assist in excluding the atmosphere, the long arc lengths used (tens of mm) allowed the weld to absorb atmospheric oxygen and nitrogen, and the steel welds were often brittle. Carbon arcs are still in use today for gouging and cutting, and also for joining platinum alloy thermocouple wires.

Carbon arc welding is a process involving a non-consumable electrode, i.e. the carbon does not melt and become part of the weld. The weld joint must be provided with extra filler metal in the form of a hand-fed rod to fill in any gaps. As a result the process is rather slow, and much work has been concentrated, and still is, on increasing the welding speed and therefore productivity. Initial attempts to replace the carbon electrode with a bare steel wire consumable electrode to arc weld steel (in the precursor of the Shielded Metal Arc Welding or SMAW process) are often attributed to a patent in 1897 by Slavyanov (a.k.a. Slawianoff). His invention was really an arc melter. A patent in 1889 by Coffin, which involved dual electrodes, one carbon and one iron, also has a strong claim for primacy.

The first welds made with these processes were often, but not always, brittle, since oxygen and nitrogen from the atmosphere reacted with the melting steel on the electrode tip and the surface of the weld pool. The brittleness described caused many structural failures in the first half of the 20th century, e.g. in steam generation plant, and delayed the acceptance of welded boilers for a few decades. The idea that the failures were caused by a *lack of strength* produced the requirement in the ASME Boiler and Pressure Vessel Code that all weld metals should at least equal, and preferably exceed, the specified minimum tensile strength in the base metal.

It was noted in early trials that the arc discharge from iron electrodes was often unstable, extinguishing without warning, or moving in unpredictable directions. Some of the problems were overcome by using a wire coated with grease, a lime (CaO) wash, or even rust. These coatings provided some protection from the atmosphere and/or stabilized the electric arc. Similar empirical improvements were the subject of work by Oscar Kjellberg in Sweden in 1907, and he was granted an American patent in 1910 for a flux-coated electrode. The company he founded, ESAB, is still in the welding supply business today. Further improvements in shielding via fluxes and inert or nearly inert gases have created the present plethora of different arc welding techniques.

Other sources of energy which evolved in the last decades of the 19th century are also used for fusion welding. The chemist, Le Chatelier, realized that the high temperatures possible from the combustion of oxygen and acetylene could melt steel, providing the basis for a welding process. The combustion produces carbon dioxide and water vapour which exclude the atmosphere from the weld area and there is no perversely unstable electric arc. The process therefore developed rapidly, with the appearance of a practical welding torch by 1903. In less than 10 years, it was in widespread industrial use, in applications as demanding as pressure vessels.

All of the mentioned welding processes are used to accomplish the apparently simple tasks needed to join materials - to provide clean surfaces and the necessary energy to bind them together. Success and failure are not widely separated, and welding personnel have the task of achieving the former while avoiding the latter. There are many processes used in industry and the American Welding Society has simplified reference to them by developing a list of initials to describe them (see table).

Table 1.1 lists the American Welding Society (AWS) initialisms for welding processes. These are accepted terminology in North America and in many, but not all, industrialized countries.

For the vast majority, the energy required for welding in this large number of processes comes from four sources:

1. Chemical reactions - typical are the oxy-fuel gas process (OFW) in which oxygen and a combustible gas, such as acetylene, burn to produce heat. Another is thermit welding (TW), in which a reaction between aluminum powder and iron oxide produces heat.
2. Mechanical action - typical processes here are the friction welding process (FW), which does not quite melt the base metal, or the ultrasonic welding process (USW). Explosive welding (EXW) is in some ways a combination of the both chemical and mechanical processes, since an explosive burns to propel one base metal on to another section at high velocity, with mechanical impact between the two causing the weld. It is probably best to classify the process on the basis of the energy acting at the actual zone of welding - in this case, mechanical energy.
3. Radiant energy - processes in this group are electron beam welding (EBW) and laser beam welding (LBW). Both use radiation concentrated on a small area to melt base metal.
4. Electricity - there are many processes which involve the use of electricity on the direct formation of welds, but only three electrical effects are used to melt the base metal. The resistance of the metal, especially at surfaces in contact, is used for resistance spot (RSW) and seam welding. High frequency induction develops eddy currents to heat surfaces by resistance in processes such as Resistance Seam Welding on pipe. Electric arcs, using a combination of radiative and resistive heating, are the most commonly used heat sources in commercial welding processes. The most typical example is the shielded metal arc welding (SMAW) process.

TABLE 1.1 AWS DESIGNATIONS FOR WELDING PROCESSES

AWS Designation	Process Name	Other Designations
Arc Welding Processes		
EGW	Electrogas Welding	
FCAW	Flux Cored Arc Welding	
GTAW	Gas Tungsten Arc Welding	TIG
GMAW	Gas Metal Arc Welding	MIG, MAG
GMAW-S	GMAW-short-circuit	Dip transfer, Short arc
GMAW-P	GMAW-pulsed	Pulse welding
PAW	Plasma Arc Welding	
SAW	Submerged Arc Welding	Sub-Arc
SMAW	Shielded Metal Arc Welding	Stick Welding, Manual Metal Arc
SW	Stud Welding	
Radiant Energy Processes		
EBW	Electron Beam Welding	Beam Welding
LBW	Laser Beam Welding	
Resistance Welding Processes		
RSEW-HF	Resistance Seam Welding	ERW
ESW	Electroslag Welding	Slag Welding
PW	Resistance Projection Welding	
RSW	Resistance Spot Welding	Spot Welding
UW-HF	Upset Welding-High Frequency	
Solid State Processes		
DFW	Diffusion Welding	Diffusion Bonding
EXW	Explosion Welding	
FOW	Forge Welding	Hammer Welding
FW	Friction Welding	
ROW	Roll Welding	Roll Bonding
USW	Ultrasonic Welding	
Chemical Energy Processes		
OFW	Oxyfuel Welding	
OAW	Oxyacetylene Welding	Gas Welding
TW	Thermit Welding	
Cutting Processes		
AAC	Air carbon Arc Cutting	
EBC	Electron Beam Cutting	
LBC	Laser Beam Cutting	
OFC	Oxyfuel Gas Cutting	
POC	Metal Powder Cutting	
LOC	Oxygen Lance Cutting	

When these welding processes are used to join base metals, a large amount of energy is put into the base metal in a very short time - fractions of a second to a few minutes, in most instances. This rapid input of heat has dramatic effects on the metallurgical structure in the weld zone, which is heated close to the melting temperature for any welding process, and beyond it in fusion welding. There are three major consequences: a very rapid heating of the base metal up to melting temperature, a superheating of molten weld metal derived from any melted base metal and any added filler metal, and lastly, a fairly rapid cooling rate. The cooling rates observed in welds are more rapid than those in most commercial heat treating processes because the passage of the welding thermal energy is directly into the relatively cold base metal, which is a highly efficient quenching medium.

The rapid heating rate has two major effects. As the time to reach elevated temperature is short, any compositional variations in the microstructure will have little time to disperse. As a result, compositional segregations will persist after heating. Thus precipitates and eutectic or eutectoid zones retain their chemistry even after dissolution or transformation takes place. A pearlite grain in a carbon steel, for instance, will turn into austenite at high temperature, but may retain its eutectoid carbon composition, even in a steel that has a very different overall carbon level. Very long times are needed to even out such compositional disparities, and welding does not allow for that to happen. The result is that the steel may behave in ways not normally associated with its nominal composition. Some areas may be more hardenable, for example, which can cause problems with cracking in the weld zone. The second consequence of rapid heating to high temperatures is that rapid grain growth takes place, with quite large grains appearing close to the fusion line. This can cause segregation of elements with limited solubility in the solid base metal to grain boundaries, where some kinds of embrittlement occur, and large grains also increase hardenability of some steels. This increases the probability of forming hard microstructures, with the attendant risk of some form of cracking.

In the presence of an arc, the molten weld metal is superheated to a few hundred degrees Celcius above the nominal fusion temperature in the bulk and to much higher temperatures at the weld pool surface. The high surface temperatures and small size of the weld pool promote rapid interaction between the molten metal and any reactive gases or slags which are present. These reactions can cause significant changes in weld metal chemistry and promote a variety of ills, from porosity to cracking. However, they may also be beneficial and a knowledge of the rate-controlling mechanisms of gas-metal and slag-metal reactions is important. The development of electric arc welding has often been concerned with controlling the reactions between weld metals and the atmospheric gases - oxygen, nitrogen, carbon dioxide and hydrogen. All of these gases can cause welding problems if dissolved in relatively small quantities. Absorbed gases form compounds (e.g. brittle iron nitrides), go into solid solution (e.g. oxygen in titanium, which embrittles the metal), or form porosity via desorption (hydrogen in aluminum and magnesium) or via a chemical reaction (carbon monoxide in steels, etc.).

GAS-METAL REACTIONS

Gas-metal reactions primarily take place at the surfaces of the weld pool and molten electrode tip. There are three basic types:

1. Absorption when gas solubility in the weld metal is *high* and effectively unlimited, e.g. oxygen in titanium.
2. Absorption when solubility is *low* and limited, e.g. hydrogen in aluminum.
3. Absorption followed by a *chemical reaction*, e.g. carbon dioxide in iron.

High rates of reaction occur in electric arc welding due to the large surface-area-to-volume ratio of weld pools and melting electrode tips and the elevated temperatures in the arc region in comparison to bulk metallurgical processes such as steelmaking. The rate of gas absorption is controlled by the rate at which the reactive gas arrives at the molten metal surface, the solubility of the gas in the molten metal, and the rate at which absorbed gases are mixed into the metal.

High Gas Solubility

The most simple reaction occurs in a static GTAW system where oxygen in an inert argon shielding gas stream reacts with titanium. The very high solubility of oxygen in titanium (30 at%) allows all of the oxygen which reaches the molten pool surface to be absorbed. The reaction rate is controlled by the diffusion of oxygen through an effectively *static* boundary layer of gas at the gas-metal interface over a high-temperature *active area*. The effective thickness of the boundary layer is controlled by the velocity of the arc plasma jet. The size of the active area on the weld pool surface increases with the current and arc length, and is influenced by the thermal properties of the base metal which dictate the rate of heat removal from the weld zone. The active area is usually a small fraction of the area of the molten weld pool at short arc lengths and high currents, and is limited to a maximum size of 100% of the surface area of the weld pool at long arc lengths and/or low current levels.

The total amount of oxygen absorbed is a linear function of arcing time at constant gas composition and welding conditions (current, voltage, etc). The *rate* of absorption varies linearly with increasing current, with increasing oxygen partial pressure in the shielding gas, with arc length up to a maximum, and is virtually independent of gas flow rate. Intuition suggests that the thickness of the boundary layer should vary with the velocity of shielding gas flow, but it does not, since the plasma jet velocity dominates (velocity of 100 m/sec versus 1 m/sec for shielding gas).

If a consumable electrode is used, the same conditions apply at the pool surface, and an extra 50% absorption takes place at the electrode tip during droplet formation. Virtually no absorption takes place during metal transfer. Reaction rates at the electrode tip are very high due to the high temperature and high surface/volume ratio, but time is relatively short. In moving pools, i.e. weld runs, increasing welding speed decreases absorption.

Low Gas Solubility

When solubility is limited, the weld pool content approaches the maximum possible solubility in the metal, which occurs in the molten phase at temperatures well above the melting point. For hydrogen absorption, maximum solubility is 50 times the solubility at the melting point for aluminum and magnesium. The ratio is about 1.5 times for iron and nickel. The same criteria for the high solubility case govern the rate of hydrogen arrival at the molten metal surface. Since the surface layer rapidly saturates due to the low gas solubility in the molten metal, the overall absorption rate is limited by the transport of saturated metal away from the active area and its replacement with unsaturated metal by weld pool motion. The saturated active area is swept away to cooler regions within the molten pool, where the absorbed gas can be rejected as porosity.

In the cooler regions, rejection is slower than the active area absorption rate and the overall concentration approaches the saturation level. In aluminum and magnesium this supersaturation causes porosity, but there is evidence to suggest that the supersaturation can be retained in ferrous alloys and does not cause significant porosity. Absorption is proportional to the square root of the hydrogen partial pressure (Sievert's Law) and is unaffected by arc length.

In fused zone runs, porosity tends to decrease with increases in welding speed. Consumable electrodes absorb hydrogen at the electrode tip in amounts which increase with droplet size and lifetime at the electrode tip.

CHEMICAL REACTION

In the third situation, absorption in the active area is immediately followed by a chemical reaction in the surface region, e.g. absorption of carbon and oxygen by iron from carbon dioxide shielding gas. The rate controlling step is the diffusion of oxygen through a static boundary layer, followed by absorption in the active area. The oxygen then achieves equilibrium at a high effective reaction temperature (about 2100°C or 3800°F) in the active area. Carbon reacts with the active zone oxygen to achieve equilibrium as:

$$\underline{\%C} + \underline{\%O} = CO$$

Both are mixed into the pool and replaced by fresh metal at the surface. Recombination of oxygen and carbon can take place in the cooler outer regions of the pool, which occasionally results in CO porosity. Deoxidants such as silicon reduce the active area oxygen level, thus increasing the carbon level and eliminating porosity. The effective reaction temperature is above 2000°C (3660°F), causing oxygen levels to approach 1.4%, which are equivalent to a carbon level of 0.02% in pure iron. This is close to a reaction with carbon dioxide rather than carbon monoxide due to the constant supply of carbon dioxide at the gaseous diffusion boundary. Increasing welding speeds drop the oxygen level, due to shorter reaction times and lower effective reaction temperatures. Carbon level rises slightly. With high levels of strong deoxidants in the metal, e.g. chromium in stainless steels, up to 0.10%C can be absorbed. This can lead to carbide sensitization and corrosion problems. Thus there may be a limit on the amount of carbon dioxide that can be present in a shielding gas for welding stainless steels with solid wire electrodes.

Nitrogen absorption in steels is primarily a function of the oxidizing potential of the shielding gas stream. Absorption is inversely proportional to the deoxidant level in the weld pool, and is thus similar in some aspects to carbon pickup. The carrier gas is very important - oxygen maximizes absorption, followed by carbon dioxide, argon and hydrogen. In pure nitrogen, the maximum absorption is about 400 ppm, which is the equilibrium level for steels at 1600°C (2900°F). The peak absorption in an oxygen carrier gas is 1400 ppm at a nitrogen partial pressure of 0.6 atm. The peak value in carbon dioxide is 900 ppm, also well above the level in pure nitrogen. Absorption increases rapidly with oxygen potential, but reaches a plateau level. When a flux is involved in the shielding, as in the SAW process, the peak level drops to about 75 ppm. This comes from air trapped among the flux particles, and can be reduced by argon gas displacement. Flux chemistry has no measurable effect. Oxygen pickup in the SAW and other flux-shielded processes is controlled both by gas-metal and slag-metal reactions.

SLAG-METAL REACTIONS

Throughout this discussion, reference will be made to both fluxes and slags. For clarity of description, the word *flux* will be used for material which *has not* been melted in a joining process, and "slag" will be used for the liquid phase formed during fusion and the solidified glassy product left in the joint area after joining. Fluxes made from mineral components, (for example silicates, fluorspar, chalk) are used to shield arc welding processes (and others) from atmospheric gases. The fluxes themselves may react to some extent with the molten weld metal. Slag-metal reactions in welding processes are difficult to analyze fundamentally due to the chemical complexity of the metal phase, and particularly, the slag phase. Commercial flux formulations are chemically complex due to the variety of demands placed upon them during the joining of metals. Some of the needs may even conflict. For example, arc stability, weld metal toughness, bead shape and ease of solidified slag detachability from the joint must all be satisfied at once, at least to an acceptable degree. Reactions between slags and metals during welding can transfer elements from the slag to the metal or vice-versa. The discussion of slag-metal interaction involves reactions of the following types:

$$(MnO) + [Fe] = \%\underline{Mn}_{Fe} + (FeO)$$
$$(SiO_2) + 2[Fe] = \%\underline{Si}_{Fe} + 2(FeO)$$
$$3(CuCl_2) + 2[Al] = 3\%\underline{Cu}_{Al} + 2\{AlCl_3\}$$
$$3(CuO) + 2[Al] = 3\%\underline{Cu}_{Al} + <Al_2O_3>$$

where   [ ] = liquid metal phase
$\%\underline{X}_y$ = wt% element 'X' dissolved in metal 'y'
( ) = liquid in slag phase
{ } = gas in slag phase
< > = solid in slag phase

If reactions of the types outlined above are conducted at a suitably elevated temperature (to provide enough energy for the reaction to proceed rapidly) and for a long enough time (for the reaction to finish completely), the reaction reaches its *final equilibrium* state. This is often the case during steelmaking operations, from which most of our knowledge of pyrometallurgy comes. In welding operations, temperatures can be very high, but the duration is very short, in the order of seconds or even fractions of a second. An increase in temperature will normally make a reaction proceed more quickly. A short time is likely to reduce the extent of the approach toward equilibrium. Thus welding processes have two prime conditions which conflict, one increasing the reaction rate and the other effectively reducing it. We have therefore to consider two aspects of slag-metal reactions: the *equilibrium* condition for a given reaction, which represents the maximum achievable extent of the reaction, and the *kinetics* of the physical situation, which will determine how close to equilibrium a given reaction condition is likely to get. As we will see, mass transfer processes are critical in moving reaction products to and away from reaction sites, and thus have a crucial role to play in the achievment of equilibrium conditions.

The equilibrium condition can be assessed by calculating an *effective equilibrium temperature* for a variety of alloying elements and associate slag components in a weld. The temperature is effective because the actual temperature at which the reactions take place is difficult to assess, and may vary from one given welding procedure to another. Thus an average, or effective temperature is an appropriate assessment, acknowledging that it may alter for the same reactants in differing welding circumstances. If the calculation is done for several alloying elements individually, it is often seen that a wide range of equilibrium temperatures occur for the different reactions in the same weld deposit, as found by Slaughter in 1942. The conclusion is inescapable. The temperature experienced at the slag-metal interface must be the same for all metal and slag phases in a single weld deposit. Therefore, since the calculated equilibrium temperatures for individual reactions differ, equilibrium was not achieved for all, if any, of the reactions.

Looked at another way, if each possible reaction was assessed at a fixed temperature corresponding to that of the slag-metal interface, say 1800°C, an equilibrium concentration for all of the alloy elements in the metal could be found if all of the necessary thermodynamic data were available. In our deposited weld metal, we would then find that the actual composition varied from the equilibrium projections, so that the reactions would be incomplete in most cases, i.e. did not reach equilibrium. For example, if the equilibrium level of Mn in a steel weld is calculated to be 1.5%, and the weld contains 0.3%, the reaction can be crudely described as 20% completed.

All of this suggests that there is interference of varying magnitude with some or all of the kinetic processes which control the mass transfer away from the slag-metal interface into the bulk slag and metal phases.

Slag-metal reactions in electroslag processes are fundamentally controlled by two factors:

1. the *thermodynamic driving force*, defined by the net free energy favouring the reaction and the chemical activity of the reacting phases. These determine the state of equilibrium achieved at the slag-metal interface - the active area of the discussion on gas-metal reactions.
2. the *rate of mass transport* in the slag and metal phases, which determines the degree to which the bulk metal and slag compositions approach the interface composition.

The following discussion will look first at the situation where reactive compounds in the slag transfer elements *into the metal phase*, i.e. slag to metal transfer. This can occur deliberately, for example in SAW procedures where Mn and Si in the weld metal are provided largely from the flux. The second section will assess the situation where the reactive compounds in the slag phase remove alloying elements *from the metal phase*, transferring them as compounds into the slag phase, i.e. metal to slag transfer. This can cause alloy losses and force the need for compensatory action, such as providing Cr in SAW fluxes for welding stainless steels, which ensures that Cr lost by slag oxidation is replaced.

Slag To Metal Transfer

When the thermodynamic driving force is high, transfer is controlled by diffusion of reactants through a boundary layer in the slag phase. The reaction rate is so high that a complete exchange takes place between the reacting phases, resulting in a 100 wt% concentration of reaction product at the interface. Three sub-divisions exist, based on the *physical nature* of the reaction product:

1. *Liquid* - The boundary layer is thin, about 1 to 10 micrometres thick, and reactions go to completion (reach equilibrium) in most cases. Examples are the reactions between Fe and slags containing oxides or halides of Ni.

2. *Gaseous* - The evolution of gaseous reaction products at the slag-metal interface increases the effective boundary layer thickness to about 100 micrometres and interferes with diffusion of reacting phases to the slag-metal interface. The reaction rate is therefore slowed down, and a true equilibrium is not achieved in the time available for welding processes. Examples are the reactions between Fe and slags containing Ni chlorides to form Fe chlorides, or between Al and slags containing Cu chlorides, which forms Al chloride. Fe chlorides boil at less than 700°C (1290°F), while Al chloride boils at less than 200°C (390°F).

3. *Solid* - In this case, the effective active area available for the slag-metal reaction is reduced by the formation of a solid phase at the interface. The reaction can proceed only at the gaps where the solid phase is temporarily absent or lifted by slag motion. An example of this is the reaction between Al and slags containing Cu oxide. Al oxide melts at 2045°C (3710°F), while Al melts at 660°C (1110°F).

When there is a very limited thermodynamic driving force, the overall reaction rate is controlled by the ability (or lack of it) of stirring forces in the liquid metal to remove reaction products from the slag-metal interface and mix them into the bulk metal. The equilibrium state at the interface is characterized by concentrations of reaction products of a few wt% or less, compared to the virtual 100 wt% achieved where a high driving force exists. The *concentration gradient* from the interface into the metal is therefore much reduced, as is the resulting rate of mass transfer via diffusion away from the interface into the bulk metal where stirring forces distribute the reaction product throughout the molten metal. The mass transfer rate drops as the bulk concentration approaches the interface concentration because the concentration gradient is continuously decreasing. True equilibrium is therefore not achieved. Examples are the reactions between Fe and the oxides of Si, Mn and P, where the reaction products are liquid and no kinetic barrier exists. Deoxidants in the Fe can increase the effective rate of mass transfer by increasing the rate of removal of part of a reaction product from the interface, thus increasing the rate of ingress of the other part, e.g. removal of O from the breakdown of an oxide such as MnO, increasing the rate of Mn transfer.

Metal To Slag Transfer

Transfer of alloying elements out of a metal phase by reaction with a slag is ultimately limited by the rate of supply of the element in question to the slag-metal interface by the stirring forces in the metal and diffusion through the static boundary layer next to the slag-metal interface. As the bulk concentration in the metal drops, the concentration gradient is lowered and the process of mass transfer becomes very slow. Equilibrium is not achieved, but is closely approached if the thermodynamic driving force is high. Sub-divisions are based on the physical nature of the reaction product:

1. *Liquid* - Transfer is rapid at first, but slows down as bulk concentration in the liquid is reduced. An example is the reaction between Fe-Si alloys and slags containing FeO. Si is lost to the slag as SiO_2.

2. *Gaseous* - Gases formed in the reaction between alloyed metals and reactive slags occur initially at the slag-metal interface and then within the bulk metal as well, when stirring forces distribute the reaction products, i.e. alloy elements and associated atoms from compounds in the slag. This gas evolution causes fragmentation of the molten metal which in turn increases the reaction rate by exposing more metal surface to the slag. The resulting variations in localized reaction rate causes a large scatter in final metal composition. The scatter is a function of the large variation in surface area/volume ratio caused during fragmentation. An example is the reaction between Fe-C alloys and FeO in the slag. O is dissolved in the Fe-C alloy at the slag-metal interface during the thermodynamic breakdown of FeO in the slag, and swept into the Fe-C alloy by stirring forces. C is then lost from the metal as CO gas when the dissolved O reacts with the dissolved C within the Fe-C alloy. Therefore true equilibrium exists only at the slag-metal interface for the general case, and is achieved in the bulk metal only when the thermodynamic driving force is high and the physical nature of the reaction product is suitable (liquid). In general, kinetic barriers prevent true bulk equilibrium.

These reactions take place primarily at the electrode tip, since the surface area to volume ratio is very high and stirring forces are strong due to sudden changes in current density. The extent to which equilibrium is achieved dictates the amounts of alloying elements gained or lost and the simultaneous absorption of oxygen or other tramp element can strongly affect toughness and other mechanical properties.

Chapter

2

BASIC METALLURGY FOR WELDING

HEAT INPUT

The rapid heat input and high energy density involved in welding causes very different thermal effects than do conventional heat treatments. Very rapid heating cycles (fractions of a second to a few seconds), high peak temperatures (in excess of the melting temperature) and relatively rapid cooling rates to ambient temperature (fractions of a second to many minutes) are all involved. A typical thermal cycle for an arc welding process is shown in Figure 2.1.

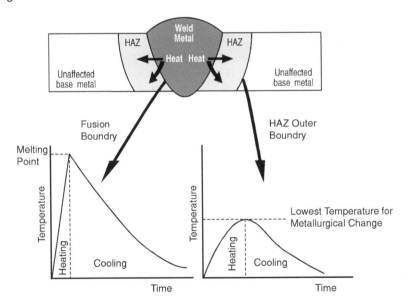

Figure 2.1 Thermal cycles in weld zones

The heat input of a given welding process is most easily defined for electric arc processes. It is simply arc voltage times arc current divided by welding speed. Since arcs lose some energy to the surrounding space by radiation, convection and conduction through gas and slag shielding media, less than 100% efficiency of energy transfer to the weld zone occurs.

$$\text{HEAT} = \frac{x \bullet \text{Volts} \bullet \text{Amps}}{\text{Welding Speed}}$$

where x = efficiency factor (typically between 40% and 95%).

However, for most welding procedures it is difficult to measure "x" reliably and all normal welding procedures are written without using a "x" factor, i.e. it is assumed to be 100%. In the SI system of units, the heat input is determined by measuring the welding speed in millimeters/second. Then the result is in Joules/mm, or if divided by 1000, kJ/mm. This is the most universally used description of heat input. Typical values for arc welding range from 0.1 to 10 kJ/mm (2.5-250 kJ/in).

Heating Rate

The effects of the rapid temperature rise during the initial heating cycle in welding have not been analyzed extensively, but do have some important implications. The most important is that any existing segregation of alloy elements or impurities in the base metal will not be evened out by diffusion, because the time available is too short. This can lead to a local region in the base metal right next to the fusion line where localized melting can occur. This is most likely where previous segregation in the base metal exists, for example at grain boundaries. These regions may contain higher than average concentrations of alloying elements and/or impurities, which cause a lowering of the melting temperature. The region is therefore called the Partially Melted Zone (PMZ). This localized, restricted area of melting may simply solidify again as the temperature falls after welding, or it may open up under stress to form a small crack, usually known as a liquation crack.

Peak Temperature

The high peak temperatures reached in welding (near or above the melting temperature of the base metal) are far in excess of temperatures in conventional heat treating processes. There are two major results:

1. Grain sizes in the heat-affected zone (HAZ) are large and are largest next to the fusion line at the weld metal. This increases hardenability in transformable steels and may concentrate impurities which are segregated to grain boundaries. This may lead to a form of hot cracking.
2. All traces of work hardening and prior heat treatment (e.g. tempering, age hardening) will be removed in at least part of the HAZ and the size of the region affected will be larger as the heat input rises.

Cooling Rate

The rapid cooling cycles for welding after solidification is complete are caused by the quenching effect of the relatively cold parent metal, which conducts heat away from the weld zone. Models involving the Laplace equation have been developed to predict cooling rates, but are beyond the scope of this book. Discussion of this equation and the developed models can be found in other publications.

However, the cooling rate must be evaluated, since important weld zone properties such as strength, toughness and cracking susceptibility are affected. As the cooling rate increases, i.e. as the temperature falls more rapidly, several important welding effects occur. In transformable metals, such as ferritic-martensitic steels or some titanium alloys, the high temperature phase in the weld metal, and particularly in the HAZ, is quenched. Since the actual cooling rate experienced at a given point varies with location and time after welding, some characterization is necessary. The most used technique is assessment of welding procedure tests, but this is expensive and not particularly useful in predicting the effect of changing variables such as heat input. In the welding of steels, the time spent dropping through a temperature range characteristic of transformation of austenite to martensite or other transformation product (e.g. bainite, ferrite + carbide) is becoming a standard test. The characteristic time is measured in seconds for a given weld metal or HAZ region to drop from 800 to 500°C (1470-930°F) and is often written as Dt_{8-5}. A large number signifies a slow cooling rate.

If weld zone properties are not as desired, it is possible, in some cases, to improve them by altering the cooling rate in the welding procedure. Two main methods are used. The first is heat input - higher heat input in a given situation slows down the cooling rate. Preheating, by warming the base material before welding, also slows down the cooling rate, with higher preheat leading to slower cooling rates. Combining preheat and heat input variations is called *Procedure Control*. Further control of cooling rate effects can be included in a welding procedure by defining a value for interpass welding temperature in multipass welding (effectively a preheat value for each succeeding pass), in order to ensure a cooling rate within a given range or 'window'. This idea will be revisited in discussions on welding various grades of steel.

METALLURGICAL EFFECTS OF THE WELDING THERMAL CYCLE

Weld Metal Solidification

Solidification of weld metal takes place in an unusual manner in comparison with castings. The molten weld pool is contained within a *crucible* of solid metal of similar chemical composition. Since the melting temperature of the base metal is reached, the fusion line and adjacent HAZ experience very high temperatures for a short time. The result is rapid grain growth. These large grains act as nuclei and the weld metal solidifies against them and assume the crystallographic orientation of the atomic structure in the large HAZ grain. Only the grains with an orientation giving rapid growth into the weld metal produce dendrites, and the result is called *epitaxial growth*.

The Metals Blue Book

There is a very thin region of melted base in the weld metal right next to the fusion line which, although 100% molten, is not stirred into the bulk of the molten pool during welding, due to viscous effects during mixing. This region is called the Melted Unmixed Zone (or MUZ) and is usually very narrow, typically 0.1 mm (0.004 in.) wide.

Since diffusion has not enough time to even out any differences with the weld metal chemistry, any segregations in the base metal will persist in this region. It is also a region in which any segregation produced in the weld metal may collect in a static region and also persist. Initial growth during solidification into this region is often planar, and cellular growth occurs as the growth rate increases. True columnar growth and dendrites form as faster growth continues in the bulk of the weld metal. As growth of dendrites proceeds into the molten weld pool, segregation of some of the alloy and residual elements in the weld pool occurs as the forming solid rejects some of the elements into the liquid due to a lower solubility in the solid state. As a result, the remaining liquid near the end of solidification can contain significantly higher levels of some elements. This can lead to the formation of cracks and fissures in the weld metal. One example is the concentration of S, P, O and C at grain boundaries in many ferrous and nickel alloys, which alone, or together with themselves and other elements, can form complex eutectics of low melting temperatures.

Growth of dendrites tends to follow the direction of heat flow, generally perpendicular to the freezing isotherm at a given instant and moving toward the welding heat source as it traverses the joint line. Consequently, at slow welding speeds, dendrites curve to follow the advancing solidification front in the elliptical weld pool. At high welding speeds, the pool is an elongated teardrop shape, and dendrites tend to grow straight toward the weld centreline. Thus any hot cracks formed at high welding speeds tend to be right down the centreline of the weld, while at slower speeds they are between the dendrites. The filler metals used for situations prone to hot cracking often contain elements which will mitigate the cracking problem by combining with the offending element to render it harmless. Mn is often added to ferrous alloys to combine with S. Since the chemistry of the filler metal often differs from the base metal and since the weld metal contains some melted base metal, the diluted filler metal region is called the Composite Zone. All of the regions mentioned are shown for a single pass weld in Figure 2.2.

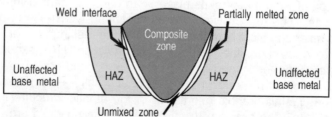

Figure 2.2 Metallurgical designations in weld zones

Heat Affected Zone

Several zones are formed within the HAZ by the combination of heating rate, peak temperature and cooling rate, depending on the alloy system. The *subcritical* HAZ occurs when the peak temperature is below an important temperature for the metal being welded. This could be, for example, the recrystallization temperature in a work-hardened metal, the solvus temperature for a precipitation-hardened alloy or the allotropic phase change temperature for a transformation hardening alloy. Above the critical temperature, the grain size increases as the temperature increases. Especially in transformation-hardening alloys, two regions exist in this high temperature zone:

1. *Fine Grained Heat Affected Zone* (FGHAZ) - this region exists when an allotropic phase change decreases the average grain size in comparison with the grain size in the subcritical region. This occurs, for example, in the intercritical temperature range in steels.
2. *Coarse Grained Heat Affected Zone* (CGHAZ) - this region occurs at high temperatures when all phase changes are completed. Grain size tends to increase up to the fusion temperature.

STRENGTHENING MECHANISMS

There are three fundamental strengthening mechanisms used in metals and alloys which can be affected significantly by the welding thermal cycle in the HAZ: cold working, precipitation (age) hardening and transformation (martensitic) hardening. All are influenced by the heating, melting and cooling aspects of rapid thermal cycles experienced in the weld zone. Metals and alloys may have one, two or all three mechanisms of strengthening in place before welding, and the reader will have to consider the overall effects of the following comments, all of which may be applicable in a single weld.

Work Hardening

Work hardened alloys are annealed during the weld thermal cycle. Dislocation density is reduced, grains are recrystallized and may grow larger, depending on the peak temperature reached. In the non-visible HAZ, in the unaffected base metal, the lower peak temperatures reached usually affect only physical properties such as electrical conductivity. The visible HAZ will have lowered strength compared to the base metal. This will occur in virtually all cases, since the grain growth in parts of the HAZ will lower the strength of even an annealed, but fine grain, metal. The visible changes are shown schematically in Figure 2.3.

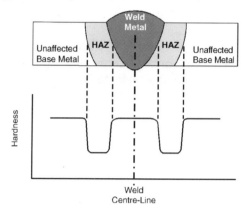

Figure 2.3 HAZ hardness in work hardened metals

If the heat input is low, two factors may minimize the loss in strength: the anneal may be incomplete, leaving the final HAZ strength somewhat above the handbook *anneal* value, and the size of the HAZ may be small compared to the thickness of the material, providing the annealed region with support from surrounding stronger material. This increases effective joint strength.

Precipitation (Age) Hardening

Precipitation hardened alloys go through more complicated changes. Age hardening is also used to describe this process, which was first noticed as a hardening of aluminum alloy rivets containing Cu for aircraft construction, in which hardness increased as a function of time delay before use after the rivets were manufactured. The most important industrial alloy systems which utilize precipitation hardening are aluminum, nickel, titanium and copper. The phase diagram of a binary alloy suitable for precipitation hardening needs two important features: a significant solubility of an alloy element in the primary metal (maximum solubility limit of several percent), and a rapid solubility drop from the maximum limit as temperature drops. Figure 2.4 is an example, with C as a typical composition used for an alloy.

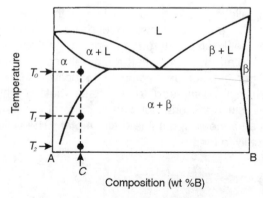

Figure 2.4 Binary phase diagram

A secondary virtue of the chosen composition is a reasonable temperature range of single phase solid solution at elevated temperature to simplify heat treatment and allow some variation in temperature for solution treatment. Usually the composition of an alloy has the precipitating element added to less than the maximum solubility, but this is not always so. Alloy elements added to levels in excess of the solubility limit will solution strengthen the alloy in addition to precipitation effects. Single elements may be involved, as may two or more. Cu is an element which precipitation hardens Al alloys. Mg and Si together do the same. The solvus line at the chosen composition should be at a reasonably high temperature, so that a range of precipitation hardening temperatures, e.g. T_1, can be used. The range from T_1 to the quench temperature, T_2, is the important parameter.

Precipitation hardening results from the precipitation of particles of a fine, dispersed second phase at low temperatures after all precipitates are dissolved by heating to the single phase region and quenching, a treatment called a solution heat treatment. Some reheating to a temperature below the single phase temperature is often used to cause diffusion to increase and promote second phase precipitation. This is called a precipitation heat treatment or artificial aging, and the whole heat treatment process is shown schematically in Figure 2.5.

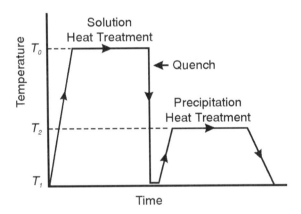

Figure 2.5 Typical age hardening heat treatment

In some cases, the precipitation heat treatment is not necessary, and the alloy hardens at room temperature over a period of time. This is known as natural aging. The progress of changes in hardness (also strength) under isothermal heating as a function of time is shown in Figure 2.6.

When an alloy capable of precipitation hardening is welded, the thermal cycle imposes further heat treatment on the weld zone. The thermal cycle of a weld is a thermal 'bump' compared to a heat treatment and has variable temperatures and times at temperatures. The result is complex in terms of the response of the alloy in the weld metal and HAZ. In addition, the alloys can be supplied in two conditions: fully precipitation hardened or solution annealed in preparation for precipitation hardening after fabrication.

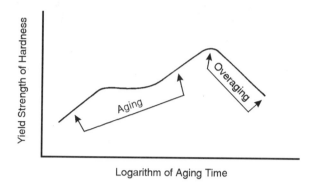

Figure 2.6 Typical strength changes during aging

Let us consider the effects of a single pass weld using two welding procedures on both fully hardened and solution treated materials. One procedure will be high heat input, leading to very slow cooling rates. The other will be a low heat input, leading to rapid cooling rates. We will also consider the effect of a post weld heat treatment (PWHT) consisting of a conventional precipitation hardening treatment. A full solution treatment after welding is usually not possible due to the probability of distorting the structure at high temperature, perilously close to its melting temperature.

Consider fully precipitation hardened metal first, joined in a conventional butt weld configuration. The weld profiles in Figure 2.7 use a horizontal dotted line to indicate the hardness of the heat-treated base metal across the zone to be welded. The solid line shows the effect of the welding cycle on the hardness across the HAZ on one side, the weld metal and the HAZ on the other side. The dashed line shows the resulting hardness after a precipitation hardening heat treatment. In Figure 2.7(a), the procedure has been low heat input. Far removed from the weld centreline, at the right side of the diagram, the peak temperature reached from the thermal cycle is too low to affect the precipitates and the hardness does not change.

As the peak temperature rises, a phenomenon takes place called reversion, which does not appear in conventional discussions of heat treating. The smallest precipitates have a large surface/volume ratio and are relatively unstable compared to the larger particles. When hit with the thermal bump from the weld cycle, the small ones disperse locally to form a highly saturated solid solution, thus softening the local region. The larger, more stable, particles begin to grow and overage, becoming larger and lowering hardness progressively as the peak temperature rises. The graph shows a continuing decrease of hardness. Once the solution treatment temperature is reached, all precipitates are dissolved, and hardness is very low. As the fusion line is approached, grain size increases, lowering hardness marginally, but no other changes take place. In the weld metal, assuming it is the same chemical composition, solidification produces a solution treated cast metal.

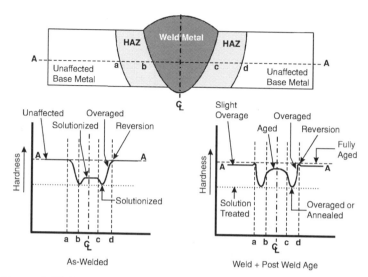

**Figure 2.7(a) Hardness changes in fully age hardened alloys
(low heat input weld)**

When these structures undergo the cooling cycle after welding with a low heat input, the material is effectively quenched, thus retaining the solute atoms in solution. Submitting the structure to a conventional precipitation heat treatment after welding (no overall solution heat treatment) will have the following effects:

1. The previously unaffected base metal will be somewhat overaged, assuming that it was at maximum hardness before welding.
2. The area of reversion will reharden to approximately maximum levels.
3. The overaged region will age further, softening it.
4. The solution treated zone in the HAZ will reharden.
5. The solution treated weld metal will reharden if it is of the same chemical composition (autogenous weld or matching filler metal). If it is of differing composition, it will partially respond or not respond to the precipitation heat treatment, depending on the final weld metal composition. Weld metals often retain as-cast properties in precipitation hardened alloys, or partial values of the base metal properties due to dilution effects.

Changing to a high heat input on precipitation hardened material has two major effects: the high heat input extends the size of the HAZ, and instead of a solution treatment occurring in the weld metal and the HAZ near the fusion line, the alloy is annealed due to the slow cooling rate. This means that the final equilibrium precipitate is present and no response to precipitation heat treatment will occur in those regions. The result is shown in Figure 2.7(b).

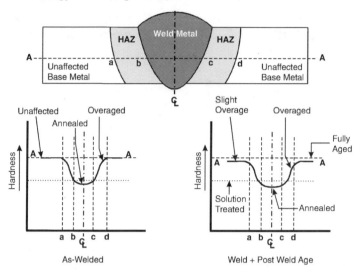

Figure 2.7(b) Hardness changes in fully age hardened alloys (high heat input weld)

These two examples show clearly that using a low heat input is more likely to retain the original precipitation treated properties in more of the weld zone.

If solution treated material is welded, the heat input starts the precipitation process. For a low heat input, hardness rises as the peak temperature increases, until the solvus temperature is achieved. Then the HAZ close to the fusion line and the weld metal are solution treated as before and kept that way by a rapid cooling rate. A post-weld precipitation heat treatment now has somewhat different effects, as shown in Figure 2.7(c):

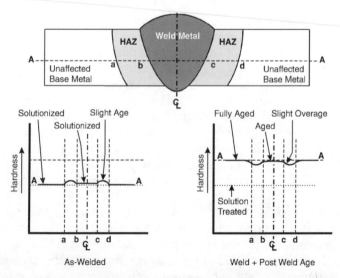

Figure 2.7(c) Hardness changes in solution treated alloys (low heat input weld)

The unaffected base metal is fully hardened.   Then the following effects occur:

1. The slightly aged zone overages, lowering hardness.
2. The solution treated zones in the HAZ and weld metal fully harden, assuming again an autogenous weld metal or matching filler metal.

If a high heat input is used, the previous comments on extent of thermal effects and the production of annealed zones in the weld metal and near HAZ still apply.   Then the post-weld precipitation treatment will have the following effects, shown in Figure 2.7(d):

1. The unaffected base metal will fully harden, as with the low heat input case.
2. The aged region will overage, but its size will be larger.
3. The annealed regions in the weld metal and HAZ will not respond and will stay soft.

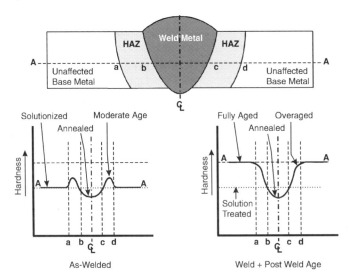

Figure 2.7(d) Hardness changes in solution treated alloys (high heat input weld)

These four scenarios show that the most effective way to retain heat treated properties in precipitation hardened alloys is to weld with a low heat input on solution treated material, followed by whichever precipitation heat treatment is appropriate for the application.

Transformation Hardening

Transformation hardened alloys experience an allotropic phase change at elevated temperatures.   Cycling the material through the phase change temperature or temperature range alters the grain size, which is small at the time of new phase appearance, and rapid cooling after the phase change can produce metastable transformation products, which are usually harder, stronger and more brittle than the equilibrium product formed on slow cooling. The visual changes in grain size are shown schematically in Figure 2.8.

No phase changes after heating

Allotropic transformation after heating

Figure 2.8 HAZ alterations in allotropic transformation alloys

The major change from the work-hardened alloys is the appearance of two fine-grained regions, one for recrystallization of the ambient temperature phase and one for the nucleation of the high temperature phase. The most common industrial alloy examples of transformable alloys are C-Mn and some alloy steels, which form austenite at high temperatures in excess of about 700°C (1290°F). On fast cooling, this may transform to martensite or bainite, rather than the equilibrium product of ferrite + carbide. Details of the metallurgy of steels and the effects of alloying elements is given in *The Metals Black Book*. Other metals of industrial importance which experience similar martensitic behavior are some titanium alloys and aluminum bronzes.

During welding, a transformable metal is exposed to every temperature from ambient to melting under rapid heating conditions, is held at a peak temperature for a short time, and is then cooled at rates between fast and exceptionally fast. This experience is entirely different from conventional heat treatments which feature slow heating rates, isothermal peak temperatures held for long and controlled periods of time, followed by controlled cooling rates (quenches) and possible isothermal follow-up heat treatments such as stress relief or tempering. Responses of weld zones to thermal cycles are therefore more complex metallurgically than would be expected from heat treating knowledge. The metallurgically important zones produced in a single pass weld in a transformable metal, for example a hardenable alloy steel, are shown in Figure 2.9.

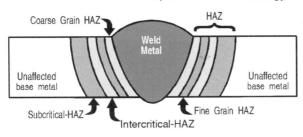

Figure 2.9 Metallurgical designations in heat affected zones

The *unaffected* base metal is defined at temperatures below the lower critical temperature, i.e. the temperature above which austenite starts to form or the A_3, which is about 720°C (1330°F) in C-Mn steels. In alloy steels, the following equations can be used to estimate transformation temperatures in °C:

Transformation on Slow Cooling
Ar_3= 868 - 396%C -68.1%Mn - 36.1%Ni - 20.7%Cu - 24.8%Cr + 24.6%Si
or
Ar_3 = 868 - 396%C - 58.7%Mn - 50%Ni - 35%Cu + 24.6%Si + 190%V

Close inspection of the unaffected zone reveals that carbides are somewhat broken up and spheroidized before the lower critical temperature is reached, so that the visible HAZ starts at temperatures about 50°C (120°F) below the critical temperature. These temperatures are sufficient to cause stress relief, tempering and precipitation hardening, hence the appearance of the subcritical HAZ as a metallurgically important zone, especially in high heat input welds.

The next zone is the intercritical zone, where the temperature is between the temperature where austenite starts to form, A_1, and the higher A_3 , where all ferrite + carbide is gone and only fine grain austenite remains. For eutectoid steel, this temperature 'range' is a single temperature, the A_3. In the intercritical zone, the austenite formed is often close to the eutectoid composition, so it is quite hardenable and may form martensite on cooling.

The fine grained HAZ is just above the upper critical temperature (A_3) and extends to peak temperature regions up to 100°C (210°F) above it. The fine austenite grain size reduces hardenability in this range, although some carbon gradients from carbide dissolution still exist because of lack of time for diffusion, especially in banded microstructures.

In the coarse grain HAZ, grain growth of the austenite takes place right up to the fusion line temperature (>1450°C or 2640°F). Austenite composition is more homogeneous than it is in the fine grain zone, but gradients still exist. Coarse initial microstructures, for example hot rolled rather than normalized steel, slow down carbide dissolution and diffusion, exacerbating gradient removal and increasing local hardenability. Higher heat inputs promote diffusion, but also cause larger grains. Large austenite grain size increases hardenability for a given composition, increasing the chance of forming martensite for a given cooling rate. Martensite is hard, strong and brittle and is susceptible to Hydrogen Assisted Cracking (HAC), also called Cold Cracking.

This varied microstructure in the HAZ, involving varying grain sizes, two inhomogeneous phases and a mixed phase region have complicated responses to the cooling rates following welding. Most weld cooling rates are relatively fast, inducing a quench in conventional heat treating terms. When austenite survives to low temperatures due to high hardenability, transformation to low temperature products (bainite or martensite) will occur. In alloy steels, the temperatures at which martensite and bainite start to form under isothermal conditions can be estimated by:

1. Martensite $M_s(^\circ C) = 561 - 474\%C - 33\%Mo - 17\%Ni - 17\%Cr - 21\%Mo$
2. Bainite $B_s(^\circ C) = 830 - 270\%C - 90\%Mn - 37\%Ni - 70\%Cr - 83\%Mo$

In continuous cooling circumstances, these temperatures are depressed. Therefore the coarse grain zone and some of the fine grain and intercritical zones may form brittle martensite or bainite, reducing toughness. Tempering with a postweld heat treatment is necessary to restore toughness in those areas.

In multipass welds in transformable alloys, the above effects are complicated by the overlapping of the thermal cycles in a given area, causing reheating and cooling effects on an already varied microstructure. These effects will be considered in more detail in the following Chapters devoted to individual alloy systems.

Chapter

3

FLAWS & DEFECTS IN THE WELD ZONE

The two main weldment zones are the weld metal, which is composed of melted base metal and (if present) filler metal and the other is the Heat-Affected Zone (HAZ), which is solely base metal which has been heated sufficiently to affect the properties of the base metal. These can be mechanical properties (strength, toughness), metallurgical (microstructure, grain size) or physical properties (hardness, electrical conductivity). The visible HAZ is made visible by metallurgical etching. The peak temperature experienced by the HAZ increases as the fusion line at the weld metal is approached. At distances well removed from the fusion line, outside the visible HAZ, there is also a region affected by lower peak temperatures experienced in the thermal cycle. This is called the non-visible HAZ, in which more subtle effects, such as changes in some physical properties, occur. Problems associated with welds in the weld metal and HAZ are caused by procedural (welding process) or metallurgical shortcomings. We will be concerned primarily with metallurgical problems, but some definition of the differences between process and metallurgical induced flaws is necessary, as is some definition of what a defective weld is.

Codes of fabrication and other welding documents use several words interchangeably to describe unacceptable areas in welds. The four most common are *imperfection, discontinuity, flaw* and *defect*. Before fracture mechanics were available and the concept of Fitness-for-Purpose did not exist, the use of any of these words to describe a problem region was acceptable. Now, however, such imprecision can lead to difficulties in interpretation, and some rationalization is called for. Let us try the following definitions of these words:

Imperfection/Discontinuity

Metallurgically, an imperfection is any lack of regularity in a three-dimensional lattice of atoms. In a metal, anything from a vacancy or dislocation to larger regions such as a grain boundary would qualify. The term discontinuity is better applied to a collection of imperfections (for example a grain boundary) which is normally undetectable by conventional inspection and non-destructive testing techniques. Neither an imperfection nor a discontinuity should be of concern in a material or fabrication. Many fabrication Codes and Standards use these words to describe large problems, such as cracks, which must be repaired.

Flaw

A flaw is a discontinuity detectable via destructive or nondestructive testing which is not likely to cause failure of the structure in the prevailing conditions. It can therefore be left in the structure without repair, if validated by fracture mechanics analysis.   Some NDT techniques (RT,UT) can detect such innocuous flaws as grain boundaries in certain alloys, so some care is necessary in interpretation.

Defect

A defect is a flaw which is likely to cause structural failure under prevailing or anticipated operating conditions.   It is also any flaw prohibited by a fabrication code or a customer specification. A given discontinuity may be a flaw in one structure, but a defect in another. ECA (Engineering Critical Analysis) involving fracture mechanics can help distinguish between flaws and defects. Otherwise, arbitrary workmanship criteria are used (a fixed size and/or type of flaw is a defect in all circumstances). Preventive maintenance is cheaper than catastrophic failure or replacement, so regular inspection is a required backup to ECA.

A somewhat cynical, but useful, engineering view is that all materials and fabrications are collections of imperfections. An acceptable material or fabrication is an organized or accidental collection of flaws suitable for a specific application (fitness-for-purpose).   Scrap material, a fabrication requiring repair, or a scrapped fabrication, are unfortunate or unlucky collections of defects for a particular application.

Flaws and defects in a weld region can be two dimensional (e.g. cracks) or three dimensional (e.g. porosity). As a general rule, two dimensional flaws are at once more dangerous and more difficult to detect and are therefore more important, despite what some codes say about porosity. However, keep in mind that both two and three dimensional flaws cause stress concentrations (important for dynamic loading) and that extensive porosity can be indicative of a poor weld which may contain other, more dangerous, flaws.

Typical flaws caused by welding process and procedure problems are the following:

Two dimensional

Lack of Penetration	inadequate heat input for joint configuration
Lack of Fusion	inadequate melting of base metal

Three dimensional

Porosity	from poor shielding (atmospheric interaction), poor quality shielding gases
Undercut	from poor welding technique
Excess reinforcement	too much weld metal
Excess penetration	high heat input

Typical flaws caused by welding metallurgy problems are the following:

Two dimensional

Cracks	from solidification and high temperature ductility problems and/or low temperature problems

Three dimensional

Porosity	from reactions with shielding gases or slags
	from rapid decrease in gas solubility as liquid cools or liquid freezes to solid

Residual Stress

Because the formation of a weld zone takes place in a small region of a welded structure, the heating of the material is non-uniform. Solidification of the weld metal and the cooling of the weld zone to ambient temperature causes shrinking for nearly all metals, and virtually all commonly welded metals. Shrinkage during freezing is about 4-7% and shrinkage during cooling to ambient is lower, but of similar magnitude. Since the shrinkage is localized, residual stresses are developed in the completed structure. These residual stresses are at the yield point in tension in both the axial and transverse directions. The actual magnitudes and distribution depend to some extent on thickness and restraint, especially in the transverse orientation. These levels of tensile stress can contribute to flaw development, such as cracking, and often need to be countered by stress relieving postweld heat treatment (PWHT). After PWHT, residual stresses are reduced to about 15% of the original value transverse to the weld and 30% in the axial direction.

CRACKING IN WELDS

The combination of gas and slag-metal reactions, segregation effects, including solidification and grain growth, and residual stress can lead to the formation of flaws in the weld metal and HAZ. Porosity can result from solubility changes or chemical reactions as temperature drops, but the most dangerous flaws are two-dimensional cracks. These can be formed at elevated temperatures, loosely associated with the solidus temperature of the bulk alloy, or at roughly ambient temperature or slightly above. These phenomena are called hot cracks and cold cracks respectively.

Hot Cracking

Hot cracking occurs at elevated temperatures. Most discussions of hot cracking in weld zones concentrate on cracking involving some molten component, and is often referred to as solidification cracking. However, it is more scientific to refer to hot cracking as a flaw which occurs when temperatures are high enough to affect the visible microstructure, and we will take this as the recrystallization temperature or a phase transformation temperature. Recrystallization usually occurs at 0.3 to $0.5T_m$, where T_m is the melting temperature of the metal in degrees absolute. This may be high or low compared to ambient temperature, depending on the metal. In lead alloys, the recrystallization temperature is below 20°C (68°F), while in refractory metals, for example tungsten, it is well over 1000°C (1830°F). Thus *hot* cracking is a relative term.

While most hot cracks form at or near the solidus of the bulk metal in either the weld metal or the HAZ, there are other types of cracks which form at lower temperatures, down to about $0.5T_m$, which are also hot cracks. Cracks which occur in the weld metal as it solidifies are called *solidification cracks*. These may involve elements deliberately added (intrinsic elements) or impurities (extrinsic elements). They are called macrocracks if visible to the naked eye, microcracks if smaller. Cracks may also form in the weld metal and the HAZ at temperatures just below the bulk solidus in each region, primarily due to segregation of alloy or impurity elements which depress the solidus locally. The resulting liquid films on the dendrite or grain boundaries cause cracks to form in a region of otherwise solid metal. In the weld metal these are often called *hot tears or fissures*, and in the HAZ *liquation cracks*. At lower temperatures, local embrittlement can cause *ductility dip cracks* and certain heating conditions after welding can lead to *reheat cracks* in the HAZ. *Temper embrittlement* is also brought on by certain elevated temperature and time conditions.

Solidification Cracking

Present knowledge of solidification cracking is based on work started in the late 1940's by Pumphrey et al on welds and castings in aluminum alloys. Consider the weld solidification of a 20% X alloy in an eutectic system, see Figure 3.1.

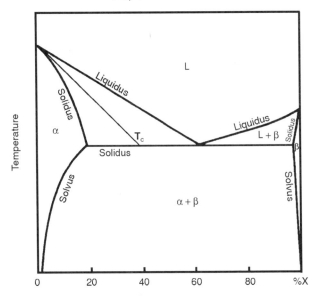

Figure 3.1 Solidification zones according to Pumphrey et al

In the early analysis, as the temperature drops below the liquidus, solid dendrites with a lower solute content begin to grow and the remaining liquid reduces in volume and increases in solute content. Any gaps between the freely growing dendrites are filled immediately with liquid and no cracking occurs. At T_c, the dendrite mass becomes coherent, and a relatively rigid mass develops. As the temperature continues to drop, thermal strains may cause gaps to appear between the dendrites, and refilling via remaining liquid becomes less and less likely as the liquid volume continues to decrease. This may leave cracks in the solidified structure. The cracking is proportional to the difference between the coherent temperature and the solidus, which peaks at the maximum solid solubility of the solute.

This approach corresponds fairly well to experimental results, but is applicable only to super-solidus cracking. It does not account for sub-solidus cracking, the effects of residual elements (extrinsic), or offer a mechanism for the formation of the cracks. Pellini approached the sub-solidus cracking problem by proposing that segregated intrinsic and/or extrinsic elements produce pockets of liquid between dendrites at temperatures below the bulk solidus. Solid to solid bridges are formed between the pockets of trapped liquid, which are separated from any remaining bulk liquid. If cracking occurs by the rupture of the solid-solid bonds, no extra liquid is available for repair. The key to this theory is the proposal that the cracking mechanism is the failure of the solid-solid bridges due to thermal strain (liquids cannot support strain). Thus, if the bridges occupy a small fraction of a dendrite or grain boundary, failure is likely. If they occupy a large portion of the boundary, the thermal strains can be resisted and the liquid solidifies as an inclusion or grain boundary precipitate.

Finally, Borland combined the two approaches and added an analysis of the interphase tension existing between any remaining liquid and the solid mass surrounding it as cooling and shrinkage take place. In this theory, the solidification range has three zones, as shown in Figure 3.2:

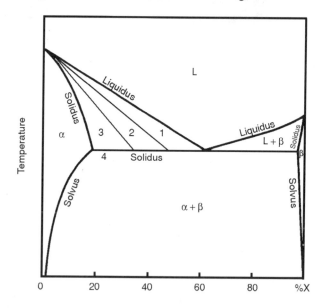

Figure 3.2 Solidification zones according to Borland

In "1", just below the liquidus, the dendrites grow freely without contact in a sea of liquid - no cracking is possible. In "2", the dendrites touch and interact, but there is enough remaining liquid to repair any cracks that appear. Region "3" corresponds to the temperature below Tc in the Pumphrey model. Region "4" corresponds to the Pellini model. Borland also suggested a concept to account for differing cracking tendencies in the sub-solidus region. The interphase tension between the remaining liquid and the surrounding solid will have an effect on how well (or how poorly) the liquid spreads on the interdendritic boundary. The interphase tension between the solid and remaining liquid produces a horizontal force tending to spread the remaining liquid along the solid-liquid interface. Borland described the horizontal force resolution as "τ", and suggested that a low value (0.5) produces complete wetting and therefore a high cracking risk. As the value rises to about 0.6, only the grain edges are coated. At a value of 0.7, the corners only are coated and cracking risk is minimal. This concept is illustrated schematically in Figure 3.3.

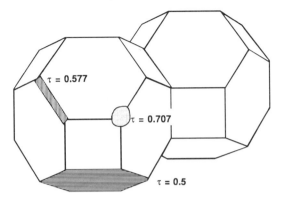

**Figure 3.3 Remaining liquid wetting characteristics
according to Borland**

Liquation Cracking

Liquation cracking is the result of microsegregation at grain boundaries remelting to become liquid at temperatures just below the bulk alloy solidus in the HAZ. The tensile stresses induced by the cooling and subsequent shrinkage of the weld zone pull the grain boundaries apart. This type of cracking typically occurs near the weld metal-HAZ fusion line, often within 10 grain diameters of the fusion line. It usually happens in the base metal HAZ, but can occur in the HAZ of previously deposited weld beads. Grain growth tends to concentrate the available impurities on reduced grain boundary area, thus exacerbating the problem. Liquation cracking susceptibility has been assessed with an *index* which is the difference between the temperature of zero ductility on heating minus the temperature of recovery (non-zero) ductility on cooling. The larger the gap in temperature, the worse the cracking.

Solidification and liquation cracks have certain characteristics which aid in identifying them. The cracks are intergranular or interdendritic, and have rounded edges and a blunt appearance. Their aspect ratio (length/width) is low. When an allotropic phase change is involved, great care must be taken to assess the cracking, since it may have occurred intergranularly in a high temperature phase, which may appear to be transgranular cracking in a low temperature phase nucleated later. For example, cracks formed at prior austenite grain boundaries in the HAZ of a steel may appear in the light microscope as transgranular cracks in ferrite. These are termed *ghost boundary cracks* since the original grain boundaries may be invisible in etched surface. Liquation cracks also have solidified traces of the liquid which promoted the cracking on the crack surfaces, which, when analyzed, can suggest avoidance remedies. For example, if sulphur is mixed into a weld metal from contaminated base metal and sulphides are found on crack surfaces, a higher manganese filler wire can be used to avoid cracking on repair.

Ductility Dip Cracking

Ductility dip cracking appears in a hot tension test as a reduction in ductility to low levels at a temperature in the vicinity of the recrystallization temperature or an allotropic phase change temperature. The mechanism is not well understood at present, nor is remedial action, but some characteristics help to identify them and differentiate them from liquation cracks. Since they tend to occur close to $0.5T_m$, their appearance in the HAZ is further away from the fusion line. In a multipass weld metal, they will appear in an area of similar kind, away from any existing fusion lines. The cracks are also intergranular and blunt, but have two distinguishing characteristics: the crack surfaces do not have solidified liquids on them, and often display thermal striations on the exposed surface, which is indicative of a free surface existing at high temperatures when dislocation climb is possible. The mechanism may be associated with the production and migration of new grain boundaries during recrystallization or allotropic phase change.

Reheat Cracking

Reheat cracking occurs in Cr-Mo, Cr-Mo-V and some austenitic stainless steels during service at high temperatures (e.g. steam piping) or occasionally during stress relief heat treatment. The mechanism is the same in both cases. All of the steels susceptible to reheat cracking contain secondary hardening elements (V, Ti, Cb) and are also susceptible to temper embrittlement (P, Sn, As, Sb segregation to prior austenite grain boundaries promoted by Mn, Cr and Ni). There are many alloy carbides which can form during tempering or high temperature service. These carbides can cause secondary hardening, which is partly responsible for reheat cracking. The carbides in the table are those formed due to the major alloying elements Cr and Mo. The minor alloying elements (V, Ti and Cb) form smaller intragranular carbides or carbonitrides which produce an even more significant increase in grain interior strength, thus promoting reheat cracking. This effect is emphasized in the empirical equations above by the weighting factor attached to these elements, which in every case exceeds the weighting factors assigned to Cr and Mo. In austenitic stainless steels, Types 321 (Ti) and 347 (Cb) are prone to reheat cracks, but Type 316 (Mo) is not.

During heat treatment or high-temperature service after welding, the grain boundaries of the large austenite grains formed near the fusion line may have higher concentrations of P, Sn, etc. in them due to the lowering of available grain boundary area per unit volume. This weakens the grain boundary cohesion to some degree, but classic temper embrittlement does not occur if the temperature is in excess of 500°C (930°F). However, the dissolved C and N may recombine with V, Ti or Cb to form small precipitates within the grains (rather than large precipitates at grain boundaries, which are used to restrict grain growth in alloy steels). These precipitates strengthen the grain interior with respect to the grain boundary. The combination of strong grain interiors with possibly weakened grain boundaries causes intergranular failure when thermal strains are induced in the structure by thermal gradients due to heat treatment or service. The failure occurs when the the grain boundary strength drops below the strength of the grain interiors.

Cold Cracking

Cold cracking occurs near, at or below ambient temperature and is largely confined to a particular form of cracking in transformable martensitic-ferritic steels called *Hydrogen Assisted Cracking* or HAC. Cold cracking is caused by four conditions, which *must all be present* for cracking to occur:

1. Low Temperature - < 200°C (390°F) in any alloy grade, < 100°C (210°F) for most steels
2. Sufficient Tensile Stress - total from applied, residual and transformation stresses
3. Hydrogen - < 5 ppm for very good low hydrogen (basic) electrodes
   5 - 10 ppm for good low hydrogen
   10 - 20 ppm for medium hydrogen
   > 20 ppm for high hydrogen
4. Susceptible Microstructure (Hardness) -
   martensite
   bainite
   ferrite/carbide

The first two conditions are inevitable in any welded structure, and effective control is confined to the two latter criteria. Hydrogen is absorbed as atomic hydrogen and is very mobile in the liquid and solid metal. Diffusible atomic hydrogen appears to cause cracking by producing fracture at low global stress values in distorted ferrite structures by acting as interstitial atoms on particular crystal planes. It may also collect at traps such as inclusions, recombining as molecular hydrogen and increasing local pressure at the inclusion-matrix interface. This type of hydrogen pressure can also assist in other forms of cracking, for example Lamellar Tearing, or Sulphide Stress Cracking (SSC). While 5 ppm appears to be the lower limit to establish good conditions, some modern high strength steels may be susceptible at 2 ppm and modern consumables exist that will achieve such low levels on a reliable basis.

Diffusible hydrogen is measured in many ways, but only two methods are reliable. The IIW mercury test collects diffusible hydrogen at about ambient temperature from specific size specimens by allowing it to diffuse out of a welded sample into a closed glass container filled with mercury. The virtue of mercury is that hydrogen gas is virtually insoluble in the metal and all hydrogen which escapes from the sample over a period of 72 hours is measured. Other media used in the past, such as glycerine, have significant solubility for hydrogen and are therefore unreliable for a routine test. Recently, a test has been developed in which a similar hydrogen charged specimen is heated to about 650°C (1200°F) in a vacuum, thus forcing out the diffusible hydrogen in a very short time.   Readings with gas chromatography (AWS A4.3) are automated and very reliable.   If the specimen is taken to its melting temperature, all of the hydrogen, including trapped molecular hydrogen, can be measured.   Diffusible hydrogen was originally measured as millilitres/100 g of deposited weld metal.   This is a rather unwieldy method scientifically and parts per million (ppm) is better.   As it happens, for iron and iron alloys, ppm differs only by +10% from the older values, and determinations of hydrogen content are rarely that accurate.

The hydrogen present in the weld zone is absorbed from gases and fluxes used in welding and from any moisture or hydrocarbons, such as oil, grease or paint, in the vicinity of the weld.   Thus normal precautions include dry base metal, electrodes, gases and fluxes, along with degreasing and the removal of any moisture or oil traps, for example rust or heavy oxide, on the base metal.   A preheat is often useful to remove condensation from the weld area and is also used in Procedure Control of cracking, as noted below.   Vigilance must be used to select and control hydrogen in welding consumables and processes, e.g. use of low hydrogen electrodes and suitable steps to ensure dry electrodes are used.   Storage of electrodes and fluxes must ensure that low hydrogen conditions are maintained.   Electrodes and fluxes are usually shipped in hermetically sealed containers to maintain low hydrogen characteristics.   If the package is intact on receipt, the consumables should be stored in heated ovens at 120°C (250°F) to avoid moisture absorption during storage.   If the seal is broken, or consumables have been exposed to moisture pickup for periods of more than 24 hours, they should be baked at 350-400°C (660-750°F) for a few hours before use to restore low hydrogen characteristics.   Manufacturers of particular consumables should be consulted for exact procedures.

The control of hydrogen cracking via microstructure and hardness manipulation is necessary if low hydrogen procedures are not completely reliable or if the base metal is susceptible.   It has been found that HAC is more likely in hard microstructures, so that martensite is most susceptible, bainite less so and ferrite + carbide least.   For a given steel, there is a relationship between peak hardness and cooling rate, shown schematically in Figure 3.4.

If HAC occurs at a critical hardness, there will be a critical cooling rate corresponding to this hardness.   If we can weld in such a way that the maximum cooling rate is slower that the critical cooling rate, then peak hardness should be below the maximum and cracking will be avoided.   The most effective method involves definition of the heat input and the use of preheat to control the cooling rate in the weld zone.   PWHT can also be used when necessary, if the preheat and heat input manipulations are inadequate. The method is called *Procedure Control*.

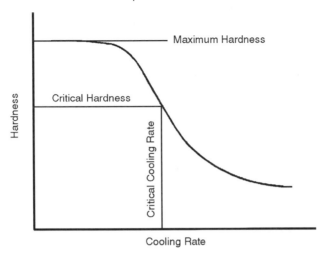

Figure 3.4 Hardness characterization curve for hardness control

Three types of control exist, depending on the chemical composition of the steel, Figure 3.5.

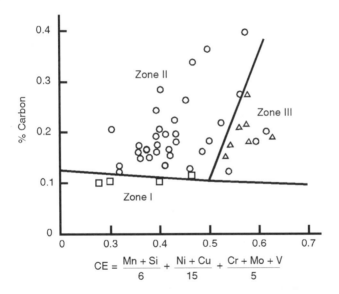

$$CE = \frac{Mn + Si}{6} + \frac{Ni + Cu}{15} + \frac{Cr + Mo + V}{5}$$

Figure 3.5 Classification zones of steels according to hardening characteristics

For low carbon conditions (Zone I), little risk is assumed, since peak hardness is low regardless of alloy content. For moderate carbon and alloy levels (Zone II), procedure control is used - a combination of preheat and/or heat input which causes the cooling rate to moderate enough to avoid hard structures. With higher carbon and alloy levels, hardness and hardenability are too great to allow control of hardness by procedure control alone (Zone III), and postweld heat treatment (PWHT) must also be used.

If carbon is less than 0.11%, then martensitic lattice distortion and hardness is relatively low even when alloy levels (and hardenability) are very high, and in general only minimal precautions are necessary, such as avoiding moisture, etc. The carbon equivalent on the X axis of the graph is a simple way to describe hardenability - the alloy level is empirically related to an equivalent amount of carbon which would give an identical hardenability, with maximum hardness determined by actual carbon level. There are many ways to calculate Carbon Equivalent (CE). All of the formulae available were derived for a particular batch of steel chemistries and in a particular test - hence the varying influence of a given alloying element on the carbon equivalent. In general, lower carbon steels show a lesser influence of alloying elements on the CE. The critical hardness to cause cracking tends to decrease as CE decreases, regardless of the formula used. These formulae are all of similar general form, in that the effect of a given alloying element is considered to be some fraction of its actual weight percentage in the steel:

$$\text{Carbon Equivalent (CE)} = \%C + \Sigma\ \%X/Y$$

where X = a given alloying element, for example Mn or Cr and Y = a whole number, usually >1.

A typical example of a simple carbon equivalent formula is:

$$CE = \%C + \frac{\%Mn}{4} + \frac{\%Si}{4}$$

Examples of CE formulae will be given in each Chapter as appropriate.

Steels of low hardenability in Zone II can be hardness controlled by preheat temperature and heat input. The preheat and heat input are set to ensure that the cooling rate in the HAZ produces a maximum hardness which is characteristic of a microstructure not susceptible to hydrogen cracking. The risk is deemed to be limited (not eliminated) by restricting the maximum hardness to between 350 Hv10 for *high* hydrogen levels, relaxed to 450 Hv10 for *very good* low hydrogen conditions. A nomogram relating combined metal thickness in the joint (quench severity), preheat, heat input and hydrogen potential for arc welding C-Mn steels has been developed by the Welding Institute in England, Figure 3.6.

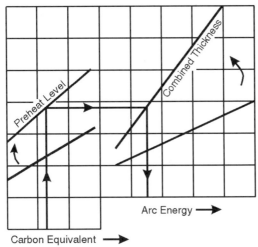

Arc Energy ⟶

Carbon Equivalent ⟶

Figure 3.6 Generic nomogram for procedure control of HAZ hardness

This is now the standard method promoted by the IIW and is available in software packages from several sources worldwide. The nomogram uses the concept that the risk of hydrogen assisted cracking (HAC) is directly related to the peak hardness in the HAZ, which in turn is a function of the carbon level and the hardenability. The Carbon Equivalent is the device to combine hardness and hardenability. The cooling rate which causes the peak hardness allowable to prevent cracking is determined by the preheat temperature/heat input combination derived from a particular test. The IIW formula was determined by tests of several steels and electrodes using the CTS (Controlled Thermal Severity) test. For a given steel chemistry (CE), process hydrogen level and joint configuration (primarily material thickness effects), the nomogram will predict required preheat for a nominated heat input or a required heat input for a nominated preheat. The use of the nomogram will be discussed in more detail in Chapter 4.

Higher carbon and alloy levels produce steels which are highly hardenable and cannot be limited in hardness via preheat and procedure control alone. These steels, Zone III of Figure 3.5, need to be postweld heat treated (tempered) in addition to having controls on hydrogen levels, preheat and heat input if cracking is to be avoided. Some steels used in structures, such as ASTM A514, are in this category. The PWHT diffuses hydrogen out of the weld zone as well as tempering the microstructure and relieving stresses due to the welding thermal cycle.   Preheat >100°C (210°F) also promotes hydrogen diffusion out of the weld zone and is a more effective method of avoiding HAC than simply relying on the PWHT.   In many cases, a hardenable steel will crack before the initiation of PWHT if Procedure Control involving preheat is not used. Preheat maintenance until PWHT is initiated is a good practice. If PWHT is not possible, holding after welding at 200-250°C (390-480°F) will diffuse hydrogen out of the weld zone.   However, this process is time dependent and not as reliable as procedure control.

Modern microalloy steels can have HAC and even weld metal problems at low carbon levels giving steel compositions in Zone I. This is apparently a result of the type of microstructure, not hardness per se. Martensitic/bainitic structures can be crack prone, even at low Hv10 hardness levels. Cracking has occurred at hardnesses below 250 Hv10 in microstructures which are bainitic, low in carbon content but high in hardenability (CE). There can also be a problem with very clean steels, with few inclusions and low oxygen and sulphur levels. This effect is due to the nucleating role of inclusions for the thermal decomposition of austenite. Many inclusions induce high temperature transformation products (e.g. ferrite and pearlite) while fewer inclusions cause delayed transformations at lower temperatures (acicular ferrite, bainite, martensite) due to a reduction in the number of transformation nucleation sites.

Coarse microstructures associated with hot-rolled steels take more time to even out carbon gradients after pearlite dissolution, leading to zones of higher hardenability near the fusion line. Required preheats can be up to 75°C (165°F)higher for hot-rolled steels than for normalized steels of identical chemistry. This can be significant for both structural welding and pressure vessel fabrication and repair.

Mixed Mode Cracking

There are also forms of cracking which involve elements of both hot and cold cracking. One is *Lamellar Tearing* and another is *Chevron Cracking*. Both occur primarily in ferritic steels and contain elements of cracking at elevated temperature followed by additional cracking assisted by hydrogen.

Lamellar tearing, as shown in figure 3.7, occurs just outside the visible HAZ in certain structural and pressure vessel steels >20 mm (0.75 in.) thick which are exposed to through-thickness stresses as a result of welding. Typical joint designs which are prone to this problem are edge joints, for example in box section headers, or in flange to web joints involving fillet welds, for example in girder construction. In pressure vessels, nozzle to shell and flange to shell welds are possible sites. Susceptible steels are likely high in S and O and therefore have large inclusion populations. The likely peak temperature in the cracked region is about 500°C (930°F).

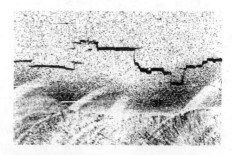

Figure 3.7 Lamellar tears

The cracking happens only in steels which contain aligned rows of flattened inclusions, usually due to hot rolling when manganese oxy-sulphides are present, although alumino-silicate inclusions can be involved as well. The aligned, or lamellar, inclusions do not constitute a lamination and are not considered a defect prior to a lamellar tearing problem. The mechanism of cracking is as follows:

1. Welding solidification and cooling shrinkage induces tensile stress perpendicular to the inclusion line.
2. The inclusion-matrix interface, possibly aided by hydrogen poisoning, parts under stress to form a small crack. It is also possible that the inclusion itself breaks if the matrix bond is stronger than the inclusion material.
3. The cracks along the inclusion lines join up, extending the crack length. These leave only a small zone of metal between the cracked regions to support any stress.
4. The matrix material between different cracked regions shears under the remaining load. This often occurs after a time delay and may be assisted by hydrogen.
5. The resulting two-stage crack is thus stepped in profile and has a characteristic *woody* appearance when the cracked surface is viewed directly or in the SEM. To avoid this kind of flaw, there are several proposed remedies, some of which are more efficient or cheaper than others. Details are in Chapter 4.

Chevron cracking appears in the weld metal of structural steels, mainly those deposited by the SAW process. This form of cracking is also stepped. The cracks have one component which occurs in the dendritic boundaries as intermittent cracks, and a component across the dendrites, which act to join up with the boundary cracks. The result is a V shaped crack (or series of cracks) pointing along the length of the weld metal, hence the name, chevron. Initial analysis of these cracks suggested that the dendritic boundary component formed first, at high temperature, as a ductility dip crack. Evidence of clean crack surfaces containing thermal striations supported this analysis. The transverse dendritic cracks were hydrogen induced in shear, similar to the connecting cracks in lamellar tearing.

Other work suggested that the thermal striations were caused by the heating effect of subsequent passes in multipass welds and were not present in single pass welds. In this scenario, both cracking components were hydrogen induced. To test the validity of both ideas, single pass welds deposited at low temperatures to retain hydrogen showed that the thermal striations did indeed appear in single pass welds, confirming the original analysis. The thermal striations were difficult to observe at the time and may have been easily missed in some investigations. Subsequent work showed that the cracking was reduced by ensuring that low hydrogen procedures were used and properly carried out.

TEMPER EMBRITTLEMENT

Temper embrittlement is caused by the segregation of P, and to a lesser extent by Sn, As and Sb, to prior austenite grain boundaries during heat treatment or high-temperature service at temperatures <500°C (930°F). Cr and Ni tend to promote segregation, while Mo tends to minimize it. In the HAZ region of welded structures, it also appears that Mn and Si promote segregation in the HAZ. In weld metals, Mn is also a problem at levels in excess of only 0.10%. The suggested method to minimize cracking is to keep Mn and Si as low as possible. Mn can be held to about 0.5% in Cr-Mo steels, but not less. Si is also a problem to limit in weld metals, due to the need for deoxidation. The preferred method in modern steels is to limit P to less than 0.010%, or even 0.005%. It is not metallurgically feasible to reduce Sn and especially As and Sb, to levels lower that the ppm levels at which they appear now. Scrap containing Sn can be limited to other steels which are not exposed to conditions which promote temper embrittlement. There have been a few empirical rules derived to minimize the probability of temper embrittlement in weld zones, for base plates (HAZ) and the weld metal in Cr-Mo and Cr-Mo-V steels, which will be given in detail in Chapter 5 on Alloy Steels.

CARBON STEELS

CARBON STEELS

Carbon steels, or Carbon-Manganese (C-Mn) steels are alloys of carbon, manganese and iron dominated by the solid state eutectoid reaction at 0.8%C. General metallurgical information on these alloys is given in *The Metals Black Book* - Ferrous Metals. Related AWS Filler Metal Specifications are A5.1, 5.2, 5.17, 5.18, 5.20, 5.25 and 5.26. Carbon levels are in the range up to about 1.2%C maximum. Manganese is always present to control sulphur and prevent hot cracking. Steels may contain up to 2%Mn before it is considered a separate alloying element.

STRUCTURAL STEELS

Modern structural steels are largely Carbon-Manganese (C-Mn) or low alloy steels. C-Mn steels used for structures and machinery typically contain up to about 0.50% carbon, 1.7% manganese, 0.6% Silicon and occasionally a small amount of aluminum for grain size refinement. No other deliberate alloying elements are added, but chromium, nickel, molybdenum, copper and columbium can be present as residual elements.

In a given steel, the maximum achievable hardness in a severe quench (high cooling rate) is determined almost exclusively by the carbon content. The ability of the steel to achieve that maximum hardness is dependent on the cooling rate and the alloy content. Most alloying elements tend to diminish the rate of diffusion of carbon, which tends to promote lower temperature transformation products and thus increases hardenability, which describes how easily the steel can achieve maximum hardness in a section. The Jominy endquench hardness test is often used to assess this behaviour. Jominy data is often correlated with an Isothermal Transformation diagram (or Time-Temperature-Transformation diagram), which shows which transformation products occur for a steel cooled rapidly from about 900°C (1650°F) to a specified lower temperature, which temperature is then held until a transformation occurs. This does not happen often in practice, especially in welding, where cooling takes place from close to the melting point in the weld Heat-Affected Zone (HAZ) and cooling is continuous, not isothermal. In these conditions, Continuous Cooling Transformation diagrams are used to determine transformation behaviour.

All of the discussion above assumes that any austenite present is homogeneous. In welding, this is usually not the case due to the variety of heating rates and peak temperatures experienced by the HAZ, hence one is dealing with a variety of compositions of austenite in a small region, all of which behave differently.   Once this is understood and transformation behaviour for the variety is determined, welding problems can be approached with some degree of confidence. Two regions are important - the heat affected zone (HAZ) and the weld metal.

Heat Affected Zone

In the HAZ, the microstructure developed is controlled by the three primary aspects of the thermal cycle - heating rate, peak temperature and cooling rate. The thermal cycle experienced by a weld zone varies widely from process to process. Short, rapid thermal cycles experienced at low heat input promote hardened structures and tend to retain hydrogen. Extended thermal cycles experienced at high heat inputs (or with preheat) minimize hardening and promote diffusion of hydrogen away from the weld zone, but also promote very large grain sizes and often produce low fracture toughness and high tough (ductile) to brittle transition temperatures. Varying hydrogen levels, chemical composition and microstructures produce large differences in the toughness of structural steels.

Welding causes a great many changes in microstructure and properties to take place in a very small region in a fabrication. We have all of physical metallurgy and a great deal of pyrometallurgy in a zone 1 to 100 mm (0.040-4 in.) wide. To study this, we will start at a point outside the HAZ and work our way in toward the weld metal, observing the changes as the peak temperature rises, initially for a single pass weld (one thermal cycle). Multipass welds simply impose another rapid heating and cooling regime on an already complicated microstructure.

Heating rates primarily have an effect on the dissolution of carbide and ferrite to form austenite and on the homogeneity of the austenite after full dissolution. Coarse microstructures, such as the pearlite in a hot rolled steel, do not dissolve and homogenize within the HAZ as quickly as do finer microstructures, such as pearlite in normalized steel.   Steels with coarse microstructures therefore have higher effective hardenability than would be suggested by chemical composition alone and this must be taken into account when welding them.

Peak temperature begins to have an effect at about 100°C (210°F) below the lower critical temperature. Pearlite begins to spheroidize slightly, producing an effect best described as *fuzzy carbide*. The temperatures are in the recrystallization level for ferrite and mechanical properties usually do not suffer, unless the steel is not killed and strain ageing is possible. This is rare in most modern steels.

Outside the visible HAZ, just before the onset of fuzzy carbide, the main problem is Lamellar Tearing. There are several methods available to avoid or repair lamellar tearing.

For repairing an existing structure or fabricating with existing poor material:

1. Alter the joint design to avoid placing problem regions in through-thickness tension. This works best with corner joints on box sections, as illustrated below in Figure 4.1:

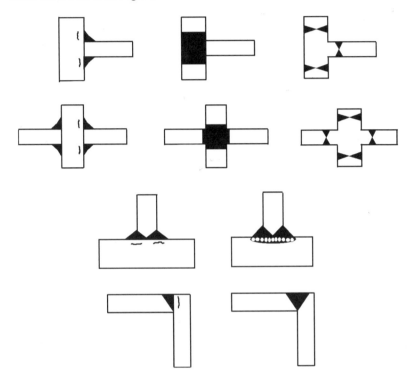

Figure 4.1 Location of lamelar tears and joint redesign for avoidance

2. Arc gouge the surface at the location of the proposed weld joint (5-10 mm (0.2-0.4 in.) deep) and butter the area with weld metal. This changes the inclusions in the susceptible region to spherical, randomly dispersed ones not prone to lamellar tearing. It is also postulated that using *low strength* weld metal absorbs strains to avoid tearing. This is highly unlikely, since the yield strength of most SMAW deposits exceeds the yield strength of steels prone to lamellar tearing by a substantial margin.
3. Insert soft wire inserts, especially in flange-web joints, to absorb strain. This is a marginal technique, although effective in some circumstances.

For preventing problems in a proposed structure:

1. Order clean steels with a total %oxygen + %sulphur <500 ppm.
2. Weld only with low hydrogen procedures.
3. Scan plate with ultrasonic NDT techniques to find problem areas. This is an inefficient and costly technique and is not always able to locate lamellar inclusion arrays even when present.

Once the lower critical temperature (A_1) is exceeded in the HAZ, some austenite is formed. It is formed within or at the edge of pearlite colonies. Near the A_1, the austenite is eutectoid composition. As the temperature rises toward the upper critical temperature (A_3), more austenite is formed and grain size increases. The carbon level in the austenite approaches the nominal level for the steel as the austenite grain grows outside the pearlite colony and devours the surrounding ferrite. However, complete homogenization takes a very long time and is unlikely to occur during any welding cycle, especially if banding is present in the steel.

Peak temperature is important in the determination of austenitic grain size once the upper critical temperature is exceeded. The higher the peak temperature, and the longer it is held (high heat input), the larger the austenite grain size. Therefore, as the fusion line is approached, austenite grain size tends to increase, and the largest grains are found adjacent to the fusion line. These grains are significantly larger than those found in conventional heat treating and the size contributes to HAZ problems. Hardenability of the steel is increased, making the appearance of low temperature transformation products (martensite, bainite) more likely if the cooling rate is high and carbon level is significant. The fracture toughness of the HAZ may then be low, although the strength will be high. PWHT is the only solution if an improved (lower) strength/toughness ratio is desired. A 'stress relief' treatment at 575-650°C (1070-1200°F) is usually sufficient, but for ESW procedures, pressure vessel and some other codes require full normalizing after fabrication, which is done at temperatures of 850-925°C (1560-1700°F). In large fabrications this can lead to significant thermal distortion from the effect of gravity on the structure.

Cooling rate has a major effect on hardness in the HAZ and cold cracking behaviour. HAC can occur in many locations and orientations, as shown in Figure 4.2.

Figure 4.2 Location of hydrogen assisted cracks in weld zones
The Metals Blue Book

Weld metals are also susceptible, but to a lesser extent. Cold cracking is associated with tensile stress, low (near ambient) temperature, hydrogen and a susceptible microstructure.

The first two are inevitable in any welded structure, and control is confined to the two latter criteria of hydrogen level and microstructure. Diffusible atomic hydrogen appears to cause cracking by producing low stress fracture in distorted ferrite structures by acting as interstitial atoms on particular crystal planes. Modern low hydrogen consumables such as SMAW electrodes are dried and hermetically packed to ensure low hydrogen (<5 ppm or so) in the weld deposit. Care must be taken to avoid breaking the seal during handling. Once the package is open, storage must be in heated holding ovens at about 120°C (250°F) to ensure that no moisture is absorbed. Until recently, exposure to the atmosphere in non-heated circumstances would cause low hydrogen electrodes to deteriorate within a few hours. Now, for electrodes of 550 MPa (80 ksi) or lower tensile strength, usually an exposure for a whole shift (8-10 hours) is permissible as long as the electrodes are returned to the holding oven. If overexposure to moisture, in the form of humid air, does occur, drying at 350-400°C (660-750°F), depending on the manufacturer, is necessary before reuse. If contact with water, oil or grease occurs, the electrodes should be discarded.

The control of hydrogen cracking via microstructure and hardness manipulation is necessary if low hydrogen procedures are not completely reliable or if the material is quite susceptible. Three types of control exist, based on the Graville Diagram, Figure 4.3. If carbon is less than 0.11%, the lattice distortion and hardness is relatively low even when alloy levels (and hardenability) are very high, and in general only minimal precautions are necessary. These steels are located in Zone I.

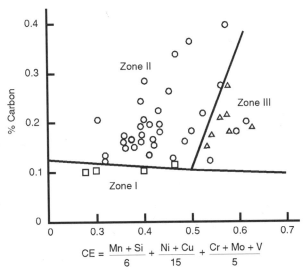

$$CE = \frac{Mn + Si}{6} + \frac{Ni + Cu}{15} + \frac{Cr + Mo + V}{5}$$

Figure 4.3 Classification zones of steels according to hardenability characteristics

The Metals Blue Book

The carbon equivalent on the X axis of the graph is a simple way to describe hardenability - the alloy level is empirically related to an equivalent amount of carbon which would give an identical hardenability. Maximum hardness is determined by actual carbon level. Steels of low hardenability in Zone II can be hardness controlled by variation of the welding procedure, notably preheat temperature and heat input. The preheat and heat input are set to ensure that the cooling rate in the HAZ produces a maximum hardness which is characteristic of a microstructure not susceptible to hydrogen cracking, often taken at 350Hv10. If hydrogen can be controlled to <5 ppm, hardnesses of up to 450Hv10 can be tolerated. A nomogram relating plate thickness (quench severity), preheat, heat input and hydrogen potential for arc welding C-Mn steels has been developed by the Welding Institute in England, Figure 4.4.

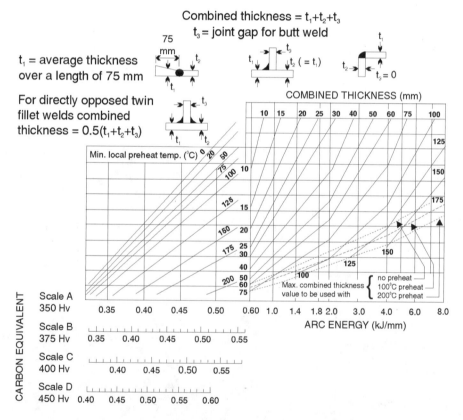

Figure 4.4 Nomogram for procedure control

The nomogram is now available on computer software, but a knowledge of how it works is useful to interpret the recommendations. In addition to the *combined thickness* of the material in the weld joint area, the nomogram uses two fixed or known *process related inputs* to predict possible combinations of *two procedural variables* to weld with minimal risk of HAC. The combined thickness for a butt weld is the sum of the thickness of the material on each side of the joint plus the joint gap. This represents the quenching and restraint potential of the mass of material around the weld bead. The two process-related inputs are the *diffusible hydrogen level* put into the weld zone by the welding process chosen, and the *carbon equivalent*, or CE, of the base metal. The process-related inputs are the combination of *welding heat input* and *preheat temperature* that will provide a cooling rate slow enough to avoid hard microstructures which may cause HAC at the diffusible hydrogen level involved.

The diffusible hydrogen level is assessed in four steps mentioned in Chapter 3.

1. Very good low hydrogen, the best available, is < 5 ppm.
2. Good low hydrogen conditions are 5-10 ppm
3. Medium levels are 10-20 ppm
4. High hydrogen conditions are > 20 ppm.

The maximum levels produced are in the order of 60 ppm for cellulosic SMAW electrodes. The maximum solubility of atomic hydrogen in ferrite at ambient temperature is about 6 ppm, so significant supersaturation is easily obtained by the time the weld zone has cooled.

The carbon equivalent for C-Mn steels uses the IIW equation for entering the nomogram. This is a slightly modified version (1967) of the original equation determined earlier by Dearden and O'Neill:

1. IIW

$$CE = \%C + \frac{\%Mn}{6} + \frac{\%Cr + \%Mo + \%V}{5} + \frac{\%Cu + \%Ni}{15}$$

Because the original equation was determined on a relatively limited set of steels which were characteristic of the steelmaking processes of the time (relatively high carbon, sulphur and oxygen), some modern steels may not conform well to predictions based on the IIW equation. Conservative estimates may be calculated by adding a term for silicon:

2. IIW (current)

$$CE = \%C + \frac{\%Mn}{6} + \frac{\%Si}{6} + \frac{\%Cr + \%Mo + \%V}{5} + \frac{\%Cu + \%Ni}{15}$$

The IIW equations work well for steels of moderate to high carbon levels, above about 0.15%. For lower carbon steels there is no agreed format to determine CE.

Once the combined thickness, hydrogen level and carbon equivalent are determined, the nomogram is entered at the appropriate point on the CE lines at the lower left. There are two ways to proceed. A heat input may be chosen at the lower right, and a vertical line taken to the combined thickness. A horizontal line is then traversed to the left, to intersect with a vertical line drawn upward from the appropriate CE/diffusible hydrogen entry. The intersection determines the required preheat. Alternatively, if a particular preheat is desirable (to provide welder comfort), the intersection of the vertical line from the CE entry with the appropriate preheat is traversed horizontally to the right to intersect with the combined thickness. A vertical drop then determines the required heat input, which may be achieved by any appropriate combination of voltage, current and welding speed. The use of a derived preheat/heat input combination is called *Procedure Control*. Steels with a CE below 0.40 are considered weldable without preheat. A CE between about 0.40 and 0.60 requires procedure control. Higher carbon (and alloy) levels produce steels with a CE above 0.60 which are highly hardenable and cannot be limited in hardness via procedure control alone. These steels, Zone III of Figure 2, need to be post-weld heat treated (tempered or stress relieved) in addition to having controls on hydrogen levels, preheat and heat input if HAC is to be avoided.

At higher peak temperatures near the fusion line, sulphur and phosphorus are the prime cause of any liquation cracking in the HAZ near the fusion line. High heat inputs tend to make this worse, since large austenitic grain size in the HAZ minimizes total grain boundary area, thus concentrating residual elements on the remaining grain boundaries. The best solution is to use low S and P steels (best is below 0.010% of each, < 0.015% adequate in most circumstances).

Weld Metal

The electrodes used in a variety of processes for welding C-Mn steels are usually lower in carbon and higher in manganese and silicon than are the steels being welded. There is a variety of reasons: low carbon tends to maximize toughness and lower hardenability and porosity, higher manganese ties up sulphur from dilution and higher silicon deoxidizes the molten metal to minimize porosity and high inclusion populations, which improves toughness. Toughness is also improved by keeping the amount of grain boundary ferrite (proeutectoid) low, which maximizes the amount of acicular ferrite in the weld metal. Manganese also promotes the formation of acicular ferrite by increasing hardenability, causing transformation temperatures to decrease. Acicular ferrite is a microstructure first observed in C-Mn steels. As its name suggests, it is needle-like in shape, usually in interlocked patterns with a hint of Widmanstätten orientation. The net result is a microstructure of fine effective grain size and random orientation. It is an austenite transformation product formed at intermediate transformation temperature, below that required for coarse ferrite and carbide and above that required for martensite. It has been observed to be nucleated by inclusions, and the size, number and chemistry of the inclusions are all important, although the exact relationship of nucleation efficiency to inclusion properties is not known with certainty.

The low carbon levels in modern electrodes caused some early concern over strength, which is required to match or exceed the base metal ultimate strength in many fabrication codes. Fine austenitic grain size in most processes, excepting ESW and high heat input SAW, tend to keep strength and toughness high. In addition, weld metals have high dislocation densities, which keeps the yield/ultimate strength ratio high, around 0.75 or more.

In the weld metal, a common problem is solidification cracking, which occurs due to segregation of low-melting point constituents, primarily associated with sulphur. High restraint increases the strain involved in solidifying the metal, and the last material to freeze is often called upon to fill larger void than its volume permits - result, hot crack. High welding speeds and small beads tend to produce tear drop shaped welds, Figure 4.5(b), which concentrate segregates at the centreline and promote cracking - a better pool shape is Figure 4.5(a), which tends to reject segregates in to the molten weld pool (at the ends of the dendrites), thus delaying or preventing centreline cracks.

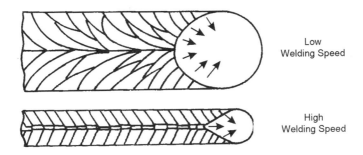

Low
Welding Speed

High
Welding Speed

Figure 4.5 Weld metal solidification patterns at low and high welding speeds

Centreline cracking is also more likely in large, deep penetration narrow welds, which can be minimized by multipass welding with small stringer beads. Manganese is used to tie up sulphur in a sulphide with less propensity to cause cracking than iron sulphides. Sulphur segregation is increased by higher carbon levels and the Mn:S ratio must increase with carbon, Figure 4.6.

This is largely a result of the effect of carbon on stabilizing austenite during solidification. When iron-carbon alloys solidify, there is a peritectic reaction involving ferrite and austenite:

$$\text{Liquid } (0.53\%C) + \delta \,(0.09\%C) \rightarrow \gamma \,(0.17\%C) \text{ at } 1495°C \,(2720°F)$$

This reaction, shown in figure 4.7, is not often discussed with regard to welding metallurgy, or even conventional steel metallurgy. It deserves more attention.

The Metals Blue Book

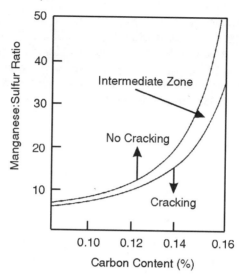

Figure 4.6 The effects of Mn/S ratio and carbon content on weld metal hot craking in C-Mn steels

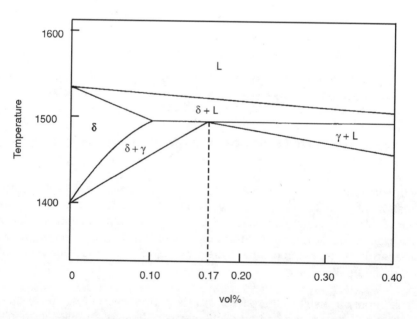

Figure 4.7 Peritectic reaction involving ferrite and austenite

The typical carbon level of structural steels is up to about 0.25%. Any weld metal with 0.10%C or less will solidify as 100% ferrite, then transform to austenite as temperature decreases. Higher carbon levels will lead to mixed ferrite-austenite solidification. At 0.25%C, just above 1495°C, there will be low carbon solid ferrite and liquid of higher carbon level. The ferrite will be 64% of the total and the liquid 36%. If the carbon content is higher, say

0.33%, there will be 45% ferrite and 55% liquid. The remaining liquid solidifies as austenite and/or is involved in the peritectic reaction. At 0.53%C or more, solidification will be 100% austenite. Austenite has a significantly lower solubility than ferrite for both sulphur and phosphorous. As solidification continues, both residual elements will tend to segregate from the austenite and promote hot cracking.

Hot Cracking Susceptibility (HCS) for carbon-manganese steels:

$$HCS = \frac{\%C\left(\%S + \%P + \dfrac{\%Si}{25} + \dfrac{\%Ni}{100}\right)}{3\%Mn + \%Cr + \%Mo + \%V} \times 10^3$$

There is no lower limit in this equation. For a given grade of steel, the tendency toward hot cracking will increase as the HCS increases. As a guide only, a number of 4 or more indicates a possible problem.

MICROALLOYED STEELS

Microalloyed steels were first developed in the 1970's and contain less than 1 wt% of alloying elements outside of manganese and silicon. Usually there are additions of one or more of molybdenum, columbium, vanadium, titanium, boron and aluminum in various combinations. These steels were developed to achieve high strength and high toughness at the same time, with improved weldability in comparison to higher carbon steels. Carbon levels are less than 0.12%C and may be lower, for example in pearlite reduced (PR) steels (<25% pearlite) and pearlite free (PF) steels (<5% pearlite). The steels are more hardenable than carbon steels. There are four main application areas: offshore drilling and production platforms, high pressure linepipe, structural steel for bridges and buildings, and sheet metal for car and truck bodies and frames. Microalloyed steels improve strength, toughness and weldability by reducing carbon levels and replacing the strengthening effects of carbon by a combination of alloy elements which provide solid solution and precipitation strengthening. Controlled rolling procedures in the temperature range for metastable austenite followed by accelerated cooling on thin materials produce a high dislocation density and a fine grain size. Normalizing and tempering heat treatments are used on thicker material. The result in both cases is strong and tough alloys. The steels often come with labels describing the treatment applied, for example CROLAC, an acronym for Control Rolled On Line Accelerated Cooling. These alloys have strength defined in terms of yield strength, rather than the more conventional ultimate strength. The usual terminology is SMYS, for Specified Minimum Yield Strength. The range of SMYS is 300-700 MPa (44-100 ksi).

Microalloyed steels often have controlled levels of impurity elements to minimize inclusion content and to alter the shape of existing inclusions. Low sulphur and phosphorous contents are provided by modern steelmaking processes, as are controlled oxygen and nitrogen levels. These actions improve the toughness, limit susceptibility to Sulphide Stress Cracking and influence austenite transformation characteristics in the HAZ and the weld metal. Low inclusion populations decrease nucleation sites for phase transformation, increasing effective hardenability.

A measure of the effect of desulphurizing on inclusion shape is:

$$ESSP = \frac{\%Ca(1-\%O)}{1.25\%S} > 1.0$$

where ESSP = Effective Sulphide Shape control Parameter.

With a low oxygen content, a high calcium to sulphur ratio ensures the formation of spherical, rather than elongated, sulphides. This improves through-thickness ductility and improves toughness.

General data on these alloys is given in *The Metals Black Book* - Ferrous Metals. Related AWS Filler Metal Specifications are A5.5, 5.23, 5.28 and 5.29.

Heat Affected Zone

The HAZ in microalloyed steels has two areas of concern, Hydrogen Assisted Cracking susceptibility and an adequate strength/toughness balance. Some older steels have a problem with lamellar tearing in thick sections, but any steels with sulphide shape control and low levels of residual elements should not cause concern. There are several differences in the behaviour of microalloyed HAZs in comparison with carbon steels. Since hardenability is higher due to the alloy content, too rapid a cooling rate will cause martensite/lower bainite to form, reducing toughness. On the other hand, too slow a cooling rate will lead to coarse, high transformation temperature products which have low strength and toughness. High peak temperatures near the fusion line increase austenitic grain size, leading to increased hardenability. The microalloying elements produce precipitates which pin austenite grain boundaries, restricting grain growth. Some are more effective than others, the effect depending on precipitate dissolution temperature. AlN is relatively ineffective, V(C,N) is better, Cb(C,N) resists dissolution until about 1050°C (1920°F), TiN is even better, and TiO_2 is the best of all. TiO_2 precipitates are now included in some microalloyed steels specifically to minimize austenitic grain growth in the welding thermal cycle. Since all precipitates will dissolve at very high temperatures, some grain growth is inevitable, but the maximum grain size varies according to which grain refining precipitates are present.

Therefore, control needs to be applied to both the upper and lower limits of cooling rate for a given steel. There is thus a *window* of heat input/preheat combinations for any given steel which will produce a good strength/toughness balance. Inclusions also influence austenite transformation behaviour, so determining the window is not a straightforward task based on alloy content alone. The most common assessment is a measure of the time (in seconds) to cool from 800 to 500°C, known as Dt_{8-5}. The maximum and minimum values are used to set the welding procedure parameters.

The precipitates used to restrict grain growth ultimately dissolve as the peak temperature in the HAZ approaches the fusion temperature. The dissolved elements often stay in solution due to the rapidity of cooling after welding. This increases effective alloy level in the austenite, increasing hardenability, and also may lead to precipitation hardening if a PWHT is used. The resulting precipitation and tempering of any martensite can have different results in different steels. Some improve in toughness after PWHT, some deteriorate. Typical temperatures for PWHT are in the range 575-625°C (1065-1155°F). Reheat cracking is also a possibility in these steels during PWHT, but is not common.

The general approach to controlling HAC with the IIW nomogram applies to microalloyed steels. However, there are some complication factors. The carbon equivalent is not adequately determined by the IIW equation, which tends to overestimate the hardenability and recommend higher than necessary heat inputs/preheats. Many alternatives exist, of which several are listed for illustration:

Carbon Equivalent for microalloyed steels.

1. Stout - pipeline steels

$$CE = \%C + \frac{\%Mn}{6} + \frac{\%Cr}{10} + \frac{\%Mo}{10} + \frac{\%Cu}{40} + \frac{\%Ni}{20}$$

2. Bersch & Koch - pipeline steel

$$CE = \%C + \frac{\%Mn + \%Si + \%Cr + \%Mo + \%V + \%Cu + \%Ni}{20}$$

3. Ito-Bessyo - for linepipe steels with carbon levels <0.15%

$$P_{cm} = \%C + \frac{\%Mn}{20} + \frac{\%Si}{30} + \frac{\%Cr}{20} + \frac{\%Mo}{15} + \frac{\%V}{10} + \frac{\%Cu}{20} + \frac{\%Ni}{60} + 5\%B$$

4. Yurioka CEN - combines IIW and P_{cm} formats

$$CEN = \%C + A(C) \times \{\frac{\%Mn}{6} + \frac{\%Si}{24} + \frac{\%Cr + \%Mo + \%V}{5} + \frac{\%Cu}{15} + \frac{\%Ni}{20} + \frac{\%Cb}{5} + 5\%B\}$$

where $A(C) = 0.75 + 0.25\{tanh[20(\%C - 0.12)]\}$.

It can be seen from these formulae that the alloying elements have a lesser effect on carbon equivalent in microalloyed steels, since the divisor for most alloying elements is less than the value used in the IIW equation.

The Ito-Bessyo equation is the most popular for pipeline steels. The Yurioka equation is intended to combine the IIW and Ito-Bessyo equations to allow for the use of one equation for all carbon and microalloyed steels.

Microalloyed steels, although not prone to HAC, are somewhat more susceptible to HAC for a given hardness level than are C-Mn steels. This shows up as a problem with cracking at well below the 350 Hv10 level used as a minimum limit for C-Mn steels. Cracking can occur at hardness levels of 250 Hv10 or even less. The reasons for this behaviour are not simple. Residual stresses from cooling are higher, due to the higher yield stresses in both base and weld metals. Low sulphur levels have been connected to an increase in HAC and the reason initially proposed was that the increased number of inclusions as sulphur levels rose acted as hydrogen traps, reducing the risk of HAC. Some investigators advocated resulphurizing the steels, which would have compromised the gains in fracture toughness. Later it was shown that the reduced numbers of inclusions in low residual steels reduced the number of nucleation sites available to initiate austenite transformation during the cooling regime of the welding thermal cycle. As a result, low temperature transformation products were more likely in low residual steels, leading to martensite/bainite structures known to be more prone to HAC. The solution is procedure control based on slowing down the cooling rate. There is still the question of cracking at low hardness levels to address.   While this problem is not entirely solved, it is possible that microstructures prone to HAC cannot be defined solely by hardness, as they can in C-Mn steels.   HAC is more likely in a distorted ferrite lattice (bainite/martensite) and although low carbon levels in microalloyed steels reduce maximum hardness, the increased hardenability may cause a shear component to be present in the austenite transformation products, which is susceptible to HAC regardless of actual hardness level. The only solution is to define, for each steel, a window of cooling rates which will address the problem of HAC in addition to the problem of strength/toughness balance. The best range of cooling rates is the one which satisfies both.

Weld Metal

The weld metal of microalloyed steels has a more complicated microstructure than C-Mn steels. The columnar dendrites formed during solidification may be mostly ferrite to start with, since the carbon content is low, but this becomes austenite as cooling continues. When the austenite transforms at lower temperatures, there are several possible outcomes:

1. If the cooling rate is slow and the transformation temperature high, grain boundary ferrite nucleates at the dendritic boundaries and polygonal ferrite and a coarse carbide phase form within the dendrites.
2. At higher cooling rates, acicular ferrite dominates the grain interiors and less grain boundary ferrite forms.
3. Microsegregation during solidification may result in the formation of aligned martensite - austenite-carbide (MAC) phases at ferrite grain

boundaries, leading to paths of low fracture toughness. The segregates are combinations of microalloy elements and residual elements, especially carbon and nitrogen. The specific composition at a segregated site will determine whether a particle is martensite (low concentration of residuals and alloy), austenite (high alloy and residual level to form stable austenite) or carbide (high carbon segregation).

4. At high cooling rates, martensite or bainite may form within dendritic grains, mixed in with ferrite.

The best microstructure is nearly all acicular ferrite, with as little grain boundary ferrite as possible, to maintain an adequate strength/toughness ratio, and a minimum of MAC phase. To achieve this, the alloying used in consumables, especially for the SAW process, often differs significantly from the base metal alloying. For strength in the lower end of the range and applications at ambient temperature, simple Mn-Si electrodes are used. As strength increases and/or application temperatures decrease, electrodes using Mo, Ti and B are used. These require strict adherence to using particular fluxes and particular procedural conditions, best obtained from the supplier. They are often not suited to multipass welding. In the most arduous conditions, especially at low temperatures, Ni is used to improve toughness, usually at the level of 1-2%. For very high strength steels for pipelines (SMYS > 550 MPa or 80 ksi), up to 5%Ni is used. There are some specifications for sour gas and oil service which restrict Ni to < 1%, so these consumables are not always an option.

The influence of the SMYS of the steel on electrode choice must be kept in mind. With the SMAW process, the usual way is to select an electrode with the next higher level of ultimate tensile strength. Thus a pipeline steel with an SMYS of 480 MPa (70 ksi) would use an electrode of a 550 MPa (80 ksi) ultimate tensile strength, e.g. an E80xx electrode. This is especially important since the root pass of many pipeline fabrication welds uses a low strength deposit E6010 to minimize cracking.

Hot cracking is not a general problem with microalloyed steels, primarily because of the low residual element levels. If a concern exists, the base metal can be assessed with the following equation:

Weld Crack Susceptibility (WCS) for microalloy steels:

$$WCS = 230\%C + 190\%S + 75\%P + 45\%Cb - 12.3\%Si - 5.4\%Mn - 1$$

A level of 10 to 20 is good, > 20 is likely to lead to hot cracking.

Chapter

5

ALLOY STEELS

Alloy steels contain deliberate additions of one or more elements to modify the corrosion resistance, creep resistance, wear resistance, strength and/or toughness. Low alloy steels contain less than 5-6 wt% total alloy additions (again not including manganese and silicon), usually of chromium, nickel, molybdenum and vanadium. Some information states an upper level of 6%, but many modern alloys exceed that level and are still included in the low alloy classification. Examples are the high end of the Cr-Mo range (9%Cr-1%Mo) and the cryogenic Ni steel range (9%Ni). Alloy steels usually contain < 10% total alloy element content, not including C and Mn, but some exceed this level, for example, the austenitic manganese Hadfield steel, which contains 13%Mn, or the maraging steels, which contain 18%Ni plus deliberate additions of Mo, Ti and Cb. General metallurgical information on these alloys is given in *The Metals Black Book* - Ferrous Metals. Related AWS Filler Metal Specifications are A5.2, 5.5, 5.23 and 5.25, 5.26, 5.28 and 5.29.

The approach taken in this chapter will be to discuss the welding metallurgy of a few alloy steel systems, since it is not possible to give details for all of the alloy steels which are welded. The systems chosen will illustrate the fundamental aspects of welding metallurgy which apply in total or in part for any alloy steel.

Alloy steels are more hardenable than C-Mn and microalloyed steels. The main concerns in the HAZ are toughness, which can be reduced by a combination of large grain size and martensitic transformations, and HAC. Many alloy steels are in Zone III of the Graville Diagram, and as a result, need Procedure Control and PWHT to avoid HAC and optimize the strength/toughness ratio. In many cases, especially if carbon content is high, it is advisable to maintain the preheat until PWHT can be carried out, so that martensite is not formed in the weld area before PWHT.

Weld metals have their major problems with strength/toughness balance, HAC and, in some cases, hot cracking. As strength level increases, toughness tends to decrease. There are three ways to improve fracture toughness at a given strength level: reduce grain size, reduce inclusion population, and add nickel.

Average grain size is difficult to reduce in cast materials without using nucleants of some kind, and these are often inclusions. Therefore, if enough suitable inclusions are available to refine as-cast microstructures, the weld metal inclusion population is sufficient to decrease toughness. One method used to overcome this conundrum is *Temper Bead Welding*, which is a procedural device to maximize the amount of weld metal and prior HAZ which is grain refined by the heat input of the following passes in a multipass welding procedure. The procedure requires strict control of the heat input and size of each layer of weld (which may be one or more individual passes). The aim is to grain refine at least 80% of the previously deposited weld metal. The last pass is placed at ½ a pass width inside the fusion line to grain refine as much as possible of the final layer. This technique is used on repair welding of steam piping in power stations, where no PWHT is possible.

To reduce the inclusion population, welding processes involving non-inert gases and most fluxes cannot be used on high strength steels. Oxidation or reactions between weld metal deposits and gases or slags causes inclusion populations to rise. The processes of choice for the highest strength alloy steels are therefore GTAW and GMAW, with some use of EBW and LBW, with either vacuum or inert gas shielding.

HAC in weld metals is more of a problem in alloy steels than it is in C-Mn steels, especially High Strength Low Alloy (HSLA) steels. One reason for this is the higher susceptibility of as-cast microstructures, with large grain sizes and inherent alloy element segregation. Another is the failure of electrode compositions to keep pace with advances in base metal chemistry and the nature of producing the alloy steel weld metal deposits. The best situation exists for GTAW and GMAW processes, with some advances being made in FCAW with metal powder cores. These processes use electrode wires of similar alloy composition and cleanliness as the base metal, and do not oxidize or contaminate the weld metal with inclusions. Processes such as SMAW often add alloying elements as ferroalloys in the flux, using core wires made of plain carbon steels. Microsegregations and impurity elements are not as well controlled, which leads to the weld metal in some of these steels being more prone to HAC than is the HAZ, the opposite of the situation with C-Mn steels.

Nickel additions are made to many low alloy welding electrodes to improve toughness. Up to 5% is used in HSLA grades, and cryogenic steels have electrodes containing up to 12%. The exact reason for the improvement of ferritic (martensitic) alloy toughness with Ni used as a substitutional alloying element is not known, but it is exploited in a variety of procedures. In cryogenic steels, where the Ni segregates into pockets of austenite, the reason is understood.

For low alloy steels, a revised Hot Cracking Susceptibility equation is available:

$$HCS = \frac{\%C\left(\%S + \%P + \dfrac{\%Si}{25} + \dfrac{\%Ni}{100}\right)}{3\%Mn + \%Cr + 2(\%Mo + \%V)} \times 10^3$$

Note the increased problem with Ni and the beneficial effect of Mo and V compared to the similar equation for C-Mn steels in Chapter 4. Ni is an austenite stabilizer and promotes the segregation of S, and Mo and V tend to tie up carbon in carbides, removing it from solution, where it also stabilizes austenite.

CHROMIUM-MOLYBDENUM STEELS

Chromium-Molybdenum (Cr-Mo) steels are used in steam piping and gas and oil processing. At elevated temperatures in service, Cr provides oxidation resistance and Mo provides secondary hardening for creep resistance. Grades range from 0.5%Cr-0.5%Mo to 9%Cr-1%Mo. They are hardenable under quite slow cooling rates and are sensitive to HAC in both the weld metal and HAZ. Preheats recommended are therefore fairly high, see Table 5.1.

Table 5.1 Cr-Mo Steels - Preheat and Postheat Conditions

Pipe ASTM Gr	Nominal Composition	Preheat °C 13 mm max	Preheat °C 60 mm max	Preheat °C Over 60 mm	Postheat[a] °C	Anneal[b] °C
A335-P2	½Cr-½Mo	20	100	150	590-700	850-910
A335-P12	1Cr-½Mo	125	150	150	650-730	850-910
A335-P11	1¼Cr-½Mo	125	150	150	675-750	850-910
A369-FP3b	2Cr-½Mo	150	150	150	675-760	850-910
A335-P22	2¼Cr-1Mo	150	150	150	675-760	850-910
A335-P21	3Cr-1Mo	150	150	150	675-760	850-910
A335-P5	5Cr-1Mo	150	150	150	700-760	850-910
A335-P5b	5Cr-½MoSi	150	150	150	700-760	850-910
A335-P5c	5Cr-½MoTi	150	150	150	700-760	850-910
A335-P7	7Cr-½Mo	200	200	200	720-760	850-910
A335-P9	9Cr-1Mo	200	200	200	720-760	850-910

a. Postheat 1 hour per 25 mm (1 in.) thickness, air cool.
b. Anneal 1 hour per 25 mm (1 in.) thickness, cool at maximum rate of 30°C (85°F) per hour to 540°C (1000°F), then air cool.

The range of PWHT temperatures for each grade allows for choice in the final strength/toughness balance. At the low end, strength is emphasized, while at the high end of the range, toughness is high and strength is low.

Weld metals should have similar chemistry to the base metal being welded. If differing compositions are welded, use the higher alloy content as a guide. If many Cr-Mo alloys are welded, some restriction on the number of filler metal compositions specified is useful to minimize storage problems and avoid mistakes. A good combination for all grades is 1.25%Cr-0.5%Mo for all grades up to and including itself, 2.25%Cr-1%Mo for up to 3%Cr-1%Mo, 5%Cr-1%Mo for itself, and 9%Cr-1%Mo for the rest. Temper bead techniques are useful to optimize weld metal properties if PWHT is impossible. Occasionally austenitic stainless steel electrodes are used for repair welding without PWHT, using E309 or E310 electrodes. E310 is less preferable given the choice, and in some circumstances, repairs with E309 should be regarded as a temporary expedient rather than a true repair. While oxidation resistance of the austenitic weld metal is satisfactory, its thermal fatigue and creep resistance is dissimilar, and premature failure may occur.

Temper embrittlement is possible in Cr-Mo steels and even more likely in Cr-Mo-V and Ni-Cr-Mo-V steels used in steam piping and other creep resisting applications. The following empirical equations can be used to assess base metals and weld metal compositions (*not* nominal electrode composition) for temper embrittlement tendency:

1. Watanabe "J" Factor (Base Plate HAZ)
    $J = (\%Mn + \%Si)(\%P + \%Sn)$
    For Cr-Mo steels, J > 100, 10^{-4} embrittles
    For Ni-Cr-Mo-V steels, J > 10, 10^{-4} embrittles

2. Brascato Factor (Weld Metals)
    $X = 10\%P + 5\%Sb + 4\%Sn + \%As$
    X > 5, 10^{-2} embrittles if Mn > 0.10%

3. Kohno Combined Factor
    $K = (\%Mn + \%Si), X$
    Limit indeterminate

Cr-Mo and particularly Cr-Mo-V alloys are susceptible to *Reheat Cracking*, especially in service. The following empirical equations have been developed to aid in determining if a metal is susceptible to reheat cracking:

1. Nakamura
    $P = \%Cr + 3.3\%Mo + 8.1\%V - 2$
    Cracking is possible if P > 0
2. Ito
    $P = \%Cr + \%Cu + 2\%Mo + 5\%Ti + 7\%Cb + 10\%V - 2$
    Cracking is possible if P > 0

3. Bonizewski - "Metal Composition Factor" for 2.25%Cr - 1%Mo steels
    $MCF = \%Si + 2\%Cu + 2\%P + 10\%As + 15\%Sn + 20\%S$

Note that some similar elements appear in both these equations and those for temper embrittlement. Temper embrittlement and reheat cracking both involve a weakening of grain boundaries relative to grain interior strength.

CRYOGENIC NICKEL STEELS

Cryogenic nickel steels are used to store liquified gases of many kinds in large tanks, with applications down to the temperature of liquid nitrogen, -196°C (-319°F). Nickel levels vary from 2.25% to 9%. The main competitors are austenitic stainless steels, which are more expensive for large fabrications, and aluminum alloys, which are difficult to find in thick sections for large fabrications.

Table 5.2
Liquefaction Temperatures of Gases and Recommended Grades of Cryogenic Steels

Gas	Liquefaction temperature, °C
Carbon steel	
Ammonia	-33.4
Fine grain Al-killed steel	
Propane (LPG)	-42.1 to-45.5
2.25%Ni steel	
Propylene	-47.7
Carbon disulphide	-50.2
3.5%Ni steel	
Hydrogen sulphide	-59.5
Carbon dioxide	-78.5
Acetylene	-84
Ethane	-88.4
5-9%Ni steel	
Ethylene (LEG)	-103.8
Krypton	-151
Methane (LNG)	-163
Oxygen	-182.9
Argon	-185.9
Fluorine	-188.1
Austenitic stainless steel	
Nitrogen	-195.8
Neon	-246.1
Al alloys	
Heavy hydrogen	-249.6
12%Ni fine grain steel	
Hydrogen	-252.8
Helium	-268.9
Absolute zero	-273.18

The high Ni levels in most of these steels lowers the ferrite-austenite transformation and martensite start temperatures remarkably. These steels are hardenable and are used in the normalized (quenched) and tempered condition. Reheating to moderate temperatures, especially in the higher Ni steels, results in the formation of reverted austenite, to which the austenite stabilizers C and Ni congregate. Since C and Ni levels in this austenite are high, it may be thermodynamically stable down to very low temperatures. If it is also mechanically stable (does not form martensite when stressed), and it occupies about 10-20% of the matrix, reverted austenite improves toughness by two effects: C is removed from the martensitic matrix, and the stable austenite acts as a crack blunting phase.

In order to produce stable austenite, especially in the high Ni alloys, complicated heat treatments may be necessary. The 9%Ni grade easily forms stable austenite within the matrix. The 5%Ni grade is more difficult to stabilize due to the lower nickel content. A three step treatment may be necessary. Austenitizing at 850-920°C (1560-1690°F) to get homogeneous austenite is followed by quenching to under 150°C (300°F), which produces untempered martensite. 'Tempering' at 690-760°C (1270-1390°F), with another quench to under 150°C (300°F), is next. This tempers the martensite and forms some austenite. Then there is a *reversion anneal* at 620-660°C (1150-1220°F) followed by quenching. This is analogous to an intercritical treatment in a carbon steel, where both austenite and ferrite co-exist, and it produces a stable austenite of higher C and Ni content. Similar results are obtained in the 9%Ni grade by double normalizing and tempering at 900°C (1650°F), 800°C (1470°F) and 570°C (1060°F) respectively.

In the HAZ regions, HAC is a possibility, but not as common as might be expected from the alloy level. Following low hydrogen practices and preheats in the order of 100-200°C (210-390°F) is usually sufficient, even in the 9%Ni steel. Liquation cracking is a potential problem only if sulphur bearing compound, e.g. grease and oil, contaminate the weld zone. Even slow cooling rates will produce martensite/bainite in the HAZ. PWHT in the range 620-730°C (1130-1350°F) is common for the 2.25 and 3.5%Ni grades. The more elaborate heat treatments mentioned above for the 5%Ni steel complicate the HAZ metallurgy, because the optimization is lost during the welding thermal cycle. Heat treating again is the only sure cure. As a result, the 5%Ni steel is not as popular as the 9%Ni grade for the lowest temperature applications. In the 9%Ni grade, the residual austenite content decreases as the peak temperature in the HAZ exceeds 700°C (1290°F). Toughness decreases as a result up to peak temperature of 900°C (1650°F). When the peak temperature is high enough to austenitize the HAZ, further decreases in toughness occur at high heat inputs due to a coarsening of the fully austenitic grain size. An ASME Code case, #1308, permits the use of 9%Ni steel in the as-welded condition.

In similar composition weld metals, proeutectoid or grain boundary ferrite is very rare and the weld metal is virtually all acicular ferrite and martensite.

Heat inputs generally range from 2.0-4.5 kJ/mm (50-110 kJ/in.). The presence of nickel in nearly all cryogenic steels increases susceptibility to sulphur induced solidification cracking. Grain refined C-Mn steels are often welded with 1%Ni electrodes in the SMAW process, E55018-C3 (8018-C3) grade. The use of this electrode may cause a lot of pointless objection from some in the sour gas and oil patch, but evidence suggests that sulphide stress cracking (SSC) is not enhanced in 1%Ni weld metals in ferritic cryogenic steels. Ni bearing base metals are welded with matching electrode compositions with both gas and flux shielded processes up to the 3.5%Ni grade. The 5%Ni and especially the 9%Ni grades require more care, especially with flux shielded processes. To maintain a similar strength/toughness balance to the base metal, three methods are available:

1. Use matching composition filler wire with an inert gas shielded process.
2. Use high nickel SMAW electrodes (about 80%Ni-20%Cr).
3. With the SAW process, the 9% Ni grade requires basic flux and ERNiMo-7 (Alloy B-2) filler metal (nominally 65%Ni-28%Mo). Austenitic fillers are occasionally recommended (E309 and E310), but these achieve adequate toughness at the expense of matching strength.

HSLA AND MEDIUM CARBON ALLOY STEELS

A common low alloy structural steel is ASTM A514, a 0.2%C steel with small (< 1%) additions of Cr and Mo. Crane booms and bridge girders are common applications of this and similar steels. Low hydrogen procedures and PWHT (Graville Diagram Zone III) are required for most welding situations. PWHT is not common, especially in repair situations. Preheat/heat input combinations, and usually minimum/maximum interpass temperatures, must be devised to enter a 'window' of cooling rates, as was necessary with microalloyed steels, to optimize the strength/toughness balance.

Medium carbon low alloy steels, such as AISI 4340, are often welded for use in a variety of industries, including oil and gas exploration and refining. Early developed grades of stronger steels, for example AISI 4340, had < 2% of each alloy element to provide hardenability, and relatively high carbon levels (0.40%) to provide strength. These steels need PWHT to develop optimum properties. Weld metals can have problems with HAC and toughness if of similar compostion to the base metal. It is possible to weld these steels with high Ni alloys to avoid these problems and still retain most of the strength. The best choices are ENiCrMo-3.

Some HSLA steels were originally developed for military applications, such as submarine hulls, where strength and toughness must both be high. They are usually supplied in a quenched and tempered (QT) condition. The most common alloying elements are Cr, Mo and Ni, with minor additions of Cu, V, etc. Later developments lowered C levels and increased Ni levels to improve toughness, for example the development of HY steels. HY80 has about 0.2%C, 1.6%Cr, 3%Ni and 0.4%Mo. There is a progression through

HY 100, 130, and 150 to the stronger HY 180, which has about 0.10%C, 2%Cr, 10%Ni and 1%Mo. Gas and flux shielded processes can be used with matching strength electrodes for the lower strength materials, but cleaner low hydrogen deposits are necessary in the higher strength materials, requiring the use of inert gas shielded processes, with cooling rate controls on heat input, as strength rises above 700 MPa (100 ksi). Structural HSLA steels, e.g. ASTM A710, have quite low carbon levels (< 0.10%) and obtain strength and toughness through additions of about 1% each of Ni and Cu, which cause precipitation hardening. Appropriate strength HY consumables are used to weld ASTM A710 type steels.

SPECIALTY ALLOY STEELS

Specialty alloy steels, with more than 10% alloy content, are not often welded. They are often fully austenitic, or martensitic with very low carbon levels, so that HAC is not a problem in most cases. One type that is welded occasionally is Hadfield steel, containing 13%Mn and about 1%C. It is therefore austenitic at ambient temperature. To weld it, the one overriding consideration is the avoidance of any carbide precipitation, because the Mn carbides appear at austenite grain boundaries and severely impair toughness. No preheat can be used, and as low a heat input as possible is necessary. The 18% Ni maraging steels, used for landing gear struts, are seldom seen outside the military and aerospace industries. Strength levels in these Ni martensitic steels are in excess of 1400 MPa (200 ksi). Since the Ni martensite has virtually no carbon in it, HAC is not a problem provided that low hydrogen procedures with the GTAW or EBW processes are used (inert gas or vacuum shielding).

Chapter

6

STAINLESS STEELS

All stainless steels contain 11.5% or more Cr and often contain significant amounts of other alloying elements such as Ni and Mo. Modern 'super' stainless steels have many other alloying elements added for specific properties and may contain less than 50%Fe. General metallurgical information on these alloys is given *The Metals Black Book* - Ferrous Metals. Related AWS Filler Metal Specifications are A5.4, 5.9 and 5.22.

Stainless steels earned their name from steels evolved from early 20th century experiments on alloying, carried out mainly in England. It was observed that high Cr steels did not rust even when exposed to rain, and they were dubbed *stainless*. The early alloys contained levels of carbon sufficient to make martensitic transformation on cooling inevitable, so the early experimental steels were very brittle and essentially useless. When steels with more controlled carbon levels were made, a new family of steels was born. Although stainless due to the continuous layer of Cr oxide on the surface, these steels are not corrosion proof and can fail catastrophically in certain environments. Welding tends to make the situation worse in some instances. The alloys are also oxidation resistant at elevated temperatures. The welding metallurgy of these steels is discussed below starting with the simplest alloys, the original martensitic grades. This is followed by the increasingly more complicated grades; ferritic, austenitic, duplex, precipitation hardening and 'super' stainless steels. Since there are so many types of stainless steel, they are used for an enormous range of applications, including corrosion resistance, creep resistance, anti-magnetic applications, cryogenic applications and high strength applications. The alloying elements in stainless steels are classified as austenite stabilizers (C, Ni, Mn and N, but occasionally Cu and Co) or ferrite stabilizers (Cr, Mo, Si, Cb and many others). The balance between these competing austenite and ferrite stabilizing elements determines the microstructure of a given stainless steel.

Martensitic stainless steels contain between 11.5 and 18%Cr and from 0.12 to 1.2%C. Particular grades have some other small alloy additions, e.g. Ni and Mo. They are usually in a hardened and tempered condition, but can be supplied annealed.The most common grade is type 410, with minimum Cr and up to 0.15%C. Cutlery grades have up to 1.2%C, e.g. Type 440, but these are seldom welded. Martensitic grades have the lowest corrosion resistance among the various types of stainless steels, but are very hardenable and quite strong, with ultimate tensile strength in the order of 500-1800 MPa (72-275 ksi), depending on tempering temperature.

Ferritic stainless steels differ from martensitic grades primarily in having a lower %C, and in most cases a higher %Cr, up to nearly 30% in some cases. Enough Cr is added to restrict the austenite phase field (γ loop) to ensure that materials with more than 12%Cr do not have an allotropic phase transformation, and stay ferritic at all temperatures, see Figure 6.1.

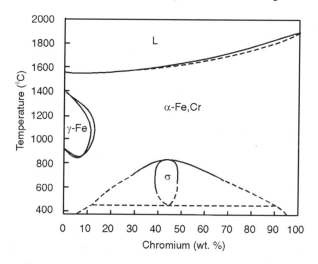

Figure 6.1 Fe-Cr phase diagram showing gamma (γ) loop

The Kaltenhauser "Ferrite Factor" describes the chemistry effect of alloying elements on the tendency to form ferrite in non-austenitic stainless steels.

$$FF = \%Cr + 2\%Al + 4\%Cb + 4\%Mo + 6\%Si + 8\%Ti - 2\%Mn - 4\%Ni - 40\%(C+N)$$

Fully ferritic structures are obtained in the following circumstances:

12%Cr steels (e.g. Types 405, 409) - FF > 13.5

17%Cr steels (e.g. Types 430, 434) - FF > 17.0

Stabilized 17%Cr steels - FF > 13.5

Since no phase change occurs, grain refinement of ferritic stainless steels is not possible during heat treatments, and grain growth may be uncontrolled at high temperatures, leading to low toughness via the formation of large 'elephant' grains. The standard alloy is Type 405, 13%Cr and up to 0.08%C, used in automobile exhaust systems. Any carbides present can create pools of austenite on grain boundaries at high temperature, so modern versions of these steels have low C + N levels (< 0.03% total) and carbide stabilizers to tie up the rest (Ti or Cb). Type 409 is 405 with added Ti. An example of the higher Cr grades is Type 430, with 18%Cr. The highest Cr level, 27%, appears in Type 446. A popular new grade is UNS S44400 (18-2), containing 18%Cr, 2%Mo, 0.02%C max and 0.4%Ti.

Ferritic steels can suffer from 475°C (885°F) embrittlement, a precipitation of a Cr rich α' phase on dislocations after exposure to temperatures between 400-540°C (750-1000°F) for long times (many tens of hours minimum), σ phase embrittlement (a Cr-Fe intermetallic) when exposed to temperatures in the range 500-800°C (930-1470°F) for long times (tens of hours minimum) and high temperature embrittlement, which can be carbide or martensite formation in grain boundaries at temperatures in the 950-1000°C (1740-1830°F) range. Small amounts of martensite in the weld metal/HAZ of these steels reduces toughness and corrosion resistance.

Ferrite has very low solubility for C and N and will reject both elements on cooling, unless other elements are present, e.g. Ti, which will form C,N compounds. General corrosion resistance of ferritic stainless steels is better than martensitic grades due to their higher average Cr content, and they are virtually immune to the chloride-induced   stress cracking that plagues austenitic grades.

Austenitic stainless steels are austenite at ambient temperatures because their Ni content (at least 8%) and Mn level is sufficient to stabilize austenite at ambient temperature without a transformation to martensite in normal circumstances. They contain 16-25%Cr and 7-20%Ni. The generic austenitic alloy is Type 302, 18%Cr, 9%Ni and up to 0.15%C. A more common grade is a lower carbon version, Type 304, 19%Cr, 9%Ni and 0.08%C max. Most heats of 304 have 0.06%C or less. Type 304L is the low carbon version, with 0.03%C or less. Carbon content is important in the control of *sensitization*, the formation of chromium carbides on grain boundaries, which reduces local corrosion resistance. Types 316, 317 and 318 contain Mo for pitting corrosion resistance, and stabilized types (to prevent sensitization) are 321 (Ti) and 347 (Cb).

In austenitic stainless steels, the M_s is well below room temperature in most cases. However, depending on the extent of solution of carbonitrides in particular, and on mechanical deformation if present, martensite can be formed. The M_s (°C) can be estimated by:

$$M_s = 502 - 810\%C - 1230\%N - 13\%Mn - 30\%Ni - 12\%Cr - 54\%Cu - 46\%Mo$$

The effect of strain can be estimated by:

$$M_{d30} = 497 - 462(\%C+\%N) - 9.2\%Si - 8.1\%Mn - 13.7\%Cr - 20\%Ni - 18.5\%Mo$$

where M_{d30} is the temperature (°C) at which 50% martensite is formed at a true strain of 0.30.

The balance between austenite and ferrite stabilizers to produce fully or partially austenitic structures was first determined by Schaeffler, see Figure 6.2 below.

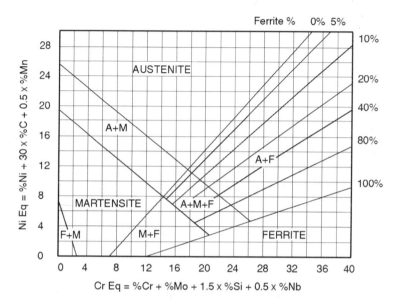

Figure 6.2 Schaeffler diagram

The Chromium and Nickel *equivalents* used to determine ferrite, martensite and austenite contents in the Schaeffler Diagram were very simple:

Ni_{eq} = %Ni + 30%C + 0.5%Mn
Cr_{eq} = %Cr + %Mo + 1.5%Si + 0.5%Cb

The diagram shows the general composition areas for each major phase or phases, and the regions is which minor phases, e.g. σ, can form in suitable conditions. *Nickel Balance* ('Ni') is an empirical parameter to assess sensitivity to σ phase formation in austenitic stainless steels in high temperature service. It is to some extent an assessment of ferrite stability, since σ and χ intermetallic phases usually form in ferrite.

'Ni' = 11.5 + 30(%C+%N) + 0.5%Mn + %Ni
    - 1.36(1.5%Si+%Cr+%Mo+0.5%Cb)

If 'Ni' is > 2.5, σ formation is unlikely. If 'Ni' > 5, it is very unlikely.

The general corrosion resistance of austenitic alloys is very good, but there are a few weaknesses.   Carbide precipitation at grain boundaries (sensitization) can occur at temperatures in the 500-800°C (930-1470°F) range. These carbides are chromium carbides, $Cr_{23}C_6$, and the local area is depleted in Cr to below 12%, which allows corrosion to occur in oxidizing acids. Low carbon grades are less prone to sensitization, as are stabilized grades. Chloride stress cracking is a problem in any stressed stainless alloy with Ni contents between 5% and 40%. When chlorides are present in very acidic conditions (low pH), pitting corrosion is enhanced. 'PRE' is Pitting Resistance Equivalent, used to assess corrosion resistance due to acid chloride attack, usually in $FeCl_3$. Mo is usually added to resist pitting, for example in Types 316, 317 and 318.

$$PRE = \%Cr + 3.3\%Mo + 16\%N$$

The PRE number should be at least 35 for good pitting resistance.

Duplex stainless steels, originally developed to replace Type 316 austenitic steel, are mixtures of austenite and ferrite, usually 50% of each. This is achieved by increasing %Cr and lowering %Ni to balance the phase content at ambient temperature. A typical composition is 25%Cr and 5%Ni. Other alloy additions also occur, e.g. Mo for protection against pitting corrosion. One early grade was UNS S31500 (3RE60), used for heat exchanger tube bundles for improved strength and higher pitting resistance. These alloys are very resistant to chloride stress cracking, as are the ferritic grades, but they are also stronger than the austenitic grades and tougher and more formable than the ferritic grades. Control of the phase balance can also be exerted by deliberate additions of N. A typical modern example is UNS S31803 (2205), containing 22%Cr , 5%Ni and 3%Mo.

'P' value is used to assess the austenite/ferrite balance in duplex stainless steels, where normally a 50-50 balance is the optimum. This ferrite forming expression is given by Gooch as:

$$P = \left\{ \frac{\%Cr + \%Mo + 3\%Si + 7\%Ti + 12\%Al}{0.7\%Mn + \%Ni + 26\%N + 30\%C} \right\}$$

Age hardening stainless steels may be austenitic, martensitic or semi-austenitic, and are typically stainless grades with addition of Ti, Cb and Al to cause precipitates of the Ni_3X variety, similar to those found in Ni alloys. X is Ti, Cb or Al, although one grade uses phosphorous as a precipitation agent (17-10P). Ultimate strength levels after age hardening are very high, up to 1500 MPa (200 ksi) in many cases.

Super stainless steels are relatively new and received the name because of their high alloy content and subsequent high corrosion resistance. The extra alloy elements used, in addition to Cr and Ni, are higher levels of Mo, Cu and Cb and N, among others. These steels are usually austenitic or duplex. Typical examples of the austenitic type are UNS S31254 (254SMO) and

UNS S08904 (904L), and a typical example of the duplex variety is 2708, with 27%Cr, 8%Ni, 3%Mo, 2.5%Cu and 0.25%N.

WELDING METALLURGY

The most common processes used to weld stainless steels are GTAW, GMAW, SMAW and SAW. SMAW electrodes have different flux coatings than those appearing on electrodes for C-Mn and alloy steels. Their AWS designation reflects this difference. EXXX-15 electrodes, where XXX is the three digit designation of the stainless (e.g. 316), have a calcium oxide-calcium fluoride based flux, which is all-positional (fast freezing slag) and uses only DC electrode positive welding current. The deposit has low inclusion content, but has a rough surface. EXXX-16 fluxes are titanium oxide based (rutile, sometimes called *acid*), use AC or DC electrode positive current, and are not for all-position welding if greater than 4 mm (5/32 in.). The deposited bead is smoother, but less crack-resistant. EXXX-17 electrodes, developed from the EXXX-16 type, have a higher silicon content. The deposit is quite smooth but needs more residual ferrite in the weld to avoid hot cracking than does EXXX-16 weld metal.

HEAT AFFECTED ZONES

Martensitic Stainless Steels

The overriding problem in the HAZ of martensitic stainless steels is HAC due to the hardenability of the alloy. Low hydrogen procedures must be used, and preheating and PWHT is needed in most cases. Cleaning via degreasing and scratch brushing with stainless steel brushes is mandatory. Type 410 can be welded without preheat or PWHT when thin sections (< 5 mm, 0.2 in.) are joined, for example in distillation trays for chemical plant. If carbon level exceeds 0.15%, preheat should be used, 200°C (390°F) up to 0.20%C and 300°C (570°F) thereafter. A subcritical anneal at 625-675°C (1150-1250°F) immediately after welding will suffice for most weld zones in martensitic grades.

Ferritic Stainless Steels

Provided that low hydrogen processes are used, ferritic grades are not prone to HAC. The HAZ in ferritic grades may experience grain growth and have 'elephant' grains which reduce toughness. No preheat and the lowest practical heat input is the best answer. If any carbides are present, either Cr or stabilized types (Ti, Cb), they may dissolve locally in the HAZ near the fusion line, forming austenite in grain boundaries which can transform to martensite on cooling. This phenomenon also reduces toughness. PWHT is the only answer to this, and the answer is in many cases inadequate. Use the lowest %C and %N steels available. Carbide dissolution may also lead to Cr carbides precipitation during cooling, causing sensitization. Again, the lowest carbon and nitrogen levels possible in the base metal are the best answer. Restricting welding to thin sections (< 7 mm or < 0.28 in.) also minimizes problems with grain growth and carbides. σ and α' precipitates are very unlikely to form during welding thermal cycles, but may be a problem during long exposures to suitable temperatures.

Austenitic Stainless Steels

Austenite is not sensitive to HAC and therefore no preheating or hydrogen control is necessary. Sensitization by the precipitation of chromium carbide is a major concern in welding austenitic grades. It is also known as *weld decay*. If corrosion resistance at ambient temperature is the only criterion, use low carbon grades, e.g Types 304L and 316L. If extreme conditions are not involved (hot oxidizing acids), most grades specifying 0.08%C maximum (e.g. 304) have < 0.06% in practice, which is low enough to avoid carbide precipitation in welding procedures using a reasonable heat input, say < 3 kJ/mm (< 75 kJ/in.). If higher temperatures are involved in service (above 500°C or 930°F), even low carbon grades are inadequate, and stabilized grades (Types 321, 347) must be used. These can suffer from a related problem known as *knifeline* corrosion, which is simply sensitization at and near the fusion line. It is caused by the stabilized carbides (Ti,Cb) dissolving at high temperatures, then having insufficient time to reform as the temperature drops during the cooling cycle. Cr carbides then form at 800-500°C (1470-930°F), using the free carbon available near the fusion line. If sensitization does occur, the complete fabrication must be solution annealed at 1000-1100°C (1830-2010°F) and quenched to remove the carbides, or stabilized at 870-900°C (1600-1650°F) for a few hours followed by air cooling to diffuse Cr back into the depleted areas. Both expedients are costly and may cause distortion, so the best cure is avoiding the problem in the first place by using low carbon/low heat input options.

If high temperature service alone is the criterion for design, strength is the controlling parameter. In this case low carbon austenitic grades, are generally inadequate above 500°C (930°F). There are several grades intended for high temperature service which have higher carbon contents. These are the H-grades, e.g. 304H.

Ductility dip cracking is occasionally observed in austenitic HAZs, but it is not common. More common is liquation cracking at or near the fusion line, caused by austenite grain growth concentrating sulphur, oxygen and/or eutectic carbides on the grain boundaries. Lower sulphur and oxygen levels in modern steels have improved the situation, but older steels often cause problems in repair welding. Low restraint and low heat input may help. Reheat cracking may be seen in Types 321 and 347 intended for creep resisting service, due to precipitation of Ti and Cb carbonitrides after dissolution at high peak temperatures. Type 316 is a viable alternative which is more resistant to reheat cracking, although more prone to σ formation in prolonged high temperature service.

Duplex Stainless Steels

In duplex stainless steels, the 50-50 austenite-ferrite balance is critical for the maintenance of corrosion resistance and mechanical properties. These steels are 100% ferrite at high temperature, and transform partially to austenite as temperature falls to ambient. The transformation is similar to the austenite to ferrite + carbide transformation at the A_1 - A_3 region on the iron-carbon phase diagram, except the phase fields are reversed. In the case of duplex steels, austenite is the low temperature phase. Since the transformation is sensitive to cooling rate, a slow cooling rate gives the desired 50-50 balance, while a faster cooling rate, typical of many welding procedures, suppresses the transformation somewhat, leaving less austenite than wanted. Very low heat inputs should be avoided. The ferrite - austenite balance may be 40-60 or even 35-65, which affects both corrosion resistance and mechanical properties. If cooling rates are excessively slow, however, intermetallics such as σ phase may form, embrittling the HAZ. Heat inputs should be < 2 kJ/mm (< 50 kJ/in.). Heat treatment to the fully ferritic range with a slow cooling rate afterward is the only way to restore the 50-50 balance. This procedure may be difficult on a finished fabrication. These steels do not suffer from the excessive grain growth seen in ferritic stainless steels, and are thus more weldable.

Age Hardening Stainless Steels

Age hardening stainless steels are best welded in the annealed or overaged condition and heat treated after fabrication to develop full properties. Most are austenitic before fabrication, and do not need preheat or low hydrogen precautions. The phosphorous bearing grades should not be welded, due to the risk of liquation cracking.

Super Stainless Steels

The super stainless steels, whether austenitic or duplex, suffer from the same problems as their lesser alloyed siblings. In addition, there is more of a problem with the formation of intermetallic phases of more kinds, which can lead to corrosion and/or brittleness. There is the familiar σ phase Cr-Fe intermetallic, but also χ, μ and a host of others. Heat inputs should be < 1.5 kJ/mm (38 kJ/in.). Cr_2N precipitation can also occur at higher heat inputs, which embrittles the steel.

WELD METALS

Martensitic Stainless Steels

Type 410 can be welded with a matching composition electrode, or a special 410NiMo grade with lower carbon that maximizes weld metal toughness. Occasionally austenitic electrodes, for example Types E308, E309 or E310, are used to join martensitic grades. The reasons for this are to use an austenitic structure in the weld metal to retain hydrogen in solution (avoiding HAC), letting the weaker weld metal absorb thermal strains on cooling (lower residual stress) and providing reasonable weld metal toughness without PWHT. However, the whole point of using the martensitic steel in the first place, high strength, is lost. Such welds should be regarded as 'glue' and not as providing mechanical properties similar to the base metal. The exception is free machining steel containing S or Se, which should be welded with great caution with electrode E312 (29%Cr-9%Ni).

Ferritic Stainless Steels

The ferritic stainless steels can be welded with the GTAW process with similar filler metal chemistry in some cases, e.g. ER409. This is especially true when the parent metal is thin, < 4 mm (< 0.16 in.) and heat input is held to less than 1.5 kJ/mm (38 kJ/in.). It is essential to protect to back of the root of the weld to avoid N absorption from the atmosphere, which will form austenite, then martensite, in the weld metal. For thicker materials, or when the SMAW process is used, it is necessary to use austenitic filler metal to ensure weld metal lack of brittleness and corrosion resistance equal to or better than the base metal. Electrodes recommended are E309 (23%Cr-12%Ni) and E310 (25%Cr-20%Ni). E309 gives a greater tolerance to dilution by the base metal and is also available in various versions which allow some fine tuning on weld metal composition, e.g., low carbon, Mo and Cb stabilized types exist.

Austenitic Stainless Steels

In general, *matching* composition electrodes are used to weld most austenitic alloys, although there are subtle differences in chemistry between base metals and weld metals. In some cases, a matching composition does not exist, or cannot be used due to weldability problems. There has been a great increase in knowledge concerning the weld metal formation in austenitic steels since the mid 1970's. The main problem has always been solidification cracking, primarily from sulphur segregation due to the low solubility in austenite. The solution to the problem has been known for some time, but there is now a solid theoretical understanding of the mechanism. Originally, it was observed that a few % of ferrite in the ambient temperature weld metal prevented solidification cracking. Reasons proposed for this effect were that the ferrite absorbed some of the sulphur because the solubility in ferrite is larger, and that the grain boundary area in the weld metal was increased by the presence of ferrite, thus spreading out any grain boundary films, minimizing cracking. These ideas, although sound in concept, do not stand up to critical analysis, because there was too much sulphur in many early steels to be neutralized by the higher solubility and greater grain boundary area of such a small amount of residual ferrite.

The Schaeffler Diagram introduced previously was used for many years to determine the composition of electrodes which, when diluted with controlled amounts of base metals, would yield predictable amounts of residual ferrite in the weld metal, thus minimizing cracking. However, a lack of an agreed standard to calibrate magnetic ferrite measuring instruments and variations in ferrite levels due to procedural variations such as arc length, induced the Welding Research Council (WRC) to define a *Ferrite Number* (FN), and a standardized method to measure it. Initially, a modified Schaeffler diagram pertaining only to *austenitic* weld metals, called the Delong Diagram, was used to determine FN up to 18. In this diagram, nitrogen appears as an austenite stabilizer, see Figure 6.3.

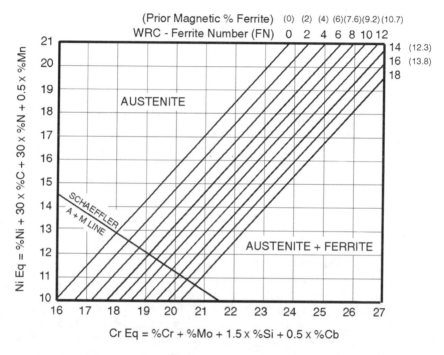

Figure 6.3 Delong diagram

Schneider and Delong added other elements to describe Ni and Cr equivalents more precisely for modern steels, so now:

Ni_{eq} = %Ni + %Co + 0.3%Cu + 0.5%Mn + 25%N + 30%C
Cr_{eq} = %Cr + 2%Si + 1.5%Mo + 5%V + 5.5%Al + 1.75%Cb
+ 1.5%Ti + 0.75%W

The WRC-1992 diagram is the most recent FN prediction diagram, using chemical composition. Note that silicon and manganese are not shown, since they have statistically insignificant effects. The WRC-1992 diagram is shown in Figure 6.4.

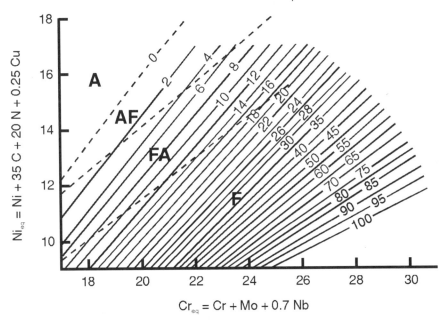

Figure 6.4 WRC - 1992 Constitution diagram for stainless steel weld metal

For solidification mode, Hammar and Svensson proposed:

Ni_{eq} = %Ni + %Cu + 0.91%Mn + 14.2%N + 22%C
Cr_{eq} = %Cr + 1.5%Si + 1.37%Mo + 2%Cb + 3%Ti

It is clear that unanimity does not exist for empirical assessments. This is because each determination of the regression coefficients was based on differing heats of steel and differing experimental techniques. Now it is known that many *austenitic* stainless steels solidify as 100% ferrite, and transform to austenite by a nucleation and growth mechanism during cooling. A small amount of residual ferrite in the ambient temperature microstructure shows that the original solidification phase was ferrite. This effect can be seen in a pseudo-binary phase diagram for Cr-Ni steels, see Figure 6.5.

It can be seen that low Ni metals on the left side of the diagram solidify as ferrite and stay ferritic to ambient temperature. These are the ferritic stainless steels, for example UNS S44400 (18-2). As composition moves to the right, some austenite is formed on cooling. These are the duplex steels, such as UNS S31803 (2205), and austenitic stainless steels are further to the right, up to close to the eutectic composition. Further to the right, the metals solidify as austenite and stay fully austenitic on cooling, e.g. Type 310. These alloys are very prone to solidification cracking. Low S and P values and control of the residual levels, whatever they may be, is essential to avoid cracking in fully austenitic alloys.

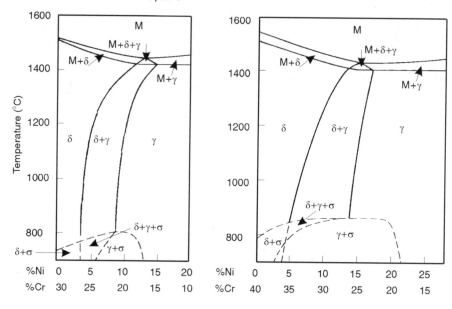

Figure 6.5 Pseudo binary phase diagrams for constant iron content

Sulphur also affects weld penetration, especially when the GTAW process is used. This has been attributed to Marangoni surface tension effects. Surface active elements induce flow in the weld pool which affects heat transfer to the weld root. Normal (historically) sulphur levels cause a flow from the edge of the weld to the centre, which pushes heat down the central axis to cause weld penetration. Low sulphur causes flow on the pool surface from the central region to the edges, producing a wide, shallow pool. Oxygen is also a surface-active element. The amount of free (uncombined) S and O is the critical parameter. Lancaster suggests the following measures of free S and O:

$$(S)_{free} = S_{total} - 0.8Ca - 1.3Mg - 0.22(Ce + La)$$

$$(O)_{free} = O_{total} - 0.9Al_{total}$$

where quantities are in ppm.

Analysis of stainless heats indicates that free Al should be less than 50 ppm for good penetration characteristics involving oxygen, but 100 ppm or more is normal. This leaves sulphur as the main agent to control. Free sulphur should be 50 ppm or more. If maximum limits of about 7 ppm are assumed for Ca, etc, then about 70 ppm is required for good penetration characteristics. This is more than available in some modern steels, which may have between 10 and 50 ppm. For reliable operation, Lancaster recommends 100 ppm S in total, with deliberate additions of Ca and rare earths avoided.

Special conditions on weld metal chemistry and welding procedure can apply for particular applications involving austenitic stainless steels. For welding Type 304, only E308 electrodes exist. E308 does not exist as a wrought or cast alloy, only as a welding electrode. When stabilized Types 321 or 347 are welded, only E347 electrodes are available, since Ti is too easily lost to oxidation in arc processes. As a general rule, if only corrosion at ambient temperature is involved in an application, use the low carbon electrode for all types of welds, e.g. E316L for 316 and 316L welding. This simplifies storage problems and minimizes mistakes, especially the use of normal carbon electrodes to weld low carbon base metals.

For antimagnetic applications, a fully austenitic weld must be used. To avoid solidification cracking, low S and P electrodes (with S + P < 0.002% for some cases), preferably allied to low C levels, should be used. This is also true for cryogenic applications, where low temperature toughness is required. Fully austenitic weld metal is essential to avoid both ferrite and the possibility of forming strain induced martensite from unstable austenite. Carbon and carbide forming elements should be kept low to minimize the number of carbide particles, which reduce low temperature fracture toughness. Low carbon electrodes should be used, with the lime flux coating for SMAW procedures, since rutile fluxes may provide Ti via slag-metal reactions.

Another application requiring fully austenitic weld metal is urea plant, where any ferrite in the weld metal will be preferentially attacked. In addition to low S and P, Mn additions are used, as are rare earths (La or Ce), to control residual sulphur. A guide from Matsuda is:

$$\%La = 4.5\%P + 8.7\%S$$

For high temperature creep and oxidation resisting applications, base metals have higher C contents (to 0.40% in HK40) as well as carbide stabilizing elements, e.g. Mo, Cb, Ti. Weld metals therefore need higher carbon levels and large grain sizes for good creep behaviour. Wrought alloys such as Types 321 and 347, or cast alloys such as HK40 (UNS J94204) can be welded with a hot wire GTAW process involving slow welding speeds and electrode oscillation to promote large grains in both the weld metal and HAZ. When SMAW electrodes are used, the lime-based types should be avoided. The rutile types provide some Ti from slag-metal reactions for stabilization of carbides.

Duplex Stainless Steels

Duplex stainless steels can suffer from an imbalance in the austenite-ferrite content due to the cooling rate effects on ferrite transformation mentioned in the section on the HAZ. This is accommodated by changing the Ni alloy balance in the electrode to promote a 50-50 balance during cooling in defined conditions, i.e. there is a restricted range of heat inputs allowed by a supplier of consumables. It is also essential to protect the root region, top and bottom, to avoid uncontrolled N pickup, which as an austenite stabilizer will alter the balance again. Type 309 and its low carbon, Mo and Cb variations is often used to weld these steels with the SMAW process. Special electrodes for specific grades are also available, often from the steel producer.

Age Hardening Stainless Steels

Age hardening stainless steels do not, in general, have matching composition electrode metals available, although the 17%Cr-4%Ni (UNS S17400 or common name 17-4 PH) grade does have a matching electrode, namely E630. An austenitic stainless steel filler metal such as E308 can be used for root passes with E630 fill passes to minimize root cracking problems. It is possible to fully weld these metals with austenitic stainless steel electrodes, notably E308, E309 or E310, but the strength of the base metal cannot be matched. If only one of these compositions is necessary, use E309. A better overall choice to maximize strength and toughness, is a Ni alloy electrode. The most useful are ENiCrFe-3 and ENiCrMo-3.

Super Stainless Steels

Due to segregation effects during solidification, weld metals in these materials need the extra levels of minor alloy elements, e.g. Mo, to maintain their enhanced pitting corrosion resistance. Otherwise, filler metals have, in most cases, similar chemistry to the base metals. Heat inputs restricted to optimize HAZ properties can lead to toughness deficiency in the weld metal, due to an unfavourable ferrite-austenite balance, with too much ferrite. Increasing the heat input limit to 2 kJ/mm (50 kJ/in.) may alleviate this problem, but will reduce HAZ properties may be compromised and fusion line corrosion resistance reduced.

Chapter

7

CAST IRONS

Cast irons are iron-carbon alloys dominated by the eutectic reaction at about 4.3%C. They contain > 2%C, significant amounts of Si, P and S in many cases, and can have large amounts of deliberate alloy additions for specific purposes, for example Cr, Ni, Si, Mo and Cu. General metallurgical information on these alloys is given in *The Metals Black Book* - Ferrous Metals. The related AWS Filler Metal Specification is A5.15.

METALLURGY OF CAST IRONS

Cast iron is essentially pig iron from the blast furnace, whose composition has been modified by remelting and adding steel scrap and/or specific alloying elements in the form of ferro-alloys, e.g. ferro-silicon. A typical cast iron contains 3.0-3.8%C, 1.5-3.0%Si, 0.3-1.0%Mn, 0.05-0.25%P and 0.02-0.15%S. The high C and Si contents are a natural result of the blast furnace process. The actual concentration of these elements is important in determining the structure and properties of cast irons. Si, and other alloy and residual elements, modify the C-Fe phase diagram in two ways:

1. The C content of the eutectic and eutectoid are reduced (to 3.7% and 0.6% respectively at 2.0%Si).
2. The eutectic and eutectoid reactions are changed from invariant reactions at a single temperature to univariant reactions over a small range of temperatures.

Cast irons are significantly different from steels in microstructure as well as chemistry. Where steels have carbon in the form of carbide or dissolved in the matrix, cast irons have these and an additional phase - graphite. Although we tend to regard iron carbide in steels as an equilibrium phase, it is not. Given slow cooling rates during solidification or extended thermal treatments of solid cast iron, graphite will form directly or Fe_3C will transform to graphite and iron, which are the true equilibrium phases.

When a cast iron is fractured, the surface may appear as shiny (white), dull (gray) or a mixture of the two. A shiny surface indicates an absence of graphite (ferrite and carbides), a dull surface has graphite showing (in a metallic matrix), while a mixture suggests some areas of graphite flakes and some with carbides from the eutectic reaction. Fracture usually takes place along either the graphite or carbide phases, ensuring that one or the other always appears on the fracture surface.

The appearance of graphite or the large amount of cementite (carbide) is controlled by the cooling and solidification of the cast iron. The eutectic reaction will occur in one of two ways:

(1) Liquid $\rightarrow \gamma$ + graphite
(2) Liquid $\rightarrow \gamma$ + Fe_3C

The particular type experienced by a casting is determined by the cooling rate and composition. The stable equilibrium iron-carbon system for high carbon levels is graphitic carbon and iron. Slow cooling rates allow the graphite to form, while rapid cooling rates suppress graphite and cause the formation of hard, metastable iron carbide. Some alloying elements promote graphitization at a given cooling rate. Others promote the formation of carbides and suppress graphite formation. Si is a very strong *graphitizer*, and > 2% causes graphite to form even at high cooling rates. Mn is a moderate carbide former, and, if present in sufficient quantity, will suppress graphite completely, regardless of Si level. S and P also act to stabilize carbides and prevent graphite formation. All carbide forming elements, e.g. Mo, Cr, V, behave the same way. Elements which do not form carbides, e.g. Ni, Cu, promote graphite formation.

The microdistribution of graphite, if present, will depend on whether the cast iron is hypereutectic (> 4.3%C) or hypoeutectic (< 4.3%C). Hypereutectic alloys form primary graphite under slow cooling rates, while hypoeutectic alloys form primary austenite. The eutectic composition can vary due to the influence on other alloying elements. This is assessed via a formula for *carbon equivalent* (CE):

Early version:
$$CE = \%C + \frac{\%Si}{3}$$
Later version:
$$CE = \%C + 0.3\%Si + 0.33\%P + 0.4\%S - 0.027\%Mn$$

Mn also controls the sulphur to minimize hot cracking due to the formation of eutectic $Fe(O,S)$, which is molten to temperatures well below the solidification temperature of the bulk of the cast iron. Mn alters the oxysulphide to $Mn,Fe(O,S)$ which does not wet the grain boundaries and is therefore present in isolated pools. The optimum level for control is:

$$\%Mn = 1.7\%S + 0.15$$

During cooling and solidification of liquid hypereutectic iron, graphite is precipitated first and is randomly distributed. In hypoeutectic iron, austenite dendrites form first (refer to the Fe-C phase diagram), and the graphite thus appears in interdendritic locations. The distribution (and size) of graphite flakes has been classified in ASTM A247.

The eutectic reaction product without graphite is called ledeburite (iron carbide-iron mixture). It may be seen in the same vicinity (macroscopically) with a ternary eutectic called steadite (iron-iron phosphide-iron carbide mixture), since eutectics tend to segregate to interdendritic areas. The amount and distinctive appearance of steadite makes it easy to observe and offers a simple way of estimating the amount of P in the cast iron. The amount of pearlite gives a similar estimate of the C level in a steel. The higher than usual levels of other residual elements, e.g. S, cause the appearance of other constituents, such as oxysulphide inclusions.

The eutectoid reaction also forms graphite in cast irons via precipitation in the solid state.

(3) $\gamma \rightarrow \alpha$ + graphite

There is also the metastable reaction.

(4) $\gamma \rightarrow \alpha + Fe_3C$

Slow cooling and a high Si level favour the first reaction. The graphite precipitation is also promoted by the presence of pre-existing nuclei of eutectic graphite, upon which it precipitates, causing further growth. However, not all of the carbon precipitation occurs on these sites, and precipitation within the austenite grains causes a variety of graphitic structures to form, particularly at high cooling rates near the eutectoid temperature.

The morphology of the graphite flakes has a strong influence on the mechanical properties of cast irons, notably on the strength, tensile ductility and fracture toughness. Gray cast iron has the graphite in the form of flakes, as observed on a metallographic section. In three dimensions, the graphite is a large nodule with petals radiating in many directions, similar to an open rose. The edges of the thin 'petals' are sharp, and start cracks easily under a tensile stress. Tensile ductility and fracture toughness are therefore low, in fact almost nil. Tensile strength can be estimated for grey iron by:

$$\sigma_{uts} = 162.37 + 16.61/D - 21.78\%C - 61.29\%Si - 10.59(\%Mn - 1.7\%S)$$
$$+ 13.80\%Cr + 2.05\%Ni + 30.66\%Cu + 39.75\%Mo$$
$$+ 14.16(\%Si)^2 - 26.75(\%Cu)^2 - 23.83(\%Mo)^2$$

Here "D" is the casting *diameter* in inches ($\frac{3}{8}$ to 2 in.), usually on a test rod.

Limits on the elements for accurate regression are:

%C	%Si	%Mn	%Cr	%Cu	%Mo	%Ni	%S
3.04	1.60	0.39	1.10	0.07	0.03	0.07	0.089
3.29	2.46	0.98	0.55	0.85	0.78	1.62	0.106

Machinability is very good, because the graphite flakes promote chip breakup ahead of a tool bit and also lubricate the tool surface.

Heat treatment of white iron was the only way, until less than fifty years ago, to form finer nodules of graphite with virtually no petals. The heat treatment involved annealing at about 900°C (1650°F) for 40-150 hours (*malleablization*). The carbide breaks down into ferrite plus fine nodular ('temper') graphite, which does not have such a severe internal notch effect. In an inert or neutral atmosphere, no carbon is lost. The initial carbon content is 2-3%, and the heat treatment is at 850-875°C (1560-1600°F) for 40-60 hours, followed by slow cooling to 690°C (1270°F), then air cooling. A pearlitic matrix, rather than ferritic, can be produced by faster cooling and/or Mn additions. A very rapid cooling rate of a pearlitic iron will produce a martensitic matrix.

If malleablization is carried out in an oxidizing atmosphere (such as packing in iron ore), surface carbon is lost, leaving a skin of ferrite. Original carbon content is higher, about 3.5%, and this is reduced to 0.5-2.0% during malleablilizing at 850-900°C (1560-1650°F) for 100-150 hours. At the surface, decarburization is nearly complete, producing a tough, ductile ferrite. In thick sections, the centre region consists of temper carbon nodules, pearlite and some ferrite. Malleable irons are machinable, have good shock absorption and relatively high strength. They are seldom made or seen today, except in repair situations on old castings.

Heat treating for 40-150 hours is neither time nor energy efficient, so for many years other ways of producing *nodules* were sought. The solution was found in the late 1940's - the interfacial energy between the graphite and the iron matrix during solidification can be altered by reducing the sulphur content to very low levels. The sulphur 'poisons' the interface between the liquid iron and the graphite nodule, producing the convoluted 'open rose' form. Removing the sulphur allows the formation of the lowest surface area/volume ratio morphology, a spheroid. This is accomplished in practice by *inoculating* the cast iron with one of two strong desulphurizers - Mg or Ce. The required amount is:

$$\%Mg_{added} = \frac{\%S_{initial} + \%Mg_{residual}}{\eta}$$

Where η is the process efficiency for the particular system used. $\%Mg_{residual}$ is best between 0.03 - 0.05%. Lower levels result in fewer nodules (more flakes), while higher levels promote carbide formation. Within the acceptable range, high cooling rates need lower $\%Mg_{residual}$ to form nodular iron. To maximize the effect of Mg additions, there are also limits on residual elements, namely a 0.1% maximum on Al, Ti & Zr, with even lower limits on Cd, Pb and elements in the S group (Se, Te) and P group (As, Sb, Bi). The resulting nodules are nearly spheroidal and produce an iron with very good tensile ductility and fracture toughness. This iron is called *nodular* or *ductile* cast iron in America and *spheroidal* cast iron in England, where it was invented by Henry Morrogh. *Compacted graphite* cast iron is similar to ductile iron in properties and has graphite nodules that area somewhat more fuzzy and less spheroidal. This is accomplished by controlling and limiting the Mg addition to avoid full spheroidization, primarily on the grounds of economy of production. The compacted graphite iron does not shrink as much as ductile iron when cast.

Alloy cast irons contain large amounts of particular alloying elements to produce material for specialty applications. Silicon irons contain 12-18%Si for use in transporting hot oxidizing acids, e.g. nitric acid. These irons are very brittle. Chromium irons contain 15-30%Cr for abrasive wear and oxidizing acids. They have better mechanical properties than the Si irons. Nickel irons contain 14-30%Ni and are fully austenitic as a result. These are used for mild acids and alkalis. Heat resistant irons contain Cr and Si, with additions of one or more of Ni, Mo and Al.

WELDING CAST IRONS

The majority of welding on cast irons is done for repair purposes, which often means welding on material of uncertain age and composition. There are several important implications. First, find out, if at all possible, the chemical composition of the iron. If this cannot be done, try to infer from the use of the iron the possible types that may have been available. White iron is extremely difficult to weld and welding is not recommended. Most ornamental (and many structural) castings are gray iron. Ductile iron is easier to weld than gray cast iron due to the lower S and P levels and the absence of crack-like graphite flakes. Second, if a crack is to be repaired, or a broken casting rejoined, ensure that the surfaces are scrupulously clean. Degrease thoroughly with solvents or steam cleaning and remove rust and any other surface scale or contaminant. There are also chemical cleaners available. Shot blasting can be used on ductile and compacted graphite irons, but not on gray irons. For gray irons, surface decarburizing with an slightly oxidizing oxy-fuel gas flame can be effective. Third, and this is very important, determine whether welding is the most appropriate joining method. Cast iron can be bolted, stitched, riveted, brazed and adhesive bonded more easily than welded, and other joining processes may be better for many applications. Finally, if a crack is being repaired, arc gouge or grind out the crack *completely* before welding.

If welding is done, there are some general precautions for all cases. Preheating is recommended, the higher the temperature, the better. Two benefits result: the formation of hard microstructures is minimized and any hydrogen introduced to the weld zone is encouraged to diffuse to less risky areas. A minimum of 300°C (570°F) is recommended for gray cast iron, and up to 700°C (1290°F) has been used for specialty irons. A full preheat is better than a local preheat, which is better than no preheat. An exception may be made for malleable iron, especially the ones with a surface layer of pure ferrite, which are easily welded with minimal or no preheat and low hydrogen SMAW electrodes.

PWHT at no higher than 650°C (1200°F) can also be used to further soften hard microstructures in the HAZ, but these will not generally help with hydrogen diffusion, because it will be too late. In many cases involving repair of large castings, however, preheat and postheating may be impractical. If that is the case, use low heat input and short welding runs (let cool before restarting) to minimize the extent of the HAZ. Peening is sometimes advocated to minimize welding stresses, but practitioners should keep in mind that the process of hammering a weld metal is very subject to the randomness of human strength levels and accuracy of location of individual blows. In short, it is not a reliable procedure.

Most of the difficulties in welding cast irons come from two sources: hard zones and lack of ductility in the HAZ and cracks in the weld metal due to contaminations from S, P and C absorbed by dilution from the melted base metal.

In the HAZ, problems arise from the high carbon content and the various forms in which it exists. Up to the eutectoid temperature of 710°C (1310°F), the metal behaves much like a high carbon steel. Above the eutectoid, ferrite changes to austenite and begins to absorb carbon from cementite. At 800°C (1470°F) graphite starts to be dissolved, supersaturating the austenite, which precipitates cementite. Graphite solubility depends on form: high surface area per unit volume of graphite encourages dissolution. This is the case for the *open rose* graphite flakes in gray iron. Spheroidal graphite is slower to dissolve. Cementite precipitation occurs intergranularly and intragranularly, forming complex carbide networks. When the HAZ cools, the network remains and the austenite transforms, high carbon areas to martensite and lower carbon areas to pearlite. There can also be extensive areas of partial melting at temperatures close to the fusion temperature due to segregation effects. This makes for a metallographic mess and results in high hardness and low ductility in most cases. Low hydrogen deposition is essential with arc welding processes to avoid HAC in this area.

Filler metals for welding cast irons are designed to accomplish two objectives. A low yield stress in the deposit is desirable to minimize the shrinkage stresses and minimize the residual stress on the HAZ, and the contamination of the weld metal by dilution must be mitigated. Historically, five types of OFW and SMAW filler metals have been available. Cast iron filler metals are available for the OFW process and for special applications with the SMAW process. For the general use of the SMAW process, there are mild steel, *pure* nickel, copper alloys, and nickel-iron alloys.

Mild steel electrodes are not recommended other than for cosmetic repair of castings or repair of malleable iron. E7018 electrodes, in low hydrogen condition, are the best ones to use. Dilution raises the possibility of both hot cracking due to S and P absorption and cold cracking due to C absorption. *Pure* nickel electrodes contain about 95%Ni, e.g. ENi-CI-A. Ni has a limited solubility for carbon and most of it absorbed from dilution is precipitated as graphite, which increases weld metal volume and reduces residual stresses. Although deposits are soft and low in strength, these are very prone to hot cracking from residual element contamination. Ni is quite intolerant of S and P, which segregate strongly during solidification. Copper alloys are usually based on Cu-Ni, Cu-Sn or Cu-Al alloys. Ni contents are at the 55-65% level and also suffer from residual element dilution. The Cu-Sn alloys are also low in strength and somewhat more tolerant of dilution. The Cu-Al alloys are much stronger. The Ni-Fe electrodes, e.g. ENiFe-CI-A, contain about 55%Ni as well, but are much more tolerant of dilution effects. These are the best electrodes for general use. The Ni level helps to keep the yield strength low and the formation of graphite high, both beneficial results.

Electrodes for specialty irons are specialty items themselves. Some require unusual procedural conditions. For example, procedures for welding Si irons with matching electrodes require preheat in the order of 700°C (1290°F) to avoid cracking in the brittle material.

One common alternate process for joining cast iron is bronze welding, a rather unfortunate choice of words, since bronze is not used and no melting of the cast iron occurs. More accurately, the process is brass brazing (with a flux), although it is also called braze welding. The thermal cycle involved is less rapid than arc welding cycles due to an oxy-acetylene heat source, and hardening in the HAZ is less severe. To begin welding, a slightly oxidizing flame is used to remove graphite at the joint surface. During joining, a slightly carburizing flame should be used to minimize zinc fuming. The brass filler metal melts below the melting temperature of cast irons, eliminating contamination by S, P and C from the base metal. In addition, the yield stress of brass is low, so that the cooling stresses are absorbed by strain in the brass, limiting the effect of residual stress on the cast iron. Skill is essential for proper use of this process.

Chapter

8

NICKEL ALLOYS

Nickel alloys are of two types: solid solution alloys that can only be work hardened and precipitation hardened alloys. They are also intended for two application areas: resistance to corrosion and oxidation in specific media, or resistance to deformation (creep) and oxidation at high temperatures, as in jet engines. General metallurgical information on these alloys is given in *The Metals Red Book* - Nonferrous Metals. Related AWS Filler Metal Specifications are A5.11 and 5.14.

Nickel and its alloys have an *austenitic* face centered cubic crystal structure and, like austenitic stainless steel, have an aversion to sulfur. Hot cracking is therefore one of the major welding problems and all sulfur bearing compounds, oil, grease, paint, etc., must be scrupulously removed from the weld area. Temperature sensitive crayons used to be a prime culprit, but new formulations are sulfur free. Low melting point metals and alloys can also be a problem. Lead and bismuth and their alloys must be excluded from the weld zone. Silicon promotes hot cracking in some alloys, and minor additions of boron and zirconium can promote liquation cracking in the HAZ, particularly in the coarse grain region, where the minor elements collect on the grain boundaries.

Cleaning should be by vapor or solvent degreasing and by brushing with stainless steel brushes (not mild steel) for 25-50 mm (1-2 in.) outside the edge of the weld preparation. Repair welding of alloys which have been in service often requires more drastic cleaning. Grit or vapor blasting may be of help in this situation. Initial degreasing is often beneficial, by minimizing the risk of embedding grease, etc., in the surface during grit blasting or minimizing the contamination of wire brushes, which may transfer contamination to other surfaces.

Since many applications of nickel alloys involve corrosion, it is imperative that weld zones, both weld metal and HAZ, have equivalent or higher corrosion resistance, i.e. are cathodic to the base metal. If they are not, corrosion will concentrate on the weld zone, with catastrophic results in most cases. This problem is usually of more importance in the weld metal. If any doubt exists for a particular application, contact the manufacturer of the alloy and/or of the proposed filler metal for information. There can be corrosion problems in addition to overall weld metal chemistry that are caused by precipitation of carbides and/or intermetallic compounds.   The carbide problem is analogous to 'sensitization' in stainless steels.   Many subtle changes in the composition of certain nickel alloys has been done over the years to minimize these problems.   Fine tuning has produced, among others UNS N10276 (Hastelloy* C-276), UNS N06022 (Hastelloy* C-22) and UNS N06455 (Hastelloy* C-4) alloys, all variations on the original C theme alloy. For example, UNS N10276 was an early low carbon version to minimize carbide formation.   Sensitization is best avoided by using low heat input and stringer passes for multipass welds in thicker material.   The less time spent in the 500-850°C (930-1560°F) range, the better.   The low carbon variations of the alloys are immune if low heat input is used.   Another, significantly less desirable solution, is to anneal the fabrication after welding at a temperature high enough to dissolve the carbides, about 900-1000°C (1650-1830°F). This carries the risk of thermal distortion of complex structures and possible reprecipitation of carbides on cooling, so low carbon and low heat inputs are preferred.

Precipitates are also a problem in the welding of precipitation hardened alloys intended for creep resistance at high temperatures.   Alloy fine-tuning has also been done on these alloys to minimize the precipitation of intermetallic phases (e.g. σ, μ & χ) which embrittle the material.   The precipitate used for hardening is $Ni_3(Al,Ti)$ for the most part.   This is a very stable precipitate which takes extensive heat treatment to form and is difficult to dissolve completely in the short time of the welding thermal cycle. Welding these alloys can cause strain-age cracking in the HAZ due to strain concentration during cooling after welding in the narrow zone which is overaged or partly solution treated.   As a result, the alloys are normally welded in the annealed (solution treated) condition and the completed fabrication is hardened with a PWHT.

The third possible welding problem is porosity, generally the result of carbon being oxidized during welding or of nitrogen absorption.   Virtually all filler metals for nickel alloys contain elements to deoxidize and denitrify the deposited weld metal to avoid porosity.   These are usually Al, Ti and Cb. Hydrogen is often added to the argon shielding gas for GTAW procedures (about $10\%H_2$), which minimizes nitrogen porosity.   GMAW procedures do not respond well to similar additions, where hydrogen can cause porosity on its own.   GMAW gas mixtures should be argon without oxidizing gas additions (CO_2 or O_2), although helium can be added to aid in smooth bead formation.   SMAW is the other common process used, in which porosity is not a problem if consumables are handled and stored according to manufacturer's recommendations, which are essentially those advocated for low hydrogen electrodes for welding hardenable steels.   Creep resistant alloys are often welded with the EBW process, which does not have porosity problems because it has a high vacuum welding environment.

In the following discussion of the welding metallurgy of the various nickel alloys, it must be kept in mind that there is a very large number of these alloys and it is not possible to discuss each one in detail. In each section, examples will be given of widely used or characteristic alloys. Welding of nickel alloys is accomplished with filler metal of similar composition to the base metal except when there are specific reasons to avoid that composition. These instances will be mentioned in the appropriate section.

Corrosion Resistant Work Hardenable Alloys

Pure nickel is the first and simplest of these, used for containing alkalis such as sodium hydroxide. UNS N0200 (Nickel 200) and UNS N02201 (Nickel 201 - low %C) are the best examples. Some carbon is present in most commercial nickels. These alloys are simple to weld with similar filler metals, modified with deoxidizers to avoid porosity. An example is ER Ni-1, which is for welding UNS N02200 (Nickel 200), but contains 1.5%Al and 2.5%Ti.

The most common grade of nickel-copper alloy is UNS N04400 (Monel* 400), containing 30%Cu and the version with added sulfur for machinability is UNS N04405 (Monel* R405). UNS N04400 is relatively easy to weld using ERNiCu-7 with precautions taken (as noted above) for hot cracking and porosity. UNS N04405 should not be welded due to the risk of hot cracking.

Nickel-Chromium and the Less Expensive Ni-Cr-Fe Alloys

These alloys were originally developed to resist oxidation at high temperatures. With additions of minor alloying elements such as Mo, Cb and Ti, they also resist corrosion in a variety of oxidizing acids at ambient or slightly elevated temperature. Examples of the oxidation-resisting type are UNS N06600 (Inconel* 600), UNS N06601 (Inconel* 601) and UNS N06690 (Inconel* 690), while examples of the corrosion-resisting type are the UNS N10276 (Hastelloy* C-276 - lower Fe) and UNS N06007 (Hastelloy* G - higher Fe). UNS N06625 (Inconel* 625) spans both application areas. Specific filler metals are available for these alloys. ENiCrMo-3 or ERNiCrMo-3 is essentially UNS N06625 in electrode form and is useful for welding a wide variety of alloys from austenitic stainless steels (Types 317 and 310) up through UNS N06007.

Iron-Nickel-Chrome Alloys

These include UNS N08800 (Incoloy* 800) and UNS N08825 (Incoloy* 825), as well as many other proprietary alloys. UNS N08020 (Carpenter* 20Cb3) alloy hot cracks in autogenous welding or with a similar composition filler metal. ERNiCr-3 (67%Ni min.-20%Cr) with low dilution procedures should be used for this alloy. The Cb content of the filler metal helps to minimize hot cracking due to silicon.

Nickel-Molybdenum Alloys

These alloys were developed to resist reducing acids and salts, and include UNS N10001 (Hastelloy* B) and N10665 (Hastelloy* B-2). They do not, therefore, resist oxidizing conditions. They are welded with similar composition filler metal and under no circumstances can other filler metals be used, nor can the electrodes suitable for these alloys be used to weld other nickel alloys. The AWS classification is ENiMo-1 or ERNiMo-1 and ENiMo-7 or ERNiMo-7. It is imperative that all Ni electrode types be stored separately and under well identified conditions in order to avoid mixing the different types in welding procedures.

Precipitation Hardened Superalloys

These alloys are of nickel base or nickel-iron base and can be simply divided into alloys which can be welded and alloys which should not be welded, on the basis of the total alloy element level which causes precipitation hardening. UNS N07080 (Nimonic* 80A) was an early example of this group of nickel base alloys. Others are UNS N07718 (Inconel* 718) and UNS N07500 (Udimet* 500). Examples of the nickel-iron alloys are UNS N09901 (Incoloy* 901) and UNS N09925. Figure 1 shows how %Al and %Ti are involved in promoting strain age cracking in various nickel alloys. Note that one of the nickel-copper alloys (Monel* K500), is on the line separating weldable from unweldable.

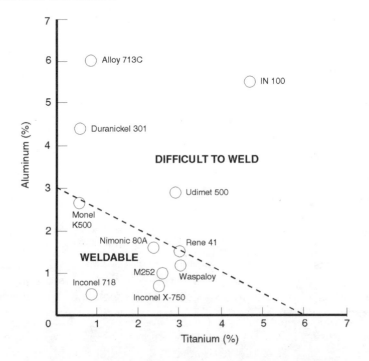

Figure 8.1 Weldability of age hardening nickel alloys

If the alloy content is within the range that can be welded, certain general rules should be followed to avoid strain-age cracking in the HAZ. Precipitation of the stable $Ni_3(Al,Ti)$ precipitate γ' takes a long time, and γ'', NiCb, is even slower. Standard heat treatment on the original alloys was a solution treatment at 1150°C (2100°F) for 2-4 hours followed by a precipitation heat treatment of 760°C (1400°F) for 12 hours. The precipitate changes little in the HAZ - just enough to have a solution treated/overaged zone or zones in which thermal strain due to cooling concentrates. Welding in the solution treated condition is best. There is a very slight aging process in the temperature range of 600-1000°C (1110-1830°F), but a PWHT to age the whole zone restores strength to consistent levels throughout. If solution treating prior to welding is not possible, a thorough overage treatment will drop the strength of the material enough to avoid cracking in most instances.

*Alloys denoted by an asterisk (*) are trade names identified with UNS numbers as listed in the Metals & Alloys in the Unified Numbering System, HS-1086 Feb. 93 and H-1086/94U, published by the Society of Automotive Engineers (SAE) and the American Society for Testing and Materials (ASTM).

Chapter

9

REACTIVE & REFRACTORY ALLOYS

Reactive metals such as titanium and zirconium react with and readily dissolve the atmospheric gases oxygen, nitrogen and hydrogen at elevated temperature and as a result they become hardened and embrittled. Refractory metals do likewise, but are differentiated by very high melting temperatures and therefore by high application temperatures. They are also brittle at room temperature: examples are tungsten, tantalum, molybdenum and columbium alloys. General metallurgical information on these alloys is given in *The Metals Red Book* - Nonferrous Metals. Related AWS Filler Metal Specifications are A5.16 and 5.24.

Reactive Alloys

Welding of the reactive alloys requires isolation and protection from atmospheric gases above all other welding precautions. This can be accomplished at several levels, depending on the limit of contamination needed. Levels must be lowest for corrosion protection and may be slightly relaxed when only mechanical strength is required. Tough weld metals also require low contamination levels. Only inert gas processes should be used, with argon as the shielding gas. The dew point should be -60°C (-70°F) maximum. High purity argon or combinations of cold traps (for moisture) and hot traps (for oxygen and nitrogen) should be used in critical welds. GTAW is the best choice in most cases. GMAW should be treated with caution, due to the extra gas-metal reaction site at the electrode tip. EBW in vacuum, or argon shielded LBW are also used for some fabrications. The extent of reaction with atmospheric gases, notably oxygen, can be assessed by the colour of the weld zone after fabrication. The color indicates the thickness of the surface oxide layer, much like an oil film on a puddle of water shows differing colors in sunlight. A shiny, bright silvery, colour is ideal, but rarely achieved. A golden or straw colour indicates minimum contamination and good corrosion resistance. Various shades of blue, from light to dark, indicate thicker oxides and lower corrosion resistance. Purple, or worse yet, dull grey, oxides indicate significant contamination. The latter situation should not be tolerated for any welds and should prompt an immediate investigation into shielding integrity.

All reactive alloys should be scrupulously cleaned and deburred before welding. Subsequent to cleaning, handling should be done only with lint-free white gloves. Cleaning is done in two stages: degreasing followed by oxide removal. Vapour degreasing or solvent degreasing are the most common methods used, although steam cleaning and alkaline solutions are available. Chlorinated and silicate solvents should be avoided due to enhanced risk of cracking. Degreasing should be followed by pickling in a nitric-hydrofluoric acid mixture ($40\%HNO_3$-$4\%HF$). Oxide removal (minimization in reality) is by two methods only: stainless steel brushing or draw filing. On no account should mild steel brushes or steel wool be used, nor should grinding wheels. Mild steel or grinding wheel residue will cause contamination problems and decreased corrosion resistance. In critical applications, a final pickling is desirable.

The first level of welding protection, for welding in an open shop area well protected from draughts, is from a combination of trailing, underbead and leading gas shields. Figure 1 illustrates these additions to inert gas shielding, which should be used in the order listed, i.e. trailing shields are vital and should always be used, underbead shielding is also vital, and leading shields are good but sometimes unnecessary.

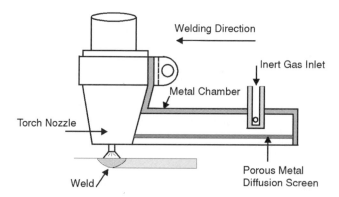

Figure 9.1 Trailing gas shielding system for welding reactive alloys

All shields should use argon gas for complete coverage and all components, including the steel wool diffuser, should be stainless to avoid rusting and subsequent shielding gas contamination.

A second level of protection can be achieved by enclosing the components near the weld zone in a plastic sheet *tent*, through which argon flows and is maintained at a slightly positive pressure. Argon also flows through the welding torch, always GTAW or PAW, and into the region of the weld underside. This method is often used on pipe fabrication in open shop areas.

The third level of protection, for components which will fit, is provided by a glove box. These are of sufficient size to hold the entire fabrication and also allow for any necessary manipulation during welding. The glove box is evacuated after insertion of the components, usually to a hard vacuum using mechanical and diffusion pumps. It is then back-filled with argon and welding is usually by the GTAW process with filler if necessary. One useful trick to minimize contamination from residual gases in the chamber or the argon supply is to have a 'strike pad' of scrap titanium in the chamber. The initial arc is struck on this scrap and held for 30 seconds or so, with the molten weld pool acting as a *getter* for residual gases. A fourth level of protection can be provided for with high vacuum EBW fabrication, since the hard vacuum needed for operation of the process has a lower residual gas level than most commercial supplies of argon.

Pure titanium undergoes an allotropic transformation (analogous to iron). It is hcp α at room temperature (c/a = 1.587, slightly less than ideal), becomes bcc β above the transus temperature of 882°C (1625°F), and melts at 1668°C (3034°F). The high T_m and hcp structure make it attractive for high temperature creep applications in the aerospace field in particular. The Concorde supersonic airliner superstructure is largely made from titanium alloys, right down to the superplastically formed washroom sinks. Titanium also has low specific gravity (4.5) and an intermediate Young's modulus E (115 GPa or 16.7 x 10^6 psi). Titanium alloys can be α, *near* α, - β, or β, depending on alloy type and level, since the high-temperature phase can be stabilized by certain elements, as is γ in iron alloys such as stainless steels. Elements with bonding electron/atom ratios of < 4 stabilize α (interstitials O, N, C and substitutional Al); a ratio of 4 is *neutral* (substitutional Zr, Sn) and elements with a ratio > 4 stabilize β (interstitial H and virtually anything else metallic). Phase diagrams produced by alloying are complicated, but are of three basic types. α stabilizers expand the α field and raise the transus temperature (Al, O, N, C, Ga). β stabilizers have two possible effects: the α field is suppressed and the transus temperature is lowered (Mo, W, V, Ta) or there is the formation of a eutectoid (Cu, Mn, Cr, Fe, Ni, Co, H). The M_s temperature for the formation of α' is also lowered at higher total β alloy levels. All titanium alloys are β phase at high temperature, and will be so in the weld metal and HAZ where the β transus is exceeded. Depending on alloy level, transformation to α phase or metastable α' and other phases will occur during cooling, with higher alloying content tending to reduce transformation, leaving more β phase in the final structure.

Strength in commercially pure titanium is determined primarily by the interstitial atom content. Each element adds between 35 and 70 MPa (5-10 ksi) to the tensile strength for each 1wt% added. The total interstitial level is expressed as an *oxygen equivalent*, probably because oxygen is the most common impurity.

$$\%O_{eq} = \%O + 2\%N + 0.67\%C$$

where 1%O_{eq} = 120 MPa (17 ksi) increase.

Hardness also rises with O_{eq}, and hardness measurement can be used to monitor shielding effectiveness. Room temperature yield strength can be predicted by the Hall-Petch relationship, for example in grade 50A:

$$s_{ys} = 231 + 10.54d^{-0.5}$$

Pure titanium (up to 1% total impurities) is used mainly for corrosion resistance, for example in chloride solutions, as is the α alloy ERTi-7 (formerly Ti-0.2%Pd). α phase stabilizers for solution hardened alloys are often factored and grouped as equivalent to aluminum, or as "Al". If a certain level of "Al" is exceeded, an ordering reaction occurs, leading to brittleness. The ordered phase is of the coherent hcp TiX type, called α_2. The danger level is 9wt% "Al_{eq}", calculated as follows:

$$Al_{eq} = \%Al + 0.33\%Sn + 0.17\%Zr + 10(\%C + \%O + 2\%N)$$

Any alloy with about 6%Al is in the danger area due to the residuals. The most common α alloy is Ti-5Al-2.5Sn. Tiny amounts of β may appear in α alloys due to some residual Fe impurity. All α alloys can be welded using pure Ti filler, preferably the low interstitial version, ERTi-1. This is essential for applications at low (cryogenic) temperatures. ERTi-2, -3 and -4 are less expensive but contain higher interstitial levels of O, N and H. There is also ERTi-7 (formerly ERTi-0.2Pd) for welding the Pd alloy to maintain the extra corrosion protection afforded by the alloy, and ERTi-6 (formerly ERTi-5Al-2.5Sn) filler metal, with an ERTi6ELI (formerly ERTi-2.5Sn-1) for low interstitial levels.

Near α alloys use small amounts of β phase to improve creep strength, as in forged compressor disks for gas turbines using alloy Ti-8Al-1Mo-1V. Creep resistance above 400°C (750°F) is very good. Near α alloys contain up to 2% of β stabilizing elements, most commonly Mo or V. Optimum creep strength is found from the Bomberger equation:

$$10 \geq 36 - 2.6\%Al - 1.1\%Sn - 0.7\%Zr - 27\%Si - 3\%Mo_{eq}$$

where $\%Mo_{eq} = \%Mo + 0.5\%Cb + 0.2\%Ta + 0.75\%V + 0.5\%W$, $\%Mo_{eq}$ must be ≥ 1.5

Near α alloys can be welded with pure Ti filler metal. There is also one Al-Mo-V filler metal, SAE/AMS 4955 (formerly ERTi-8Al-1Mo-1V) and one Al-Cb-Ta-Mo, ERTi-15 (formerly ERTi-6Al-2Cb-1Ta-1Mo) available.

α-β alloys, the most common of which is Ti-6Al-4V, are the most widely used alloys. β stabilizing elements can (as do γ stabilizing elements in iron alloys) lead to martensite formation in α-β alloys quenched from the β temperature field. This phase in titanium alloys is hexagonal α', but as solute level rises, may also be α'' or α''', both of which are apparently face-centred orthorhombic. α' is tempered to form α and intermetallic compounds such as Ti_3Al, which precipitation harden the matrix. The M_s decreases at higher solute levels, making martensite formation easier. Ti-6Al-4V can be strengthened by quenching and tempering, but only at low thicknesses of 25 mm (1 in.) or less, since it is alloyed at a low level and *hardenability* is thus marginal. Weldability of Ti-Al-V alloys is good. They are often welded in the annealed condition, then aged in a PWHT to develop full properties in the base metal, HAZ and weld metal. ERTi-5 (formerly ERTi-6Al-4V) and ERTi-5ELI (formerly ERTi-6Al-4V-1) filler metals can be used for these alloys, including low interstitial versions (ELI). Commercially pure Ti can also be used, with slight loss in strength.

More *hardenable* alloys have several β elements added at the 1-5% level. At sufficiently high alloy levels, some metastable β is retained (analogous to retained austenite in steels). During aging or tempering, α is difficult to nucleate directly in the retained β if tempering temperatures are below 500°C (930°F). Intermediate phases may then form, with ω phase occurring in medium solute levels and coherent β_1 (depleted in solute) at higher levels. β_1 phase is not considered to be important in commercial alloys. ω phase is *weird hexagonal* (c/a = 0.613) and is responsible for severe embrittlement. Particle shape is cuboid if the lattice misfit with the matrix is high, ellipsoidal if low. Composition depends on solute atom type(s), since electron/atom ratios in all ω phases is 4.2. It may therefore be an electron compound. ω is avoided during isothermal heat treatments by temperatures in excess of 475°C (880°F), or by increasing solute level. At temperatures above 500°C (930°F), when α does nucleate directly from β, two possibilities exist: Widmanstätten α in dilute alloys or high Al alloys, which reduces ductility, or a fine dispersion of α in β in alloys containing more β stabilizing elements. The Widmanstätten α can be suppressed in susceptible alloys by deformation prior to aging to promote precipitation of a more uniformly distributed a phase. These alloys, and the β alloys discussed below, are more difficult to weld if the amount of β phase exceeds 20% or the ultimate tensile strength exceeds 700 MPa (100 ksi). Pure Ti filler metal can be used to weld these alloys to limit the amount of β phase in the weld metal. Preheating to 150°C (300°F) and a PWHT also helps to minimize cracking.

β alloys naturally are alloyed enough to have the M_s well below room temperature. The first commercial alloys, developed to maintain the easily deformed bcc structure for cold forming, were Ti-13V-11Cr-3Al and Ti-15Mo. Strength is achieved by the dual means of solution hardening and precipitation hardening (α phase precipitate) at 480°C (890°F) for up to 24 hours. Embrittlement may occur by the formation of ω phase, the formation of eutectoid compounds such as $TiCr_2$, or strain induced martensite from unstable β. More modern alloys are therefore more sophisticated chemically to avoid these problems. A typical modern alloy is Ti-15V-3Sn-3Cr-3Al. Note the low Cr level to avoid $TiCr_2$ and Sn to help suppress ω. Strength/toughness combinations of 1400MPa/50MPa√m (200ksi/45ksi√in) are possible in these alloys. There is only one β type filler metal available commercially, SAE/AMS 4959. These alloys generally have low weld metal strength but good ductility in the as-welded condition. Response to PWHT differs in the weld metal and HAZ and may lead to low ductility in the weld metal. Complete heat treatment, including solution treatment and precipitation heat treatment, of the entire fabrication is necessary if the weld metal must have similar properties to the base metal.

Zirconium alloys are fewer in number. They require the same initial precautions. ERZr2 filler metal is used for commercially pure Zr. There are also filler metals specifically for alloys with Sn (ERZr3) and Cb (ERZr4). Nuclear grade Zr alloys require specialty filler metals with strict control of Hf content.

Refractory Alloys

Refractory alloys, in addition to absorbing atmospheric gases and becoming embrittled, are often brittle at ambient temperature due to a tough/brittle transition temperature well above ambient caused in many cases by a bcc crystal structure. Thus the need to protect the alloys from the atmosphere is still the main welding requirement and the same methods are used. Unalloyed refractory metals and many alloys are weldable without cracking at high or low temperatures. Tungsten and molybdenum and their alloys can crack when cool from residual stresses if welded under restraint. Part of this is caused by grain growth in the HAZ and coarse dendrites in the weld metal. Stress relief heat treatment after welding can alleviate this (900°C or 1650°F for Mo, 1400°C or 2550°F for W, both in inert atmospheres). The high melting temperatures of these alloys place an additional burden on the welding process. The only truly practical processes are GTAW, EBW and LBW.

Chapter

10

ALUMINUM & MAGNESIUM ALLOYS

General metallurgical information on aluminum and magnesium alloys is given in *The Metals Red Book* - Nonferrous Metals. Related AWS Filler Metal Specifications are A5.3, 5.10 and 5.19.

The welding metallurgy of aluminum and magnesium and their alloys is dominated by some shared characteristics. Both alloy systems melt at low temperatures, < 650°C (1200°F). This makes melting and welding fairly easy. Wrought alloys are welded best with inert gas shielded processes, GTAW, PAW or GMAW. Gases are pure Ar or Ar-He mixtures. Pure He can be used, but that is rare due to the cost and poor shielding efficiency (low density of the gas). Aluminum alloys of all compositions suffer mainly from hot cracking in the weld metal. There is some risk of liquation cracking in the HAZ, but it is quite unusual. Solidification cracks are minimized by choosing filler metal composition to avoid weld metal compositions close to the solid solubility limit of the primary alloying element(s) in the base metal. The 2xxx Cu alloys, the 7xxx Mg-Zn alloys and the 8xxx Li alloys give the most trouble.

For magnesium alloys, filler metals of similar chemical composition to the base metal are normal. Preheat is more common than it is with aluminum alloys, to help minimize residual stresses and therefore, weld metal cracking. PWHT is also more common, for the same purpose, and also for minimizing stress corrosion cracking in service.

The solubility of hydrogen is very high in the liquid compared to the solid in both aluminum and magnesium, with a ratio of about 50:1. The drop in solubility as temperature decreases is especially steep just above the melting temperature. This makes both alloy systems very prone to hydrogen porosity in the weld metal. The only remedy is to avoid hydrogen absorption during welding. It is essential to remove hydrocarbons such as oil and grease with solvent degreasing. It is also vital to avoid water on the surface of the metal, as a small amount can cause significant porosity, especially when welding outside. A preheat of about 50°C (120°F) is recommended to remove any condensation, even though aluminum alloys do not normally require it. It is also useful to run an arc on a scrap piece of metal when welding is initiated after a long shutdown, for example overnight. This helps to clear any condensation in the inert gas shielding lines, which can also contribute to startup porosity.

Both aluminum and magnesium react with the atmosphere to produce tightly adherent refractory oxides that melt at high temperatures (> 2000°C or 3630°F) and protect the underlying metal against corrosion. The oxides do not melt during welding and have to be dispersed or removed from the weld area to ensure full fusion. This is usually accomplished by the use of alternating current with the GTAW process and electrode positive polarity in the GMAW process. In both cases, when electrons pass from the base metal to the electrode on electrode positive polarity, the oxide on the weld pool surface is disrupted in a process called cathode sputtering. There may also be some oxide disruption from argon positive ion bombardment of the weld pool surface. Prior to welding, mechanical cleaning of the surface is also important. Oxide thickness should be minimized by scratch brushing, but only with stainless steel brushes. Carbon steel brushes will leave small particles of steel embedded in the surface, which will corrode. Stainless steel particles will not. Stainless brushes, manual or powered, should be designated for Al or Mg use and kept separate from other cleaning equipment to avoid contamination, especially in a shop where metals other than aluminum are fabricated. The stainless steel brushing can also help minimize porosity, since thick oxide can be hydrated, especially after long storage times.

ALUMINUM ALLOYS

Wrought alloys are welded into many types of fabricated items, from electric bus bars to railway hopper cars. Non heat treatable alloys are separated from heat treatable grades for welding. If an aluminum to be welded has an unknown history and the chemical composition is not obtainable, the filler metal of choice is ER4043, containing 5%Si. Virtually any aluminum alloy can be welded satisfactorily without cracking using this filler metal, but optimum properties (mechanical, anodizing response) may not be achieved. Peak cracking tendency in Al-Si alloys occurs at only 0.5%Si, so a large amount of dilution will have virtually no effect.

The 1xxx *pure* aluminum series have the best electrical conductivity and corrosion resistance of all Al alloys. Bus bars need filler metals of similar purity, typically 1040 or better, to maintain electrical continuity. For commercially pure alloys of 1060 type and up, ER1100 is the normal choice. 3xxx grades with Mn are used in petrochemical and food processing. They can be welded with the ubiquitous ER4043, but ER1100 also works well. The 5xxx alloys are best welded with ER 5xxx filler metals. 5xxx materials containing > 3%Mg must not be exposed extensively to temperatures above 120°C (250°F), which cause precipitation of β phase (Mg_5Al_8), leading to stress corrosion problems. This includes most of the commercial alloys. Limitations on interpass temperature of 150°C (300°F) apply. The peak hot cracking tendency in 5xxx alloys is between 1% and 3%Mg. It is useful to minimize the number of filler metals required for a fabrication shop as mistakes are minimized. For the 5xxx alloys, one filler will suffice most of the time and is readily available: ER5356 (5%Mg-0.13%Ti). ER5183 (4.7%Mg-0.12%Ti) and ER5556 (5.1%Mg-0.12%Ti) are acceptable substitutes. The titanium helps to grain refine the weld metal and reduce cracking. Although 4043 can be used as a filler and will prevent cracking, the weld has low toughness due to copious Mg_2Si precipitation.

The heat treatable alloys require great care in the choice of filler metal composition. The 2xxx alloys are difficult to weld without cracking, and cannot be welded with anything close to a similar composition filler. 2219 is the most weldable and has a specific filler metal developed for it, ER2319. ER4043 can also be used but does not respond to PWHT, leaving the weld metal as the weak link in the fabrication. ER4145 (10%Si-4%Cu) was developed to respond to PWHT, but is still weaker than the fully aged base metal. Regaining full strength in the weld metal by using a PWHT is not possible. The 6xxx series have peak cracking tendency at about 1%Mg and 1%Si, so that high Si and high Mg fillers are equally efficient. ER4043 has long been recommended for these alloys, but ER5356 gives equal strength and better ductility and toughness in the weld metal. The 7xxx series has a split personality for welding - low copper levels, for example in 7005 and 7039, are easy to weld, again with ER5356 or substitutes as noted above. High copper levels, for example in 7075 and 7178, have a wide melting range and are very hard to weld without weld metal cracking. The new Al-Li alloys can be welded with ER2319.

Casting alloys are usually welded for two reasons; repair of surface blemishes on new castings or repair of cracks in used castings. Inert gas shielded arc welding processes are again best. SMAW electrodes are available for field use, but have corrosive fluorides and chlorides in the flux which must be completely removed after welding to avoid subsequent corrosion. The most commonly used filler metals for castings are ER4043 and ER5356. For a casting of unknown chemistry or one that is extremely dirty, the near-eutectic silicon brazing alloys 4045 (BAlSi-5) and 4047 (BAlSi-4) can be very useful in avoiding weld metal cracking. ER4145 is useful for welding castings of the Al-Cu or Al-Cu-Si types.

MAGNESIUM ALLOYS

Wrought (primarily extruded) magnesium alloys can be prone to solidification cracking, especially if strength is high, which increases residual stress on the weld metal. Zn alloying additions above 2% cause problems, as do Cd additions. Thorium and rare earth (RE) additions are beneficial, probably due to grain refinement in the weld metal. The weld metal is quite fine grained in Mg alloys generally, leading to weld metal strength in excess of the base metal in most cases. This can cause cracking in the coarse grain HAZ. The use of a low melting point filler metal helps to avoid HAZ cracking. One filler metal, 6.5%Al-1%Zn, is widely used for many Mg alloys for that reason.

Cast magnesium alloys are usually preheated before welding between 300-400°C (570-750°F). Magnesium alloy castings can be readily welded if broken cleanly, or if there is a cold shut or sand inclusion. However, areas containing significant porosity or flux inclusions are difficult to impossible to repair, as is a casting that was impregnated or used as an oil receptacle.

Filler metals for both wrought and cast alloys can be limited to a very few compositions. ER AZ61A (6%Al-1%Zn) is used for Al-Zn alloys to resist weld metal cracking. ER AZ92A (9%Al-2%Zn) is also used for Al-Zn alloys and is preferred for castings. Castings for high temperature use, for example in aircraft, are welded together with ER EZ33A (3%Rare Earth, e.g. Ce and 3%Zn). If high temperature alloys are welded to other alloys, ER AZ92A is preferred.

Chapter
11

OTHER METALS AND ALLOYS

In this chapter, copper alloys and some other alloys which are not often welded will be considered. Some metals and alloys which should not be welded will also be mentioned. For these materials, other joining methods are preferable, for example bolting, adhesive bonding or brazing/soldering. General metallurgical information on these alloys is given in *The Metals Red Book* - Nonferrous Metals. Related AWS Filler Metal Specifications are A5.6, 5.7 and 5.8.

COPPER ALLOYS

For the most part, copper and its alloys can be welded, but they have some peculiar characteristics which need to be addressed. Pure copper and slightly alloyed coppers have very high thermal diffusivity and are difficult to melt, even with electric arcs. Copper dissolves significant amounts of oxygen, which forms a copper-copper oxide eutectic. If any hydrogen is absorbed during welding, large quantities of steam porosity form. Porosity is also a problem in Cu-Ni alloys. Free machining copper alloys contain a variety of additions to improve chip formation, including lead, bismuth selenium and sulphur. All cause significant problems with solidification cracking which cannot be addressed with filler metal modifications. Welding is not recommended. In brasses, zinc fuming can be a problem at levels of Zn in excess of 15%.

Pure copper comes in three types: tough pitch copper, phosphor deoxidized and oxygen free high conductivity copper. All have very high thermal diffusivity and are difficult to melt. Although oxy-acetylene procedures can be used, total heat levels can cause welder discomfort. SMAW electrodes are also available, but the flux residue is corrosive, and gas shielded arc welding (GTAW, PAW or GMAW) is preferable. ERCu fillers are appropriate. Preheat can help, as can the use of shielding gas modifications to increase heat input. In argon shielded processes, there are a few options, and one definite prohibition. Helium can be added to about 50% in argon to increase heat input, as can up to 30% of the diatomic gas nitrogen. High levels of nitrogen can cause some porosity in the weld metal. Hydrogen, which is also diatomic and is used to increase heat input for the welding of austenitic stainless steel, must not be used for tough pitch and nearly pure coppers due to the steam reaction with copper oxide. Deoxidation of the weld pool is also important to avoid porosity. Welding on phosphor deoxidized parent metal is best and filler metals often include deoxidants such as Mn, Si, Ti or Al. Ti and Al can also act as denitrifiers if Ar-N_2 shielding gases are used. High conductivity copper contains very low oxygen levels, and the main welding problem is loss of electrical conductivity in the weld metal due to the deoxidant additions. This conductivity loss can be minimized by using a special filler metal which is deoxidized with boron.

Brasses, i.e. alloys of copper and zinc, have low thermal diffusivity and can be arc welded easily with the GTAW process, as far as melting the alloy is concerned. Zinc fuming, however, is a problem that increases in severity as zinc content increases. Some relief can be obtained by using a phospher (tin) bronze (ERCuSn-A) or silicon bronze (ERCuSi-A) filler metals, or by welding in the overhead position, so that the weld metal recaptures the zinc fume. Brass filler metals (RCuZn) are available, but only for use with oxy-acetylene welding, not arc welding.

Alloys of copper and nickel are called cupronickels when copper rich. The most common compositions are 5-30% nickel. GTAW and GMAW processes are usual, using a similar filler composition, or the 30%Ni composition if not (ERCuNi). Thermal diffusivity is low, so fusion is easy. Since Ni is present, S and sulphur-bearing compounds must be excluded. Be wary of things like paint or crayon markings in the weld vicinity, and also of oil and grease. Lead contamination (from tooling) is to be avoided for the same reason.

Bronzes may be tin, silicon, or aluminum alloys of copper. All have low thermal diffusivity. Tin bronzes are rarely welded - if need be, phosphor or berylium bronze (ERCuSn-A) filler metal is used. Silicon bronzes can be easily welded using similar filler wire compositions, usually with inert gas shielded arc processes. High restraint may lead to hot cracking, so minimal restraint during fabrication is important. Aluminum bronzes are similar to Al alloys, in that the aluminum content is sufficient to form a continuous oxide film on the surface, which improves corrosion resistance. It also interferes with welding the same way, so inert gas processes, with alternating current for GTAW procedures, are mandatory. They are also similar to carbon steels, in that they can have a eutectoid reaction which can lead to martensitic transformation products and they can also be precipitation hardened. The composition for this behaviour is at approximately 9-10%Al. This alloy has low thermal diffusivity and excellent resistance to hot cracking in the weld metal using a similar composition of filler metal (ERCuAl-A3). Lower alloy levels, for example 7%Al, produce single phase alloys that are very prone to weld metal cracking. Using the 10%Al filler reduces weld metal cracking, but the alloy can crack elsewhere during fabrication and welding single phase alloys is not recommended. Berylium bronzes are welded autogenously on occasion, but only under fume hoods or in glove boxes to avoid Berylium contamination, which can be toxic to humans.

OTHER ALLOYS

Silver is very difficult to weld, especially in a commercially pure alloy, due to its very high thermal diffusivity, which exceeds that of pure copper. Silver and Ag-Cu alloys are used for cladding and for liners in vessels for the chemical process industry. Welding is accomplished with argon shielded GTAW procedures using similar composition filler to maintain corrosion resistance. Silver will absorb oxygen and reject it as porosity, so the inert gas shield is necessary. If porosity is a problem, lithium (Li) deoxidized filler metals are available.

Lead and lead alloys can be welded and were some of the first alloys to be arc welded over a century ago. They should not be arc welded now, due to the danger of toxic lead and lead oxide particulate formation. Welding can be done using oxy-fuel gas procedures with virtually no metallurgical problems. Manipulation of the low melting point weld pool is the dominant problem. The main application area is vessel linings for the chemical industry. Beryllium and its alloys are also not to be arc welded in the open due to their toxicity in particulate form. Welding can be done in glove boxes with inert gas shielding.

Zinc alloys, such as those in die castings, should not be welded due to their low melting temperatures. Instant puddles of former castings, but few welds, are produced. Zinc fuming is also a concern. Soldering and brazing are more appropriate processes to use.

Platinum and platinum group metals are usually fusion welded without filler for small applications, such as jewellery or thermocouples. Palladium requires a neutral or slightly oxidizing atmosphere, while Rhodium, iridium, ruthenium and osmium require inert atmospheres.

Chapter

12

CARBON STEEL ELECTRODES FOR SHIELDED METAL ARC WELDING

ELECTRODE CLASSIFICATION

AWS A5.1 Classification	Type of Covering	Welding Position[a]	Type of Current[b]
E6010	High cellulose sodium	F, V, OH, H	dcep
E6011	High cellulose potassium	F, V, OH, H	ac or dcep
E6012	High titania sodium	F, V, OH, H	ac or dcen
E6013	High titania potassium	F, V, OH, H	ac, dcep or dcen
E6019	Iron Oxide titania potassium	F, V, OH, H	ac, dcep or dcen
E6020	High iron oxide	H-fillets, F	ac or dcen
			ac, dcep or dcen
E6022[c]	High iron oxide	F, H	ac or dcen
E6027	High iron oxide, iron powder	H-fillets, F	ac, dcep or dcen
			ac, dcep or dcen
E7014	Iron powder, titania	F, V, OH, H	ac, dcep or dcen
E7015[d]	Low hydrogen sodium	F, V, OH, H	dcep
E7016[d]	Low hydrogen potassium	F, V, OH, H	ac or dcep
E7018[d]	Low hydrogen potassium, iron powder	F, V, OH, H	ac or dcep
E7018M	Low hydrogen iron powder	F, V, OH, H	dcep
E7024[d]	Iron powder, titania	H-fillets, F	ac, dcep or dcen
E7027	High iron oxide, iron powder	H-fillets, F	ac or dcen
			ac, dcep or dcen
E7028[d]	Low hydrogen potassium, iron powder	H-fillets, F	ac or dcep
E7048[d]	Low hydrogen potassium, iron powder	F, OH, H, V-down	ac or dcep

a.   The abbreviations indicate the welding positions as follows:
     F = Flat, H = Horizontal, H-fillets = Horizontal fillets, V-down = Vertical with downward progression
     V = Vertical, OH = Overhead: For electrodes 3/16 in. (4.8 mm) and under, except 5/32 in. (4.0 mm) and under for classifications E7014, E7015, E7016, E7018
     and E7018M.
b.   The term "dcep" refers to direct current electrode positive (dc, reverse polarity). The term "dcen" refers to direct current electrode negative (dc, straight
     polarity).
c.   Electrodes of the E6022 classification are intended for single-pass welds only.
d.   Electrodes with supplemental elongation, notch toughness, absorbed moisture, and diffusible hydrogen requirements may be further identified as shown in
     AWS A5.1 Tables 2, 3, 10 and 11.

CHEMICAL COMPOSITION REQUIREMENTS FOR WELD METAL

AWS A5.1 Classification	UNS Number[a]	Composition, Weight Percent[b]
E6010, E6011, E6012, E6013, E6019, E6020, E6022, E6027	W06010, W06011, W06012, W06013, W06019, W06020, W06022, W06027	-------- Chemical Composition Not Specified --------
E7016, E7018, E7027	W07016, W07018, W07027	C Not Specified Mn 1.60 Si 0.75 P Not Specified S Not Specified Ni 0.30 C 0.20 Mo 0.30 V 0.08 Mn+Ni+Cr+Mo+V 1.75
E7014, E7015, E7024	W07014, W07015, W07024	C Not Specified Mn 1.25 Si 0.90 P Not Specified S Not Specified Ni 0.30 C 0.20 Mo 0.30 V 0.08 Mn+Ni+Cr+Mo+V 1.50
E7028, E7048	W07028, W07048	C Not Specified Mn 1.60 Si 0.90 P Not Specified S Not Specified Ni 0.30 C 0.20 Mo 0.30 V 0.08 Mn+Ni+Cr+Mo+V 1.75
E7018M	W07018	C 0.12 Mn 0.40-1.60 Si 0.80 P 0.030 S 0.020 Ni 0.25 Cr 0.15 Mo 0.35 V 0.05 Mn+Ni+Cr+Mo+V Not Specified

a. SAE/ASTM Unified Numbering System for Metals and Alloys.
b. Single values are maximum.

TENSION TEST REQUIREMENTS[a,b]

AWS A5.1 Classification	Tensile Strength		Yield Strength at 0.2% Offset[c]		% Elongation in 2 in. (50.8 mm)[c]
	ksi	MPa	ksi	MPa	
E6010	60	414	48	331	22
E6011	60	414	48	331	22
E6012	60	414	48	331	17
E6013	60	414	48	331	17
E6019	60	414	48	331	22
E6020	60	414	48	331	22
E6022[d]	60	414	not specified	not specified	not specified
E6027	60	414	48	331	22
E7014	70	482	58	399	17
E7015	70	482	58	399	22
E7016	70	482	58	399	22
E7018	70	482	58	399	22
E7024	70	482	58	399	17[e]
E7027	70	482	58	399	22
E7028	70	482	58	399	22
E7048	70	482	58	399	22
E7018M	g	482	53-72[f]	365-496[f]	24

a.   See AWS A5.1 Table 4 for sizes to be tested.
b.   Requirements are in the as-welded condition with aging as specified in AWS A5.1 paragraph 11.3.
c.   Single values are minimum.
d.   A transverse tension test, as specified in AWS A5.1 paragraph 11.2 and Figure 9 and a longitudinal guided bend test, as specified in AWS A5.1 Section 12, Bend Test, and Figure 10, are required.
e.   Weld metal from electrodes identified as E7024-1 shall have elongation of 22% minimum.
f.   For 3/32 in. (2.44 mm) electrodes, the maximum for the yield strength shall be 77 ksi (531 MPa).
g.   Tensile strength of this weld meal is a nominal 70 ksi (482 MPa).

CHARPY V-NOTCH IMPACT REQUIREMENTS

AWS A5.1 Classification	Limits for 3 out of 5 Specimens[a] Average, min.	Single Value, min.
E6010, E6011, E6027, E7015, E7016[b], E7018[b], E7027, E7048	20 ft•lbf at -20°F (27 J at -29°C)	15 ft•lbf at -20°F (20 J at -29°C)
E6019, E7028	20 ft•lbf at 0°F (27 J at -18°C)	15 ft•lbf at 0°F (20 J at -18°C)
E6012, E6013, E6020, E6022, E7014, E7024[b]	Not Specified	Not Specified

	Limits for 5 out of 5 Specimens[c]	
	Average, min.	Single Value, min.
E7018M	50 ft•lbf at -20°F (67 J at -29°C)	40 ft•lbf at -20°F (54 J at -29°C)

a. Both the highest and lowest test values obtained shall be disregarded in computing the average. Two of these remaining three values shall be equal or exceed 20 ft•lbf (27 J).

b. Electrodes with the following optional supplemental designations shall meet the lower temperature impact requirements specified in the next table (below):

		Charpy V-Notch Impact Requirements, Limits for 3 out of 5 Specimens (Refer to Note a above)	
AWS A5.1 Classification	Electrode Designation	Average, min.	Single Value, min.
E7016	E7016-1	20 ft•lbf at -50°F (27 J at -46°C)	15 ft•lbf at -50°F (20 J at -46°C)
E7018	E7018-1	20 ft•lbf at -50°F (27 J at -46°C)	15 ft•lbf at -50°F (20 J at -46°C)
E7024	E7024-1	20 ft•lbf at 0°F (27 J at -18°C)	15 ft•lbf at 0°F (20 J at -18°C)

c. All five values obtained shall be used in computing the average. Four of the five values shall equal, or exceed, 50 ft•lbf (67 J).

MOISTURE CONTENT LIMITS IN ELECTRODE COVERINGS

AWS A5.1 Classification	Electrode Designation	Limit of Moisture Content, % by Wt., max.	
		As-Received or Conditioned[a]	As-Exposed[b]
E7015	E7015	0.6	Not specified
E7016	E7016, E7016-1	0.6	Not specified
E7018	E7018, E7018-1	0.6	Not specified
E7028	E7028	0.6	Not specified
E7048	E7048	0.6	Not specified
E7015	E7015R	0.3	0.4
E7016	E7016R, E7016-1R	0.3	0.4
E7018	E7018R, E7018-1R	0.3	0.4
E7028	E7028R	0.3	0.4
E7048	E7048R	0.3	0.4
E7018M	E7018M	0.1	0.4

a.   As-received or conditioned electrode coverings shall be tested as specified in AWS A5.1 Section 15, Moisture Test.
b.   As-exposed electrode coverings shall have been exposed to a moist environment as specified in AWS A5.1 paragraphs 16.2 through 16.6 before being tested as specified in 16.1.

DIFFUSIBLE HYDROGEN LIMITS FOR WELD METAL

AWS A5.1 Classification	Diffusible Hydrogen Designator	Diffusible Hydrogen Content, Average mL (H^2)/100g Deposited Metal, Max.[a,b]
E7018M	None	4.0
E7015, E7016, E7018, E7028, E7048	H16, H8, H4	16.0, 8.0, 4.0

a.   Diffusible hydrogen testing in AWS A5.1 Section 17, Diffusible Hydrogen Test, is required for E7018M. Diffusible hydrogen testing of other low hydrogen electrodes is only required when diffusible hydrogen designator is added as specified in AWS A5.1 Figure 16.
b.   Some low hydrogen classifications may not meet the H4 and H8 requirements.

STANDARD SIZES AND LENGTHS

AWS A5.1 Standard Sizes,[a] (Core Wire Diameter)			Standard Lengths [a,b]						
			E6010, E6011, E6012, E6013, E6022, E7014, E7015, E7016, E7018, E7018M		E6020, E6027, E7024, E7027, E7028, E7048		E6019		
in.	(in.)	mm	in.	mm	in.	mm	in.	mm	
1/16 [c]	(0.063)	1.6 [c]	9	230	-	-	-	-	
5/64 [c]	(0.072)	2.0 [c]	9 or 12	230 or 300	-	-	9 or 12	230 or 300	
3/32 [c]	(0.094)	2.4 [c]	12 or 14	300 or 350	12 or 14	300 or 350	12 or 14	300 or 350	
1/8	(0.125)	3.2	14	350	14	350	14	350	
5/32	(0.156)	4.0	14	350	14	350	14 or 18	350 or 450	
3/16	(0.188)	4.8	14 or 18	350 or 460	14 or 18	350 or 460	14 or 18	350 or 450	
7/32 [c]	(0.219)	5.6 [c]	18	460 or 700	18 or 28	460 or 700	18	450	
1/4 [c]	(0.250)	6.4 [c]	18	460 or 700	18 or 28	460 or 700	18	450	
5/16 [c]	(0.313)	8.0 [c]	18	460 or 700	18 or 28	460 or 700	18	450	

a. Lengths and sizes other than these shall be as agreed to by purchaser and supplier. b. In all cases, end-gripped electrodes are standard. c. These diameters are not standard sizes for all classifications (see AWS A5.1 Table 4).

TYPICAL STORAGE AND DRYING CONDITIONS FOR COVERED ARC WELDING ELECTRODES

AWS A5.1 Classification	Storage Conditions[a]		Drying Conditions[b]
	Ambient Air	Holding Ovens	
E6010, E6011	Ambient temperature	Not recommended	Not recommended
E6012, E6013, E6019, E6020, E6022, E6027, E7014, E7024, E7027	80 ± 20°F (30 ± 10°C) 50 percent max relative humidity	20°F (12°C) to 40°F (24°C) above ambient temperature	1 hour at temperature 275 ± 25°F (135 ± 15°C)
E7015, E7016, E7018, E7028, E7018M, E7048	Not Recommended[c]	50°F (30°C) to 250°F (140°C) above ambient temperature	500 to 800°F (260 to 427°C) 1 to 2 hours at temperature

a. After removal from manufacturer's packaging. b. Because of inherent differences in covering composition, the manufacturers should be consulted for the exact drying conditions. c. Some of these electrode classifications may be designated as meeting low moisture absorbing requirements. This designation does not imply that storage in ambient air is recommended.

AWS A5.1 TYPICAL AMPERAGE RANGES

Electrode Diameter in.	mm	E6010 and E6011	E6012	E6013	E6019	E6020	E6022	E6027 and E7027	E7014	E7015 and E7016	E7018M and E7018	E7024 and E7028	E7048
1/16	1.6	-	20 to 40	20 to 40	-	-	-	-	-	-	-	-	-
5/64	2.0	-	25 to 60	25 to 60	35 to 55	-	-	-	-	-	-	-	-
3/32ᵃ	2.4ᵃ	40 to 80	35 to 85	45 to 90	50 to 90	-	-	-	80 to 125	65 to 110	70 to 100	100 to 145	-
1/8	3.2	75 to 125	80 to 140	80 to 130	80 to 140	100 to 150	110 to 160	125 to 185	110 to 160	100 to 150	115 to 165	140 to 190	80 to 140
5/32	4.0	110 to 170	110 to 190	105 to 180	130 to 190	130 to 190	140 to 190	160 to 240	150 to 210	140 to 200	150 to 220	180 to 250	150 to 220
3/16	4.8	140 to 215	140 to 240	150 to 230	190 to 250	175 to 250	170 to 400	210 to 300	200 to 275	180 to 255	200 to 275	230 to 305	210 to 270
7/32	5.6	170 to 250	200 to 320	210 to 300	240 to 310	225 to 310	370 to 520	250 to 350	260 to 340	240 to 320	260 to 340	275 to 365	-
1/4	6.4	210 to 320	250 to 400	250 to 350	310 to 360	275 to 375	-	300 to 420	330 to 415	300 to 390	315 to 400	335 to 430	-
5/16	8.0	275 to 425	300 to 500	320 to 430	360 to 410	340 to 450	-	375 to 475	390 to 500	375 to 475	375 to 470	400 to 525	-

a.   This diameter is not manufactured in the E7028 classification.

AWS A5.1 CLASSIFICATION SYSTEM

The system for electrode classification in this specification follows the standard pattern used in other AWS filler metal specifications. The letter "E" at the beginning of each classification designation stands for electrode. The first two digits, 60, for example, designate tensile strength of at least 60 ksi of the weld metal, produced in accordance with the test assembly preparation section of the specification. The third digit designates position usability that will allow satisfactory welds to be produces with the electrode. Thus, the "1", as in E6010, means that the electrode is usable in all positions (flat, horizontal, vertical, and overhead). The "2", as in E6020 designates that the electrode is suitable for use in flat position and for making fillet welds in the horizontal position. The "4", as in E7048, designates that the electrode is suitable for use in vertical welding with downward progression and for other positions (see AWS A5.1 Table 1). The last two digits taken together designate the type of current with which the electrode can be used and the type of covering on the electrode, as listed in AWS A5.1 Table 1.

Optional designators are also used in this specification in order to identify electrodes that have met the mandatory classification requirements and certain supplementary requirements as agreed to between the supplier and the purchaser. A "-1" designator following classification identifies an electrode which meets optional supplemental impact requirements at a lower temperature than required for the classification (see AWS A5.1 Note b to Table 3). An example of this is the E-7024-1 electrode which meets the classification requirements of E7024 and also meets the optional supplemental requirements for fracture toughness and improved elongation of the weld metal (see AWS A5.1 Note e to Table 2). Certain low hydrogen electrodes also may have optional designators.

A letter "R" is a designator used with the low hydrogen electrode classifications. The letter "R" is used to identify electrodes that have been exposed to a humid environment for a given length of time and tested for moisture absorption in addition to the standard moisture test required for classification of low hydrogen electrodes. See AWS A5.1 Section 16, Absorbed Moisture Test, and AWS A5.1 Table 10.

An optional supplemental designator "HZ" following the four digit classification designators or following the "-1" optional supplemental designator, if used, indicates an average diffusible hydrogen content of not more than "Z" ml/100g of deposited metal when tested in the "as-received" or conditioned state in accordance with ANSI/AWS A4.3, *Standard Methods for Determination of Diffusible Hydrogen Content of Martensitic, Bainitic, and Ferritic Steel Weld Metal Produced by Arc Welding*. Electrodes that are designated as meeting the lower or lowest hydrogen limits, as specified in AWS A5.1 Table 11, are also understood to be able to meet any higher hydrogen limits even though these are not necessarily designated along with the electrode classification. Therefore, as an example, an electrode designated as "H4" also meets "H8" and "H16" requirements without being designated as such. See AWS A5.1 Section 17, Diffusible Hydrogen Test, and AWS A5.1 Table 11.

AWS A5.1 Table A1 shows the classification for similar electrodes from Canadian Standards Association Specification W48.1-M1980, *Mild Steel Covered Arc Welding Electrodes*.

Canadian Electrode Classifications Similar to AWS Classifications

Canadian Electrode Classification[a]	AWS Classification
E41000	-
E41010	E6010
E41011	E6011
E41012	E6012
E41013	E6013
E41022	E6022
E41027	E6027
E48000	-
E48010	-
E48011	-
E48012	-
E48013	-
E48014	E7014
E48015	E7015
E48016	E7016
E48018[b]	E7018
E48022	-
E48024	E7024
E48027	E7027
E48028	E7028
E48048	E7048

a.   From CSA Standard W48.1-M1980, *Mild Steel Covered Arc Welding Electrodes, published by Canadian Standards Association*, 178 Rexdale Boulevard, Rexdale, Ontario, Canada M9W 1R3.

b.   Also includes E48018-1 designated electrode.

Chapter
13

CARBON & LOW ALLOY STEEL RODS FOR OXYFUEL GAS WELDING

TENSION TEST REQUIREMENTS

AWS A5.2 Classification	Tensile Strength[a], min.		% Elongation in 1-in. (25 mm), min.
	ksi	MPa	
R45	Not Specified		Not Specified
R60	60	410	20
R65	65	450	16
R100	100	690	14
RXXX-G[b]	XXX[b]		Not Specified

a. Specimens shall be tested in the as-welded condition.

b. Classification designators (XXX) shall be based on minimum tensile strength of all-weld-metal tension test of the test assembly. These designators shall be limited to 45, 60, 65, 70, 80, 90, and 100.

CHEMICAL COMPOSITION REQUIREMENTS FOR WELDING RODS AND ROD STOCK

AWS A5.2 Classification	UNS Number[a]	Composition, Weight Percent[b]
R45	K00045	C 0.08 Mn 0.50 Si 0.10 P 0.035 S 0.040 Cu 0.30 Cr 0.20 Ni 0.30 Mo 0.20 Al 0.02
R60	K00060	C 0.15 Mn 0.90 to 1.40 Si 0.10 to 0.35 P 0.035 S 0.035 Cu 0.30 Cr 0.20 Ni 0.30 Mo 0.20 Al 0.02
R65	K00065	C 0.15 Mn 0.90 to 1.60 Si 0.10 to 0.70 P 0.035 S 0.035 Cu 0.30 Cr 0.40 Ni 0.30 Mo 0.20 Al 0.02
R100	K12147	C 0.18 to 0.23 Mn 0.70 to 0.90 Si 0.20 to 0.35 P 0.025 S 0.025 Cu 0.15 Cr 0.40 to 0.60 Ni 0.40 to 0.70 Mo 0.15 to 0.25 Al 0.02
RXXX-G[c]	Not Specified	

a. SAE/ASTM Unified Numbering System for Metals and Alloys.

b. Single values are maximums.

c. Designators, "XXX" correspond to minimum strength of weld metal in the nearest ksi. See AWS A5.2 Note "b" of Table 1.

AWS A5.2 CLASSIFICATION SYSTEM

The system for identifying the rod classifications in this specification follows the standard pattern used in other AWS filler metal specifications. The letter "R" at the beginning of each classification designation stands for rod. The digits (45, 60, 65, and 100) designate a minimum tensile strength of the weld metal, in the nearest thousands of pounds per square inch, deposited in accordance with the test assembly preparation section of this specification.

This specification includes filler metals classified as RXXX-G. The "G" indicates that the filler metal is of a "general" classification. It is general because not all of the particular requirements specified for each of the other classifications are specified for this classification. The intent in establishing this classification is to provide a means by which filler metals that differ in one respect or another (chemical composition, for example) from all other classifications (meaning that the composition of the filler metal, in the case of the example, does not meet the composition specified for any of the classifications in the specification) can still be classified according to the specification. The purpose is to allow a useful filler metal - one that otherwise would have to await a revision of the specification - to be classified immediately under the existing specification. This means, then, that two filler metals, each bearing the same "G" classification, may be quite different in some certain respect (chemical composition, again, for example).

The point of difference (although not necessarily the amount of the difference) referred to above will be readily apparent from the use of the words "not required" and "not specified" in the specification.

The use of these words is as follows:

"Not Specified" is used in those areas of the specification that refer to the results of some particular test. It indicates that the requirements for that particular test are *not specified* for that particular classification.

"Not Required" is used in those areas of the specification that refer to the test that must be conducted in order to classify a filler metal (or a welding material). It indicates that the test is *not required* because the requirements (results) for the test have not been specified for that particular classification.

Restating the case, when a requirement is not specified, it is not necessary to conduct the corresponding test in order to classify a filler metal to that classification. When a purchaser wants the information provided by that test in order to consider a particular product of that classification for a certain application, the purchaser will have to arrange for that information with the supplier of that product. The purchaser will also have to establish with that supplier just what the testing procedure and the acceptance requirements are to be for that test. The purchaser may want to incorporate that information (via ANSI/AWS A5.01, *Filler Metal Procurement Guidelines*) in the purchase order.

Chapter
14

ALUMINUM & ALUMINUM ALLOY ELECTRODES FOR SHIELDED METAL ARC WELDING

CHEMICAL COMPOSITION REQUIREMENTS FOR CORE WIRE (WEIGHT PERCENT)[a,b]

AWS A5.3 Classification	UNS Designation[c]	Composition, Weight Percent
E1100	A91100	Si (d) Fe (d) Cu 0.05-0.20 Mn 0.05 Zn 0.10 Be 0.0008 Other Elements Each 0.05 Other Elements Total 0.15 Al 99.00 min[e]
E3003	A93003	Si 0.6 Fe 0.7 Cu 0.05-0.20 Mn 1.0-1.5 Zn 0.10 Be 0.0008 Other Elements Each 0.05 Other Elements Total 0.15 Al Remainder
E4043	A94043	Si 4.5-6.0 Fe 0.8 Cu 0.30 Mn 0.05 Mg 0.05 Zn 0.10 Ti 0.20 Be 0.0008 Other Elements Each 0.05 Other Elements Total 0.15 Al Remainder

a. The core wire, or the stock from which it is made, shall be analyzed for the specific elements for which values are shown in this table. If the presence of other elements is indicated in the course of work, the amount of those elements shall be determined to ensure that they do not exceed the limits specified for "Other Elements".

b. Single values are maximum, except where otherwise specified.

c. SAE/ASTM Unified Numbering System for Metals and Alloys.

d. Silicon plus iron shall not exceed 0.95 percent.

e. The aluminum content for unalloyed aluminum is the difference between 100.00 percent and the sum of all other metallic elements present in amounts of 0.010 percent or more each, expressed to the second decimal before determining the sum.

TENSION TEST REQUIREMENTS[a]

AWS A5.3 Classification	Tensile Strength, min.	
	psi	MPa
E1100	12000	82.7
E3003, E4043	14000	96.5

a. Fracture may occur in either the base metal or the weld metal.

AWS A5.3 CLASSIFICATION SYSTEM

The system for identifying the electrode classifications in this specification follows the standard pattern used in other AWS filler metal specifications. The letter E at the beginning of each classification designation stands for electrode. The numerical portion of the designation in this specification conforms to the Aluminum Association registration for the composition of the core wire used in the electrode.

Chapter
15

STAINLESS STEEL ELECTRODES
FOR SHIELDED METAL ARC WELDING

CHEMICAL COMPOSITION REQUIREMENTS FOR UNDILUTED WELD METAL

AWS A5.4 Classification[c]	UNS Number[d]	Weight Percent[a,b]
E209-XX[e]	W32210	C 0.06 Cr 20.5-24.0 Ni 9.5-12.0 Mo 1.5-3.0 Mn 4.0-7.0 Si 0.90 P 0.04 S 0.03 N 0.10-0.30 Cu 0.75
E219-XX	W32310	C 0.06 Cr 19.0-21.5 Ni 5.5-7.0 Mo 0.75 Mn 8.0-10.0 Si 1.00 P 0.04 S 0.03 N 0.10-0.30 Cu 0.75
E240-XX	W32410	C 0.06 Cr 17.0-19.0 Ni 4.0-6.0 Mo 0.75 Mn 10.5-13.5 Si 1.00 P 0.04 S 0.03 N 0.10-0.30 Cu 0.75
E307-XX	W30710	C 0.04-0.14 Cr 18.0-21.5 Ni 9.0-10.7 Mo 0.5-1.5 Mn 3.30-4.75 Si 0.90 P 0.04 S 0.03 Cu 0.75
E308-XX	W30810	C 0.08 Cr 18.0-21.0 Ni 9.0-11.0 Mo 0.75 Mn 0.5-2.5 Si 0.90 P 0.04 S 0.03 Cu 0.75
E308H-XX	W30810	C 0.04-0.08 Cr 18.0-21.0 Ni 9.0-11.0 Mo 0.75 Mn 0.5-2.5 Si 0.90 P 0.04 S 0.03 Cu 0.75
E308L-XX	W30813	C 0.04 Cr 18.0-21.0 Ni 9.0-11.0 Mo 0.75 Mn 0.5-2.5 Si 0.90 P 0.04 S 0.03 Cu 0.75
E308Mo-XX	W30820	C 0.08 Cr 18.0-21.0 Ni 9.0-12.0 Mo 2.0-3.0 Mn 0.5-2.5 Si 0.90 P 0.04 S 0.03 Cu 0.75
E308MoL-XX	W30823	C 0.04 Cr 18.0-21.0 Ni 9.0-12.0 Mo 2.0-3.0 Mn 0.5-2.5 Si 0.90 P 0.04 S 0.03 Cu 0.75
E309-XX	W30910	C 0.15 Cr 22.0-25.0 Ni 12.0-14.0 Mo 0.75 Mn 0.5-2.5 Si 0.90 P 0.04 S 0.03 Cu 0.75
E309L-XX	W30913	C 0.04 Cr 22.0-25.0 Ni 12.0-14.0 Mo 0.75 Mn 0.5-2.5 Si 0.90 P 0.04 S 0.03 Cu 0.75
E309Cb-XX	W30917	C 0.12 Cr 22.0-25.0 Ni 12.0-14.0 Mo 0.75 Cb (Nb) plus Ta 0.70-1.00 Mn 0.5-2.5 Si 0.90 P 0.04 S 0.03 Cu 0.75
E309Mo-XX	W30920	C 0.12 Cr 22.0-25.0 Ni 12.0-14.0 Mo 2.0-3.0 Mn 0.5-2.5 Si 0.90 P 0.04 S 0.03 Cu 0.75
E309-MoL-XX	W30923	C 0.04 Cr 22.0-25.0 Ni 12.0-14.0 Mo 2.0-3.0 Mn 0.5-2.5 Si 0.90 P 0.04 S 0.03 Cu 0.75
E310-XX	W31010	C 0.08-0.20 Cr 25.0-28.0 Ni 20.0-22.5 Mo 0.75 Mn 1.0-2.5 Si 0.75 P 0.03 S 0.03 Cu 0.75
E310H-XX	W31015	C 0.35-0.45 Cr 25.0-28.0 Ni 20.0-22.5 Mo 0.75 Mn 1.0-2.5 Si 0.75 P 0.03 S 0.03 Cu 0.75
E310Cb-XX	W31017	C 0.12 Cr 25.0-28.0 Ni 20.0-22.0 Mo 0.75 Cb (Nb) plus Ta 0.70-1.00 Mn 1.0-2.5 Si 0.75 P 0.03 S 0.03 Cu 0.75
E310Mo-XX	W31020	C 0.12 Cr 25.0-28.0 Ni 20.0-22.0 Mo 2.0-3.0 Mn 1.0-2.5 Si 0.75 P 0.03 S 0.03 Cu 0.75
E312-XX	W31310	C 0.15 Cr 28.0-32.0 Ni 8.0-10.5 Mo 0.75 Mn 0.5-2.5 Si 0.90 P 0.04 S 0.03 Cu 0.75
E316-XX	W31610	C 0.08 Cr 17.0-20.0 Ni 11.0-14.0 Mo 2.0-3.0 Mn 0.5-2.5 Si 0.90 P 0.04 S 0.03 Cu 0.75
E316H-XX	W31610	C 0.04-0.08 Cr 17.0-20.0 Ni 11.0-14.0 Mo 2.0-3.0 Mn 0.5-2.5 Si 0.90 P 0.04 S 0.03 Cu 0.75
E316L-XX	W31613	C 0.04 Cr 17.0-20.0 Ni 11.0-14.0 Mo 2.0-3.0 Mn 0.5-2.5 Si 0.90 P 0.04 S 0.03 Cu 0.75
E317-XX	W31710	C 0.08 Cr 18.0-21.0 Ni 12.0-14.0 Mo 3.0-4.0 Mn 0.5-2.5 Si 0.90 P 0.04 S 0.03 Cu 0.75
E317L-XX	W31713	C 0.04 Cr 18.0-21.0 Ni 12.0-14.0 Mo 3.0-4.0 Mn 0.5-2.5 Si 0.90 P 0.04 S 0.03 Cu 0.75
E318-XX	W31910	C 0.08 Cr 17.0-20.0 Ni 11.0-14.0 Mo 2.0-3.0 Cb (Nb) plus Ta 6 X C, min to 1.00 max Mn 0.5-2.5 Si 0.90 P 0.04 S 0.03 Cu 0.75

CHEMICAL COMPOSITION REQUIREMENTS FOR UNDILUTED WELD METAL (Continued)

AWS A5.4 Classification[c]	UNS Number[d]	Weight Percent [a,b]
E320-XX	W88021	C 0.07 Cr 19.0-21.0 Ni 32.0-36.0 Mo 2.0-3.0 Cb (Nb) plus Ta 8 X C, min to 1.00 max Mn 0.5-2.5 Si 0.60 P 0.04 S 0.03 Cu 3.0-4.0
E320LR-XX	W88022	C 0.03 Cr 19.0-21.0 Ni 32.0-36.0 Mo 2.0-3.0 Cb (Nb) plus Ta 8 X C, min to 0.40 max Mn 1.50-2.50 Si 0.30 P 0.020 S 0.015 Cu 3.0-4.0
E330-XX	W88331	C 0.18-0.25 Cr 14.0-17.0 Ni 33.0-37.0 Mo 0.75 Mn 1.0-2.5 Si 0.90 P 0.04 S 0.03 Cu 0.75
E330H-XX	W88335	C 0.35-0.45 Cr 14.0-17.0 Ni 33.0-37.0 Mo 0.75 Mn 1.0-2.5 Si 0.90 P 0.04 S 0.03 Cu 0.75
E347-XX	W34710	C 0.08 Cr 18.0-21.0 Ni 9.0-11.0 Mo 0.75 Cb (Nb) plus Ta 8 X C, min to 1.00 max Mn 0.5-2.5 Si 0.90 P 0.04 S 0.03 Cu 0.75
E349-XX[e,f,g]	W34910	C 0.13 Cr 18.0-21.0 Ni 8.0-10.0 Mo 0.35-0.65 Cb (Nb) plus Ta 0.75-1.20 Mn 0.5-2.5 Si 0.90 P 0.04 S 0.03 Cu 0.75
E383-XX	W88028	C 0.03 Cr 26.5-29.0 Ni 30.0-33.0 Mo 3.2-4.2 Mn 0.5-2.5 Si 0.90 P 0.02 S 0.02 Cu 0.6-1.5
E385-XX	W88904	C 0.03 Cr 19.5-21.5 Ni 24.0-26.0 Mo 4.2-5.2 Mn 1.0-2.5 Si 0.75 P 0.03 S 0.02 Cu 1.2-2.0
E410-XX	W41010	C 0.12 Cr 11.0-13.5 Ni 0.7 Mo 0.75 Mn 1.0 Si 0.90 P 0.04 S 0.03 Cu 0.75
E410NiMo-XX	W41016	C 0.06 Cr 11.0-12.5 Ni 4.0-5.0 Mo 0.40-0.70 Mn 1.0 Si 0.90 P 0.04 S 0.03 Cu 0.75
E430-XX	W43010	C 0.10 Cr 15.0-18.0 Ni 0.6 Mo 0.75 Mn 1.0 Si 0.90 P 0.04 S 0.03 Cu 0.75
E502-XX[h]	W50210	C 0.10 Cr 4.0-6.0 Ni 0.4 Mo 0.45-0.65 Mn 1.0 Si 0.90 P 0.04 S 0.03 Cu 0.75
E505-XX[h]	W50410	C 0.10 Cr 8.0-10.5 Ni 0.4 Mo 0.85-1.20 Mn 1.0 Si 0.90 P 0.04 S 0.03 Cu 0.75
E630-XX	W37410	C 0.05 Cr 16.00-16.75 Ni 4.5-5.0 Mo 0.75 Cb (Nb) plus Ta 0.15-0.30 Mn 0.25-0.75 Si 0.75 P 0.04 S 0.03 Cu 3.25-4.00
E16-8-2-XX	W36810	C 0.10 Cr 14.5-16.5 Ni 7.5-9.5 Mo 1.0-2.0 Mn 0.5-2.5 Si 0.60 P 0.03 S 0.03 Cu 0.75
E7Cr-XX[h]	W50310	C 0.10 Cr 6.0-8.0 Ni 0.04 Mo 0.45-0.65 Mn 1.0 Si 0.90 P 0.04 S 0.03 Cu 0.75
E2209-XX	W39209	C 0.04 Cr 21.5-23.5 Ni 8.5-10.5 Mo 2.5-3.5 Mn 0.5-2.0 Si 0.90 P 0.04 S 0.03 N 0.08-0.20 Cu 0.75
E2553-XX	W39553	C 0.06 Cr 24.0-27.0 Ni 6.5-8.5 Mo 2.9-3.9 Mn 0.5-1.5 Si 1.0 P 0.04 S 0.03 N 0.10-0.25 Cu 1.5-2.5

a. Analysis shall be made for the elements for which specific values are shown in the table. If, however, the presence of other elements is indicated in the course of routine analysis, further analysis shall be made to determine that the total of these other elements, except iron, is not present in excess of 0.50 percent. b. Single values are maximum percentages. c. Classification suffix -XX may be -15, -16, -17, -25, or -26. See AWS A5.4 Section A8 of the Appendix for an explanation. d. SAE/ASTM Unified Number System for Metals and Alloys. e. Vanadium shall be from 0.10 to 0.30 percent. f. Titanium shall be 0.15 percent max. g. Tungsten shall be from 1.25 to 1.75 percent. h. This grade also will appear in the next revision of AWS A5.5, Specification for Low Alloy Steel Electrodes for Shielded Metal Arc Welding. It will be deleted from AWS A5.4 at the first revision of A5.4 following publication of the revised A5.5

TYPE OF WELDING CURRENT AND POSITION OF WELDING

AWS A5.4 Classification[a]	Welding Current[b]	Welding Position[c]
EXXX(X)-15	dcep	All[d]
EXXX(X)-25	dcep	H, F
EXXX(X)-16	dcep or ac	All[d]
EXXX(X)-17	dcep or ac	All[d]
EXXX(X)-26	dcep or ac	H, F

a. See AWS A5.4 Section A8, Classification as to Useability, for explanation of positions.

b. dcep = Direct current electrode positive (reverse polarity); ac = Alternating current.

c. The abbreviations F and H indicate welding positions (AWS A5.4 Figure 3) as follows:

   F = Flat

   H = Horizontal

d. Electrodes 3/16 in. (4.8 mm) and larger are not recommended for welding all positions.

ALL-WELD-METAL MECHANICAL PROPERTY REQUIREMENTS

AWS A5.4 Classification	Tensile Strength, min		% Elongation, min.	Heat Treatment
	ksi	MPa		
E209-XX	100	690	15	None
E219-XX	90	620	15	None
E240-XX	100	690	15	None
E307-XX	85	590	30	None
E308-XX	80	550	35	None
E308H-XX	80	550	35	None
E308L-XX	75	520	35	None
E308Mo-XX	80	550	35	None
E308MoL-XX	75	520	35	None
E309-XX	80	550	30	None
E309Cb-XX	80	550	30	None
E309Mo-XX	80	550	30	None

ALL-WELD-METAL MECHANICAL PROPERTY REQUIREMENTS (Contined)

AWS A5.4 Classification	Tensile Strength, min		% Elongation, min.	Heat Treatment
	ksi	MPa		
E309MoL-XX	75	520	30	None
E310-XX	80	550	30	None
E310H-XX	90	620	10	None
E310Cb-XX	80	550	25	None
E310Mo-XX	80	550	30	None
E312-XX	95	660	22	None
E316-XX	75	520	30	None
E316H-XX	75	520	30	None
E316L-XX	70	490	30	None
E317-XX	80	550	30	None
E317L-XX	75	520	30	None
E318-XX	80	550	25	None
E320-XX	80	550	30	None
E320LR-XX	75	520	30	None
E330-XX	75	520	25	None
E330H-XX	90	620	10	None
E347-XX	75	520	30	None
E349-XX	100	690	25	None
E383-XX	75	520	30	None
E385-XX	75	520	30	None
E410-XX	75	450	20	a
E410NiMo-XX	110	760	15	c
E430-XX	65	450	20	d
E502-XX	60	420	20	b
E505-XX	60	420	20	b
E630-XX	135	930	7	e
E16-8-2-XX	80	550	35	None

ALL-WELD-METAL MECHANICAL PROPERTY REQUIREMENTS

AWS A5.4 Classification	Tensile Strength, min		% Elongation, min.	Heat Treatment
	ksi	MPa		
E7Cr-XX	60	420	20	b
E2209-XX	100	690	20	None
E2553-XX	110	760	15	None

a. Heat to 1350 to 1400°F (730 to 760°C), hold for one hour, furnace cool at a rate of 100°F (60°C) per hour to 600°F (315°C) and air cool to ambient.

b. Heat to 1550 to 1600°F (840 to 870°C), hold for two hours, furnace cool at a rate not exceeding 100°F (55°C) per hour to 1100°F (595°C) and air cool to ambient.

c. Heat to 1100 to 1150°F (595 to 620°C), hold for one hour, and air cool to ambient.

d. Heat to 1400 to 1450°F (760 to 790°C), hold for two hours, furnace cool at a rate not exceeding 100°F (55°C) per hour to 1100°F (595°C) and air cool to ambient.

e. Heat to 1875 to 1925°F (1025 to 1050°C), hold for one hour, and air cool to ambient, and then precipitation harden at 1135 to 1165°F (610 to 630°C), hold for four hours, and air cool to ambient.

STANDARD SIZES AND LENGTHS

AWS A5.4 Electrode Size, (Diameter of Core Wire)[a]		Standard Lengths[b,c]	
in.	mm	in.	mm
1/16	1.6	9	230
5/64	2.0	9	230
3/32	2.4	9, 12, 14[d]	230, 305, 350[d]
1/8	3.2	14, 18[d]	350, 460[d]
5/32	4.0	14, 18[d]	350, 460[d]
3/16	4.8	14, 18[d]	350, 460[d]
7/32	5.6	14, 18	350, 460
1/4	6.4	14, 18	350, 460

a. Tolerance on the diameter shall be ± 0.002 in. (± 0.05 mm).

b. Tolerance on length shall be ± 1/4 in. (± 6.4 mm).

c. Other sizes and lengths shall be as agreed upon between purchaser and supplier.

d. These lengths are intended only for the EXXX-25 and EXXX-26 types.

AWS A5.4 CLASSIFICATION SYSTEM

The system of classification is similar to that used in other filler metal specifications. The letter "E" at the beginning of each number indicates an electrode. The first three digits designate the classification as to its composition. (Occasionally, a number of digits other than three is used and letters may follow the digits to indicate a specific composition.) The last two digits designate the classification as to usability with respect to position of welding and type of current as described in AWS A5.4 Appendix A8. The smaller sizes of EXXX(X)-15, EXXX(X)-16, or EXXX(X)-17 electrodes [up to and including 5/32 in. (4.0 mm)] included in this specification are used in all welding positions.

The mechanical tests measure strength and ductility, qualities which are often of lesser importance than the corrosion and heat resisting properties. These mechanical test requirements, however, provide an assurance of freedom from weld metal flaws, such as check cracks and serious dendritic segregations which, if present, may cause failure in service.

It is recognized that for certain applications, supplementary tests may be required. In such cases, additional tests to determine specific properties, such as corrosion resistance, scale resistance, or strength at elevated temperatures may be required as agreed upon between supplier and purchaser.

Chapter
16

LOW ALLOY STEEL COVERED ARC WELDING ELECTRODES

ELECTRODE CLASSIFICATION

AWS A5.5 Classification	Type of Covering	Capable of producing satisfactory welds in positions shown[b]	Type of Current[c]
E70 series - Minimum tensile strength of deposited metal, 70,000 psi (480 MPa)			
E7010-X	High cellulose sodium	F, V, OH, H	DCEP
E7011-X	High cellulose potassium	F, V, OH, H	AC or DCEP
E7015-X	Low hydrogen sodium	F, V, OH, H	DCEP
E7016-X	Low hydrogen potassium	F, V, OH, H	AC or DCEP
E7018-X	Iron powder, low hydrogen	F, V, OH, H	AC or DCEP
E7020-X	High iron oxide	H-fillets	AC or DCEN
E7020-X	High iron oxide	F	AC or DC, either polarity
E7027-X	Iron powder, iron oxide	H-fillets	AC or DCEN
E7027-X	Iron powder, iron oxide	F	AC or DC, either polarity
E80 series - Minimum tensile strength of deposited metal, 80,000 psi (550 MPa)			
E8010-X	High cellulose sodium	F, V, OH, H	DCEP
E8011-X	High cellulose potassium	F, V, OH, H	AC or DCEP
E8013-X	High titania potassium	F, V, OH, H	AC or DC, either polarity
E8015-X	Low hydrogen sodium	F, V, OH, H	DCEP
E8016-X	Low hydrogen potassium	F, V, OH, H	AC or DCEP
E8018-X	Iron powder, low hydrogen	F, V, OH, H	AC or DCEP
E90 series - Minimum tensile strength of deposited metal, 90,000 psi (620 MPa)			
E9010-X	High cellulose sodium	F, V, OH, H	DCEP
E9011-X	High cellulose potassium	F, V, OH, H	AC or DCEP
E9013-X	High titania potassium	F, V, OH, H	AC or DC, either polarity
E9015-X	Low hydrogen sodium	F, V, OH, H	DCEP
E9016-X	Low hydrogen potassium	F, V, OH, H	AC or DCEP
E9018-X	Iron powder, low hydrogen	F, V, OH, H	AC or DCEP
E100 series - Minimum tensile strength of deposited metal, 100,000 psi (690 MPa)			
E10010-X	High cellulose sodium	F, V, OH, H	DCEP

ELECTRODE CLASSIFICATION (Continued)

AWS A5.5 Classification	Type of Covering	Capable of producing satisfactory welds in positions shown[b]	Type of Current[c]
E100 series - Minimum tensile strength of deposited metal, 100,000 psi (690 MPa) (Continued)			
E10011-X	High cellulose potassium	F, V, OH, H	AC or DCEP
E10013-X	High titania potassium	F, V, OH, H	AC or DC, either polarity
E10015-X	Low hydrogen sodium	F, V, OH, H	DCEP
E10016-X	Low hydrogen potassium	F, V, OH, H	AC or DCEP
E10018-X	Iron powder, low hydrogen	F, V, OH, H	AC or DCEP
E110 series - Minimum tensile strength of deposited metal, 110,000 psi (760 MPa)			
E11015-X	Low hydrogen sodium	F, V, OH, H	DCEP
E11016-X	Low hydrogen potassium	F, V, OH, H	AC or DCEP
E11018-X	Iron powder, low hydrogen	F, V, OH, H	AC or DCEP
E120 series - Minimum tensile strength of deposited metal, 120,000 psi (830 MPa)			
E12015-X	Low hydrogen sodium	F, V, OH, H	DCEP
E12016-X	Low hydrogen potassium	F, V, OH, H	AC or DCEP
E12018-X	Iron powder, low hydrogen	F, V, OH, H	AC or DCEP

a. The letter suffix 'X' as used in this table stands for the suffixes, A1, B1, B2, etc. and designates the chemical composition of the deposited weld metal.

b. The abbreviations, F, V, OH, H and H-fillets indicate welding positions as follows:

F = Flat; H = Horizontal; H-fillets = Horizontal fillets.
V = Vertical; OH = Overhead: For electrodes 3/16 in. (4.8 mm) and under, except 5/32 in. (4.0 mm) and under for classifications EXX15-X, EXX16-X, and EXX18-X.

c. DCEP means electrode positive (reverse polarity). DCEN means electrode negative (straight polarity).

CHEMICAL REQUIREMENTS[i]

AWS A5.5 Classification[a]	Chemical Composition, Weight Percent[b]
Carbon-molybdenum steel electrodes	
E7010-A1	C 0.12 Mn 0.60 P 0.03 S 0.04 Si 0.40 Mo 0.40-0.65
E7011-A1	C 0.12 Mn 0.60 P 0.03 S 0.04 Si 0.40 Mo 0.40-0.65
E7015-A1	C 0.12 Mn 0.90 P 0.03 S 0.04 Si 0.60 Mo 0.40-0.65
E7016-A1	C 0.12 Mn 0.90 P 0.03 S 0.04 Si 0.60 Mo 0.40-0.65
E7018-A1	C 0.12 Mn 0.90 P 0.03 S 0.04 Si 0.80 Mo 0.40-0.65
E7020-A1	C 0.12 Mn 0.60 P 0.03 S 0.04 Si 0.40 Mo 0.40-0.65
E7027-A1	C 0.12 Mn 1.00 P 0.03 S 0.04 Si 0.40 Mo 0.40-0.65
Chromium-molybdenum steel electrodes	
E8016-B1	C 0.05 to 0.12 Mn 0.90 P 0.03 S 0.04 Si 0.60 Cr 0.40-0.65 Mo 0.40-0.65
E8018-B1	C 0.05 to 0.12 Mn 0.90 P 0.03 S 0.04 Si 0.80 Cr 0.40-0.65 Mo 0.40-0.65
E8015-B2L	C 0.05 Mn 0.90 P 0.03 S 0.04 Si 1.00 Cr 1.00-1.50 Mo 0.40-0.65
E8016-B2	C 0.05 to 0.12 Mn 0.90 P 0.03 S 0.04 Si 0.60 Cr 1.00-1.50 Mo 0.40-0.65
E8018-B2	C 0.05 to 0.12 Mn 0.90 P 0.03 S 0.04 Si 0.80 Cr 1.00-1.50 Mo 0.40-0.65
E8018-B2L	C 0.05 Mn 0.90 P 0.03 S 0.04 Si 0.80 Cr 1.00-1.50 Mo 0.40-0.65
E9015-B3L	C 0.05 Mn 0.90 P 0.03 S 0.04 Si 1.00 Cr 2.00-2.50 Mo 0.90-1.20
E9015-B3	C 0.05 to 0.12 Mn 0.90 P 0.03 S 0.04 Si 0.60 Cr 2.00-2.50 Mo 0.90-1.20
E9016-B3	C 0.05 to 0.12 Mn 0.90 P 0.03 S 0.04 Si 0.60 Cr 2.00-2.50 Mo 0.90-1.20
E9018-B3	C 0.05 to 0.12 Mn 0.90 P 0.03 S 0.04 Si 0.80 Cr 2.00-2.50 Mo 0.90-1.20
E9018-B3L	C 0.05 Mn 0.90 P 0.03 S 0.04 Si 0.80 Cr 2.00-2.50 Mo 0.90-1.20
E8015-B4L	C 0.05 Mn 0.90 P 0.03 S 0.04 Si 1.00 Cr 1.75-2.25 Mo 0.40-0.65
E8016-B5	C 0.07 to 0.15 Mn 0.40 to 0.70 P 0.03 S 0.04 Si 0.30-0.60 Cr 0.40-0.60 Mo 1.00-1.25 V 0.05
Nickel steel electrodes	
E8016-C1	C 0.12 Mn 1.25 P 0.03 S 0.04 Si 0.60 Ni 2.00-2.75
E8018-C1	C 0.12 Mn 1.25 P 0.03 S 0.04 Si 0.80 Ni 2.00-2.75
E7015-C1L	C 0.05 Mn 1.25 P 0.03 S 0.04 Si 0.50 Ni 2.00-2.75
E7016-C1L	C 0.05 Mn 1.25 P 0.03 S 0.04 Si 0.50 Ni 2.00-2.75
E7018-C1L	C 0.05 Mn 1.25 P 0.03 S 0.04 Si 0.50 Ni 2.00-2.75

CHEMICAL REQUIREMENTS[1] (Continued)

AWS A5.5 Classification[a]	Chemical Composition, Percent[b]
Nickel steel electrodes (Continued)	
E8016-C2	C 0.12 Mn 1.25 P 0.03 S 0.04 Si 0.60 Ni 3.00-3.75
E8018-C2	C 0.12 Mn 1.25 P 0.03 S 0.04 Si 0.80 Ni 3.00-3.75
E7015-C2L	C 0.05 Mn 1.25 P 0.03 S 0.04 Si 0.50 Ni 3.00-3.75
E7016-C2L	C 0.05 Mn 1.25 P 0.03 S 0.04 Si 0.50 Ni 3.00-3.75
E7018-C2L	C 0.05 Mn 1.25 P 0.03 S 0.04 Si 0.50 Ni 3.00-3.75
E8016-C3[c]	C 0.12 Mn 0.40-1.25 P 0.03 S 0.03 Si 0.80 Ni 0.80-1.10 Cr 0.15 Mo 0.35 V 0.05
E8018-C3[c]	C 0.12 Mn 0.40-1.25 P 0.03 S 0.03 Si 0.80 Ni 0.80-1.10 Cr 0.15 Mo 0.35 V 0.05
Nickel-molybdenum steel electrodes	
E1018-NM[d]	C 0.10 Mn 0.80-1.25 P 0.02 S 0.03 Si 0.60 Ni 0.80-1.10 Cr 0.05 Mo 0.40-0.65 V 0.02
Manganese-molybdenum steel electrodes	
E-9015-D1	C 0.12 Mn 1.25-1.75 P 0.03 S 0.04 Si 0.60 Mo 0.25-0.45
E-9018-D1	C 0.12 Mn 1.25-1.75 P 0.03 S 0.04 Si 0.80 Mo 0.25-0.45
E8016-D3	C 0.12 Mn 1.00-1.75 P 0.03 S 0.04 Si 0.60 Mo 0.40-0.65
E8018-D3	C 0.12 Mn 1.00-1.75 P 0.03 S 0.04 Si 0.80 Mn 0.40-0.65
E10015-D2	C 0.15 Mn 1.65-2.00 P 0.03 S 0.04 Si 0.60 Mn 0.25-0.45
E10016-D2	C 0.15 Mn 1.65-2.00 P 0.03 S 0.04 Si 0.60 Mn 0.25-0.45
E10018-D2	C 0.15 Mn 1.65-2.00 P 0.03 S 0.04 Si 0.80 Mn 0.25-0.45
All other low-alloy steel electrodes[e]	
EXX10-G[e]	Mn 1.00 min[f] Si 0.80 min[f] Ni 0.50 min[f] Cr 0.30 min[f] Mn 0.20 min[f] V 0.10 min[f]
EXX11-G	Mn 1.00 min[f] Si 0.80 min[f] Ni 0.50 min[f] Cr 0.30 min[f] Mn 0.20 min[f] V 0.10 min[f]
EXX13-G	Mn 1.00 min[f] Si 0.80 min[f] Ni 0.50 min[f] Cr 0.30 min[f] Mn 0.20 min[f] V 0.10 min[f]
EXX15-G	Mn 1.00 min[f] Si 0.80 min[f] Ni 0.50 min[f] Cr 0.30 min[f] Mn 0.20 min[f] V 0.10 min[f]
EXX16-G	Mn 1.00 min[f] Si 0.80 min[f] Ni 0.50 min[f] Cr 0.30 min[f] Mn 0.20 min[f] V 0.10 min[f]
EXX18-G	Mn 1.00 min[f] Si 0.80 min[f] Ni 0.50 min[f] Cr 0.30 min[f] Mn 0.20 min[f] V 0.10 min[f]
E7020-G	Mn 1.00 min[f] Si 0.80 min[f] Ni 0.50 min[f] Cr 0.30 min[f] Mn 0.20 min[f] V 0.10 min[f]
E9018-M[c]	C 0.10 Mn 0.60-1.25 P 0.030 S 0.030 Si 0.80 Ni 1.40-1.80 Cr 0.15 Mn 0.35 V 0.05
E10018- M[c]	C 0.10 Mn 0.75-1.70 P 0.030 S 0.030 Si 0.60 Ni 1.40-2.10 Cr 0.35 Mn 0.25-0.50 V 0.05

CHEMICAL REQUIREMENTS[i] (Continued)

AWS A5.5 Classification[a]	Chemical Composition, Percent[b]
All other low-alloy steel electrodes[e] (Continued)	
E11018- M[c]	C 0.10 Mn 1.30-1.80 P 0.030 S 0.030 Si 0.60 Ni 1.25-2.50 Cr 0.40 Mn 0.25-0.50 V 0.05
E12018- M[c]	C 0.10 Mn 1.30-2.25 P 0.030 S 0.030 Si 0.60 Ni 1.75-2.50 Cr 0.30-1.50 Mn 0.30-0.55 V 0.05
E12018- M1[c]	C 0.10 Mn 0.80-1.60 P 0.015 S 0.012 Si 0.65 Ni 3.00-3.80 Cr 0.65 Mn 0.20-0.30 V 0.05
E7018- W[g]	C 0.12 Mn 0.40-0.70 P 0.025 S 0.025 Si 0.40-0.70 Ni 0.20-0.40 Cr 0.15-0.30 V 0.08
E8018- W[h]	C 0.12 Mn 0.50-1.30 P 0.03 S 0.04 Si 0.35-0.80 Ni 0.40-0.80 Cr 0.45-0.70

a. The suffixes A1, B3, C2, etc. designate the chemical composition of the electrode classification.

b. For determining the chemical composition, DCEN (electrode negative) may be used where DC, both polarities, is specified.

c. These classifications are intended to conform to classifications covered by the military specifications for similar compositions.

d. Copper shall be 0.10% max and aluminum shall be 0.05% max for E8018-NM electrodes.

e. The letters "XX" used in the classification designations in this table stand for the various strength levels (70, 80, 90, 100, 110, and 120) of electrodes.

f. In order to meet the alloy requirements of the G group, the weld deposit need have the minimum, as specified in the table, of only one of the elements listed. Additional chemical requirements may be as agreed between supplier and purchaser.

g. Copper shall be 0.30 to 0.60% for E7018-W electrodes.

h. Copper shall be 0.30 to 0.75% for E8018-W electrodes.

I. Single values shown are maximum percentages, except where otherwise specified.

TENSILE STRENGTH, YIELD STRENGTH, AND ELONGATION REQUIREMENTS FOR ALL-WELD-METAL TENSION TEST[a,b]

AWS A5.5 Classification	Tensile strength, min[d]		Yield strength, at 0.2% offset[d,e]		% Elongation, min
	ksi	MPa	ksi	MPa	
E7010-X, E7011-X	70	480	57[f]	390[f]	22
E7015-X, E7016-X, E7018-X, E7020-X, E7027-X	70	480	57[f]	390[f]	25
E8010-X, E8011-X	80	550	67	460	19
E8013-X	80	550	67	460	16
E8015-X, E8016-X, E8018-X	80	550	67	460	19
E8016-C3, E8018-C3	80	550	68 to 80	470 to 550	24
E9010-X, E9011-X	90	620	77	530	17
E9013-X	90	620	77	530	14
E9015-X, E9016-X, E9018-X	90	620	77	530	17
E9018-M	90	620	78 to 90	540 to 620	24
E10010-X, E10011-X	100	690	87	600	16
E10013-X	100	690	87	600	13
E10015-X, E10016-X, E10018-X	100	690	87	600	16
E10018-M	100	690	88 to 100	610 to 690	20
E11015-X, E11016-X, E11018-X	110	760	97	670	15
E11018-M	110	760	98 to 110	680 to 760	20
E12015-X, E12016-X, E12018-X	120	830	107	740	14
E12018-M, E12018-M1	120[g]	830[g]	108 to 120	745 to 830	18

a. For specimens which are tested in the as-welded or stress-relieved condition, see AWS A5.5 Table 11 and Table 12 for details. The "M" classifications shall meet the requirements of equivalent strength EXX18-X classifications when tested on AC. b. See AWS A5.5 Table 3 for sizes to be tested. c. The letter suffix "X" as used in this table stands for all the suffixes (A1, B1, B2, etc.) except the M and C3 suffixes. d. Single values shown are minimums. e. Yield strength may be increased 5000 psi (35 MPa) max for 3/32 in. (2.4 mm) "M" classifications. f. For the as-welded condition, the required yield strength is 60 ksi (415 MPa). g. The minimum UTS for the E12018-M1 classification is nominally 120 ksi (830 MPa). However, the required UTS may be other than 120 ksi (830 MPa) as agreed between supplier and purchaser.

The Metals Blue Book

IMPACT PROPERTY REQUIREMENTS

AWS A5.5 Classification	Charpy V-notch impact requirement, min[a]
E8018-NM, E8016-C3, E8018-C3	20 ft•lbf at -40°F (27 J at -40°C)
E8016-D3, E8018-D3, E9015-D1, E9018-D1, E10015-D2, E10016-D2, E10018-D2	20 ft•lbf at -60°F[b] (27 J at -51°C)
E9018-M, E10018-M, E11018-M, E12018-M	20 ft•lbf at -60°F (27 J at -51°C)
E12018-M1	50 ft•lbf at 0°F (68 J at -18°C)
E7018-W, E8018-W	20 ft•lbf at 0°F (27 J at -18°C)
E8016-C1, E8018-C1	20 ft•lbf at -75°F[b] (27 J at -59°C)
E7015-C1L, E7016-C1L, E7018-C1L, E8016-C2, E8018-C2	20 ft•lbf at -100°F[b] (27 J at -73°C)
E7015-C2L, E7016-C2L, E7018-C2L	20 ft•lbf at -150°F[b] (27 J at -101°C)
All other classifications	Not required

a. The lowest value obtained, together with the highest value, shall be disregarded for this test. See AWS A5.5 paragraph 3.8 for details of test value determination.

b. Stress-relieved impact properties.

STANDARD SIZES AND LENGTHS

AWS A5.5 Standard sizes, (core wire diameter)[a]		Standard lengths[b,c,d] All AWS A5.5 classifications except E7020-A1, E7020-G and E7027-A1		Standard lengths[b,c,d] AWS A5.5 classifications E7020-A1, E7020-G and E7027-A1 classifications	
in.	mm	in.	mm	in.	mm
3/32[e] (0.093)	2.4[e]	12[f]	300[f]	12	300
1/8 (0.125)	3.2	14	350	14	350
5/32 (0.156)	4.0	14	350	14	350
3/16 (0.187)	4.8	14	350		
7/32[e] (0.218)	5.6[e]	14 or 18	350 or 450	14 or 18	350 or 450
1/4[e] (0.250)	6.4[e]	18	450	18 or 28	450 or 700
5/16[e] (0.312)	8.0[e]	18	450	18 or 28	450 or 700
				18 or 28	450 or 700

a. Tolerance on the core wire diameter shall be ± 0.002 in. (± 0.05 mm). Electrodes produced in sizes other than those shown may be classified. See AWS A5.5 footnote "d" of Table 3 for more details. b. Tolerance on the length shall be ± 1/4 in. (± 10 mm). c. In all cases, end gripping is standard. d. Other lengths are acceptable and shall be as agreed upon by the supplier and purchaser. e. These diameters are not manufactured in all electrode classifications (see AWS A5.5 Table 3 for more details). f. 12 or 14 in. (300 or 350 mm) length for EXX18-X classification only.

COATING MOISTURE CONTENT REQUIREMENTS

AWS A5.5 Classification[a]	Coating moisture content, maximum percent by weight (all low hydrogen electrodes)
E7015-X	0.4
E7016-X	0.4
E7018-X	0.4
E8015-X	0.2
E8016-X	0.2
E8018-X	0.2
E9015-X	0.15
E9016-X	0.15
E9018-X	0.15
E10015-X	0.15
E10016-X	0.15
E10018-X	0.15
E11015-X	0.15
E11016-X	0.15
E11018-X	0.15
E12015-X	0.15
E12016-X	0.15
E12018-X	0.15
E12018-M1	0.10

a.   The letter suffix "X" used in this table stands for all the suffixes (A1, B2, C3, M, etc.) except for the E12018-M1 classification.

AWS A5.5 CLASSIFICATION SYSTEM

The classification system used in this specification follows the established pattern. The letter E designates an electrode; the first two digits (or three digits of a five digit number) 70 for example, designate the minimum tensile strength of the deposited metal in 1000 psi. The third digit (or fourth digit of a five digit number) indicates the position in which satisfactory welds can be made with the electrode. Thus, the "1", as in E7010, means that the electrode is satisfactory for use in all positions (flat, vertical, overhead, and horizontal). The "2", as in E7020, indicates that the electrode is suitable for the flat position and also for making fillet welds in the horizontal position. The last two digits, taken together, indicate the type current with which the electrode can be used, and the type of covering on the electrode. In addition, a letter suffix, such as A1, designates the chemical composition of the deposited weld metal. Thus, a complete classification of an electrode would be E7010-A1, E8016-C2, etc.

Note: The specific chemical compositions are not always identified with specific mechanical properties in the specification. However, a supplier is required by the AWS A5.5 specification to include the mechanical properties appropriate for a particular electrode in classification of that electrode. Thus, a complete designation is E8016-C2; EXX16-C2 is not a complete classification.

Chapter
17

COVERED COPPER & COPPER ALLOY ARC WELDING ELECTRODES

CHEMICAL COMPOSITION REQUIREMENTS FOR UNDILUTED METAL

AWS A5.6 Classification	UNS Number	Common Name	Composition, Weight Percent[a,b]
ECu	W60189	Copper	Cu including Ag remainder, Zn f Sn f Mn 0.10 Fe 0.20 Si 0.10 Ni[d] f P f Al 0.10 Pb 0.02 Total Other Elements 0.50
ECuSi	W60656	Silicon bronze (copper silicon)	Cu including Ag remainder, Zn f Sn 1.5 Fe 0.50 Si 1.4 to 4.0 Ni[d] f P f Al 0.01 Pb 0.02 Total Other Elements 0.50
ECuSn-A	W60518	Phosphor bronze (copper-tin)	Cu including Ag remainder, Zn f Sn 4.0-6.0 Mn f Fe 0.25 Si f Ni[d] f P 0.05-0.35 Al 0.01 Pb 0.02 Total Other Elements 0.50
ECuSn-C	W60521	Phosphor bronze (copper-tin)	Cu including Ag remainder, Zn f Sn 7.0-9.0 Mn f Fe 0.25 Si f Ni[d] f P 0.05-0.35 Al 0.01 Pb 0.02 Total Other Elements 0.50
ECuNi[e]	W60751	Copper nickel (70/30)	Cu including Ag remainder, Zn f Sn f Mn 1.00-2.50 Fe 0.40-0.75 Si 0.50 Ni[d] 29.0-33.0 P 0.020 Pb 0.02 Ti 0.50 Total Other Elements 0.50
ECuAl-A2	W60614	Aluminum bronze	Cu including Ag remainder, Zn f Sn f Mn f Fe 0.50-5.0 Si 1.5 Ni[d] f Al 6.5-9.0 Pb 0.02 Total Other Elements 0.50
ECuAl-B	W60619	Aluminum bronze	Cu including Ag remainder, Zn f Sn f Mn f Fe 2.5-5.0 Si 1.5 Ni[d] f Al 7.5-10.0 Pb 0.02 Total Other Elements 0.50
ECuNiAl	W60632	Nickel aluminum bronze	Cu including Ag remainder, Zn f Sn f Mn 0.50-3.5 Fe 3.0-6.0 Si 1.5 Ni[d] 4.0-6.0 Al 6.0-8.5 Pb 0.02 Total Other Elements 0.50
ECuMnNiAl	W60633	Manganese-nickel aluminum bronze	Cu including Ag remainder, Zn f Sn f Mn 11.0-13.0 Fe 2.0-6.0 Si 1.5 Ni[d] 1.0-2.5 Al 5.0-7.5 Pb 0.02 Total Other Elements 0.50

a. Analysis shall be made for the elements for which specific values or an "f" are shown in this table. If, however, the presence of other elements is indicated in the course of routine analysis, further analysis shall be made to determine that the total of these other elements is not present in excess of the limits specified for 'total other elements' in the last column in the table.

b. Single values shown are maximum.

c. SAE/ASTM Unified Number System for Metals and Alloys.

d. Includes cobalt.

e. Sulfur shall be restricted to 0.015 percent for the ECuNi classification.

f. Those elements must be included in total of other elements.

MECHANICAL PROPERTY REQUIREMENTS

AWS A5.6 Classification	Tensile strength, min		% Elongation, in 4 X Dia. - gage length[a], min.
	ksi	MPa	
ECu	25	170	20
ECuSi	50	350	20
ECuSn-A	35	240	20
ECuSn-C	40	280	20
ECuNi	50	350	20
ECuAl-A2	60	410	20
ECuAl-B	65	450	10
ECuNiAl	72	500	10
ECuMnNiAl	75	520	15

a. See AWS A5.6 Figure 5 for specimen dimensions.

STANDARD SIZES OF ELECTRODES

AWS A5.6 Classification	Standard sizes, diameter of core wire, in. (mm)			
	3/32 (2.4)	1/8 (3.2)	5/32 (4.0)	3/16 (4.8)
ECu, ECuSi, ECuSn-A, ECuSn-C, ECuNi	X	X	X	X
ECuAl-A2	X	X	X	X
ECuAl-B	-	X	X	X
ECuNiAl	-	X	X	X
ECuMnNiAl	X	X	X	X

| HARDNESS OF COPPER AND COPPER ALLOY WELD METAL DEPOSITED USING COVERED ELECTRODES[a] | | |
AWS A5.6 Classification	Hardness	Testing Load
ECu	20 to 40 HRF	
ECuSi	80 to 100 HB	(500 kg load)
ECuSn-A	70 to 85 HB	(500 kg load)
ECuSn-C	85 to 100 HB	(500 kg load)
ECuNi	60 to 80 HB	(500 kg load)
ECuAl-A2	130 to 150 HB	(3000 kg load)
ECuAl-B	140 to 180 HB	(3000 kg load)
ECuNiAl	160 to 200 HB	(3000 kg load)
ECuMnNiAl	160 to 200 HB	(3000 kg load)

a. Hardness values as listed above are average values for undiluted weld metal deposited in accordance with this specification. This table is included for information only; hardness testing is not required under AWS A5.6.

AWS A5.6 CLASSIFICATION SYSTEM

The system for identifying the electrode classifications is as follows:

The letter E at the beginning of each number indicates a covered electrode.

The chemical symbol Cu is used to identify the electrodes as copper-base alloys, and the additional chemical symbol, such as Si in ECuSi, Sn in ECuSn, etc., indicates the principal alloying element of each classification or group of similar classifications. Where more than one classification is included in a basic group, the individual classifications in the group are identified by the letters A, B, C, etc. as in ECuSn-A. Further subdividing is done by using a 1, 2, etc. after the last letter, as the 2 in ECuAl-2.

Chapter
18

COPPER & COPPER ALLOY BARE WELDING RODS & ELECTRODES

CHEMICAL COMPOSITION REQUIREMENTS, PERCENT

AWS A5.7 Classification	UNS Number[d]	Common Name	Composition, Weight Percent [a,b,c]
ERCu	C18980	Copper	Cu including Ag 98.0 min, Sn 1.0 Mn 0.50 Si 0.50 P 0.15 Al 0.01 Pb 0.02 Total Other Elements 0.50
ERCuSi-A	C65600	Silicon bronze (copper-silicon)	Cu including Ag remainder, Zn 1.0 Sn 1.0 Mn 1.5 Fe 0.50 Si 2.8-4.0 Al 0.01 Pb 0.02 Total Other Elements 0.50
ERCuSn-A	C51800	Phosphor bronze (copper-tin)	Cu including Ag remainder, Sn 4.0-6.0 P 0.10-0.35 Al 0.01 Pb 0.02 Total Other Elements 0.50
ERCuNi[e]	C71580	Copper-nickel	Cu including Ag remainder, Mn 1.00 Fe 0.40-0.75 Si 0.25 Ni Including Co 29.0-32.0 P 0.02 Pb 0.02 Ti 0.20 to 0.50 Total Other Elements 0.50
ERCuAl-A1	C61000	Aluminum bronze	Cu including Ag remainder, Zn 0.20 Mn 0.50 Si 0.10 Al 6.0-8.5 Pb 0.02 Total Other Elements 0.50
ERCuAl-A2	C61800	Aluminum bronze	Cu including Ag remainder, Zn 0.02 Fe 1.5 Si 0.10 Al 8.5-11.0 Pb 0.02 Total Other Elements 0.50
ERCuAl-A3	C62400	Aluminum bronze	Cu including Ag remainder, Zn 0.10 Fe 2.0-4.5 Si 0.10 Al 10.0-11.5 Pb 0.02 Total Other Elements 0.50
ERCuNiAl	C63280	Nickel-aluminum bronze	Cu including Ag remainder, Zn 0.10 Mn 0.60-3.50 Fe 3.0-5.0 Si 0.10 Ni Including Co 4.0-5.50 Al 8.50-9.50 Pb 0.02 Total Other Elements 0.50
ERCuMnNiAl	C63380	Manganese-nickel aluminum bronze	Cu including Ag remainder, Zn 0.15 Mn 11.0-14.0 Fe 2.0-4.0 Si 0.10 Ni Including Co 1.5-3.0 Al 7.0-8.5 Pb 0.02 Total Other Elements 0.50

a. Analysis shall be made for the elements for which specific values are shown in this table. If, however, the presence of other elements is indicated in the course of routine analysis, further analysis shall be made to determine that the total of these other elements is not present in excess of the limits specified for 'Total other elements' in the last column in this table.

b. Single values shown are maximum, unless otherwise noted.

c. Classifications RBCuZn-A, RCuZn-B, RCuZn-C, and RBCuZn-D now are included in AWS A5.27-78, *Specification for Copper and Copper Alloy Gas Welding Rods*.

d. SAE/ASTM Unified Numbering System for Metals and Alloys.

e. Sulfur shall be 0.01 percent maximum for the ERCuNi classification.

STANDARD SIZES

AWS A5.7 Form	Diameter[a,b]		
	in.	in.	mm
Straight lengths	1/16	(0.062)	1.6
Straight lengths	5/64	(0.078)	2.0
Straight lengths	3/32	(0.092)	2.4
Coils, with or without support	1/8	(0.125)	3.2
Coils, with or without support	5/32	(0.156)	4.0
Coils, with or without support	3/16	(0.187)	4.8
Coils, with or without support	1/4	(0.250)	6.4
Wound on spools	0.020		0.5
Wound on spools	0.030		0.8
Wound on spools	0.035		0.9
Wound on spools	0.045		1.2
Wound on spools	0.062	(1/16)	1.6
Wound on spools	0.078	(5/64)	2.0
Wound on spools	0.094	(3/32)	2.4

a. Filler metal shall not vary more than ± 0.002 in. (0.05 mm) in diameter.
b. Other sizes, lengths, and forms may be supplied as agreed upon between the purchaser and supplier.

AWS A5.7 Classification	Hardness Testing		Tensile strength, min.	
	Hardness	Testing Load	psi	MPa
ERCu	25 HRF		25 000	172
ERCuSi-A	80 to 100 HB	(500 kg load)	50 000	345
ERCuSn-A	70 to 85 HB	(500 kg load)	35 000	240
ERCuNi	60 to 80 HB	(500 kg load)	50 000	345
ERCuAl-A1	80 to 110 HB	(500 kg load)	55 000	380
ERCuAl-A2	130 to 150 HB	(3000 kg load)[b]	60 000	414
ERCuAl-A3	140 to 180 HB	(3000 kg load)[b]	65 000	450
ERCuNiAl	160 to 200 HB	(3000 kg load)[b]	72 000	480
ERCuMnNiAl	160 to 200 HB	(3000 kg load)[b]	75 000	515

HARDNESS AND TENSILE STRENGTH OF COPPER AND COPPER ALLOY WELD METAL[a]

a. Hardness values as listed above are average values for an as-welded deposit made with the filler metal specified. This table is included for information only.
b. Gas tungsten arc process only.

AWS A5.7 CLASSIFICATION SYSTEM

The specification classifies those copper and copper alloy filler metals used most extensively at the time of issuance of the specification. In AWS A5.7 appendix A4, the filler metals are arranged in five basic groups. The tensile properties, bend ductility, and soundness of welds produced with the filler metals classified within this specification frequently are determined during procedure qualification. It should be noted that the variables in the procedure (current, voltage, and welding speed), variables in shielding medium (the specific gas mixture or the flux), variables in the composition of the base metal and the filler metal influence the results which may be obtained. When these variables are properly controlled, however, the filler metal shall give sound welds whose strengths (determined by all-weld-metal tension tests) will meet or exceed the minimums shown in AWS A5.7 Table A1. Typical hardness properties are also included in AWS A5.7 Table A1. When supplementary tests for mechanical properties are specified, the procedures should be in accordance with AWS B4.0, *Standard Methods for Mechanical Testing of Welds*.

The system for identifying the filler metal classification in this specification follows the standard pattern used in other AWS filler metal specifications. The letters ER at the beginning of a classification indicate that the bare filler metal may be used either as an electrode or as a welding rod.

The chemical symbol Cu is used to identify the filler metals as copper-base alloys. The additional chemical symbols, as the Si in ERCuSi, the Sn in ERCuSn, etc., indicate the principal alloying element of each group. Where more than one classification is included in a basic group, the individual classifications in the group are identified by the letters, A, B, C, etc., as in ERCuSn-A. Further subdividing is done by using 1, 2, etc., after the last letter, as the 2 in ERCuAl-2.

Chapter

19

FILLER METALS FOR BRAZING & BRAZE WELDING

CHEMICAL COMPOSITION REQUIREMENTS FOR SILVER FILLER METALS

AWS A5.8 Classification	UNS Number[a]	Composition, Weight Percent
BAg-1	P07450	Ag 44.0-46.0 Cu 14.0-16.0 Zn 14.0-18.0 Cd 23.0-25.0 Other Elements Total[b] 0.15
BAg-1a	P07500	Ag 49.0-51.0 Cu 14.5-16.5 Zn 14.5-18.5 Cd 17.0-21.0 Other Elements Total[b] 0.15
BAg-2	P07350	Ag 34.0-36.0 Cu 25.0-27.0 Zn 19.0-23.0 Cd 17.0-21.0 Other Elements Total[b] 0.15
BAg-2a	P07300	Ag 29.0-31.0 Cu 26.0-28.0 Zn 21.0-25.0 Cd 19.0-21.0 Other Elements Total[b] 0.15
BAg-3	P07501	Ag 49.0-51.0 Cu 14.5-16.5 Zn 13.5-17.5 Cd 15.0-17.0 Ni 2.5-3.5 Other Elements Total[b] 0.15
BAg-4	P07400	Ag 39.0-41.0 Cu 29.0-31.0 Zn 26.0-30.0 Ni 1.5-2.5 Other Elements Total[b] 0.15
BAg-5	P07453	Ag 44.0-46.0 Cu 29.0-31.0 Zn 23.0-27.0 Other Elements Total[b] 0.15
BAg-6	P07503	Ag 49.0-51.0 Cu 33.0-35.0 Zn 14.0-18.0 Other Elements Total[b] 0.15
BAg-7	P07563	Ag 55.0-57.0 Cu 21.0-23.0 Zn 15.0-19.0 Sn 4.5-5.5 Other Elements Total[b] 0.15
BAg-8	P07720	Ag 71.0-73.0 Cu Remainder Other Elements Total[b] 0.15
BAg-8a	P07723	Ag 71.0-73.0 Cu Remainder Li 0.25-0.50 Other Elements Total[b] 0.15
BAg-9	P07650	Ag 64.0-66.0 Cu 19.0-21.0 Zn 13.0-17.0 Other Elements Total[b] 0.15
BAg-10	P07700	Ag 69.0-71.0 Cu 19.0-21.0 Zn 8.0-12.0 Other Elements Total[b] 0.15
BAg-13	P07540	Ag 53.0-55.0 Cu Remainder Zn 4.0-6.0 Ni 0.5-1.5 Other Elements Total[b] 0.15
BAg-13a	P07560	Ag 55.0-57.0 Cu Remainder Ni 1.5-2.5 Other Elements Total[b] 0.15
BAg-18	P07600	Ag 59.0-61.0 Cu Remainder Sn 9.5-10.5 Other Elements Total[b] 0.15
BAg-19	P07925	Ag 92.0-93.0 Cu Remainder Li 0.15-0.30 Other Elements Total[b] 0.15
BAg-20	P07301	Ag 29.0-31.0 Cu 37.0-39.0 Zn 30.0-34.0 Other Elements Total[b] 0.15
BAg-21	P07630	Ag 62.0-64.0 Cu 27.5-29.5 Ni 2.0-3.0 Sn 5.0-7.0 Other Elements Total[b] 0.15
BAg-22	P07490	Ag 48.0-50.0 Cu 15.0-17.0 Zn 21.0-25.0 Ni 4.0-5.0 Mn 7.0-8.0 Other Elements Total[b] 0.15
BAg-23	P07850	Ag 84.0-86.0 Mn Remainder Other Elements Total[b] 0.15
BAg-24	P07505	Ag 49.0-51.0 Cu 19.0-21.0 Zn 26.0-30.0 Ni 1.5-2.5 Other Elements Total[b] 0.15
BAg-26	P07250	Ag 24.0-26.0 Cu 37.0-39.0 Zn 31.0-35.0 Ni 1.5-2.5 Mn 1.5-2.5 Other Elements Total[b] 0.15
BAg-27	P07251	Ag 24.0-26.0 Cu 34.0-36.0 Zn 24.5-28.5 Cd 12.5-14.5 Other Elements Total[b] 0.15
BAg-28	P07401	Ag 39.0-41.0 Cu 29.0-31.0 Zn 26.0-30.0 Sn 1.5-2.5 Other Elements Total[b] 0.5
BAg-33	P07252	Ag 24.0-26.0 Cu 29.0-31.0 Zn 26.5-28.5 Cd 16.5-18.5 Other Elements Total[b] 0.15
BAg-34	P07380	Ag 37.0-39.0 Cu 31.0-33.0 Zn 26.0-30.0 Sn 1.5-2.5 Other Elements Total[b] 0.15
BAg-35	P07351	Ag 34.0-36.0 Cu 31.0-33.0 Zn 31.0-35.0 Other Elements Total[b] 0.15

CHEMICAL COMPOSITION REQUIREMENTS FOR SILVER FILLER METALS (Continued)

AWS A5.8 Classification	UNS Number[a]	Composition, Weight Percent
BAg-36	P07454	Ag 44.0-46.0 Cu 26.0-28.0 Zn 23.0-27.0 Sn 2.5-3.5 Other Elements Total[b] 0.15
BAg-37	P07253	Ag 24.0-26.0 Cu 39.0-41.0 Zn 31.0-35.0 Sn 1.5-2.5 Other Elements Total[b] 0.15

a. SAE/ASTM Unified Numbering System for Metals and Alloys.
b. The brazing filler metal shall be analyzed for those specific elements for which values are shown in this table. If the presence of other elements is indicated in the course of this work, the amount of those elements shall be determined to ensure that their total does not exceed the limit specified.

See AWS A5.8 Table A2 for discontinued brazing filler metal classification.
See AWS A5.8 Table 6 for the following Ag classifications not included here: BVAg-0, BVAg-6b, BVAg-8b, BVAg-29 to BVAg-32.

CHEMICAL COMPOSITION REQUIREMENTS FOR GOLD FILLER METALS

AWS A5.8 Classification	UNS Number[a]	Composition, Weight Percent
BAu-1	P00375	Au 37.0-38.0 Cu Remainder Other Elements Total[b] 0.15
BAu-2	P00800	Au 79.5-80.5 Cu Remainder Other Elements Total[b] 0.15
BAu-3	P00350	Au 34.5-35.5 Cu Remainder Other Elements Total[b] 0.15
BAu-4	P00820	Au 81.5-82.5 Ni Remainder Other Elements Total[b] 0.15
BAu-5	P00300	Au 29.5-30.5 Pd 33.5-34.5 Ni 35.5-36.5 Other Elements Total[b] 0.15
BAu-6	P00700	Au 69.5-70.5 Pd 7.5-8.5 Ni 21.5-22.5 Other Elements Total[b] 0.15

a. SAE/ASTM Unified Numbering System for Metals and Alloys.
b. The brazing filler metal shall be analyzed for those specific elements for which values are shown in this table. If the presence of other elements is indicated in the course of this work, the amount of those elements shall be determined to ensure that their total does not exceed the limit specified.

CHEMICAL COMPOSITION REQUIREMENTS FOR ALUMINUM AND MAGNESIUM FILLER METALS

AWS A5.8 Classification	UNS Number[b]	Chemical Composition, Weight Percent[a]
BAlSi-2	A94343	Si 6.8-8.2 Cu 0.25 Fe 0.8 Zn 0.20 Mn 0.10 Al Remainder Other Elements[c] Each 0.05 Other Elements[c] Total 0.15
BAlSi-3	A94145	Si 9.3-10.7 Cu 3.3-4.7 Mg 0.15 Fe 0.8 Zn 0.20 Mn 0.15 Cr 0.15 Al Remainder Other Elements[c] Each 0.05 Other Elements[c] Total 0.15
BAlSi-4	A94047	Si 11.0-13.0 Cu 0.30 Mg 0.10 Fe 0.8 Zn 0.20 Mn 0.15 Al Remainder Other Elements[c] Each 0.05 Other Elements[c] Total 0.15
BAlSi-5	A94045	Si 9.0-13.0 Cu 0.30 Mg 0.05 Fe 0.8 Zn 0.10 Mn 0.05 Ti 0.20 Al Remainder Other Elements[c] Each 0.05 Other Elements[c] Total 0.15
BAlSi-7	A94004	Si 9.0-10.5 Cu 0.25 Mg 1.0-2.0 Fe 0.8 Zn 0.20 Mn 0.10 Al Remainder Other Elements[c] Each 0.05 Other Elements[c] Total 0.15
BAlSi-9	A94147	Si 11.0-13.0 Cu 0.25 Mg 0.10-0.5 Fe 0.8 Zn 0.20 Mn 0.10 Al Remainder Other Elements[c] Each 0.05 Other Elements[c] Total 0.15
BAlSi-11	A94104	Si 9.0-10.5 Cu 0.25 Mg 1.0-2.0 Bi 0.02-0.20 Fe 0.8 Zn 0.20 Mn 0.10 Al Remainder Other Elements[c] Each 0.05 Other Elements[c] Total 0.15
BMg-1	M19001	Si 0.05 Cu 0.05 Mg Remainder Fe 0.005 Zn 1.7-2.3 Mn 0.15-1.5 Ni 0.005 Be 0.0002-0.0008 Al 8.3-9.7 Other Elements[c] Total 0.30

a.   Single values are maximum, unless otherwise noted.

b.   SAE/ASTM Unified Numbering System for Metals and Alloys.

c.   The filler metal shall be analyzed for those specific elements for which values are shown in this table.   If the presence of other elements is indicated in the course of this work, the amount of those elements shall be determined to ensure that their total does not exceed the limit specified.

CHEMICAL COMPOSITION REQUIREMENTS FOR COPPER, COPPER-ZINC, AND COPPER-PHOSPHORUS FILLER METALS

AWS A5.8 Classification	UNS Number[b]	Chemical Composition, Weight Percent[a]
BCu-1[i]	C14180	Cu 99.90 min P 0.075 Pb 0.02 Al 0.01* Other Elements Total[c] 0.10[i]
BCu-1a	-	Cu 99.00 min[d] Other Elements Total[c] 0.30[d]
BCu-2[e]	-	Cu 86.50 min Other Elements Total[c] 0.50
RBCuZn-A[i]	C47000	Cu 57.0-61.0 Zn Remainder Sn 0.25-1.00 Fe * Mn * Pb 0.05* Al 0.01* Si * Other Elements Total[c] 0.50[i]
RBCuZn-B[i]	C68000	Cu 56.0-60.0[g] Zn Remainder Sn 0.80-1.10 Fe 0.25-1.20 Mn 0.01-0.50 Ni 0.20-0.80[h] Pb 0.05* Al 0.01* Si 0.04-0.15 Other Elements Total[c] 0.50[i]
RBCuZn-C[i]	C68100	Cu 56.0-60.0 Zn Remainder Sn 0.80-1.10 Fe 0.25-1.20 Mn 0.01-1.20 Pb 0.05* Al 0.01* Si 0.04-0.15 Other Elements Total[c] 0.50[i]
RBCuZn-D[i]	C77300	Cu 46.0-50.0 Zn Remainder Ni 9.0-11.0 P 0.25 Pb 0.05* Al 0.01* Si 0.04-0.25 Other Elements Total[c] 0.50[i]
BCuP-1	C55180	Cu Remainder P 4.8-5.2 Other Elements Total[c] 0.15
BCuP-2	C55181	Cu Remainder P 7.0-7.5 Other Elements Total[c] 0.15
BCuP-3	C55281	Cu Remainder Ag 4.8-5.2 P 5.8-6.2 Other Elements Total[c] 0.15
BCuP-4	C55283	Cu Remainder Ag 5.8-6.2 P 7.0-7.5 Other Elements Total[c] 0.15
BCuP-5	C55284	Cu Remainder Ag 14.5-15.5 P 4.8-5.2 Other Elements Total[c] 0.15
BCuP-6	C55280	Cu Remainder Ag 1.8-2.2 P 6.8-7.2 Other Elements Total[c] 0.15
BCuP-7	C55282	Cu Remainder Ag 4.8-5.2 P 6.5-7.0 Other Elements Total[c] 0.15

a. Single values are maximum, unless noted.
b. SAE/ASTM Unified Numbering System for Metals and Alloys.
c. The filler metal shall be analyzed for those specific elements for which values or asterisks are shown in this table. If the presence of other elements is indicated in the course of this work, the amount of those elements shall be determined to ensure that their total does not exceed the limit specified.
d. The balance is oxygen, present as cuprous oxide. Oxygen is not to be included in "Other Elements".
e. These chemical composition requirements pertain only to the cuprous oxide powder and do not include requirements for the organic vehicle in which the cuprous oxide is suspended, when supplied in paste form.
f. The total of all other elements including those for which a maximum value or asterisk is shown, shall not exceed the value specified in "Other Elements, Total".
g. Silver residual is included in Cu analysis.
h. Includes cobalt.

CHEMICAL COMPOSITION REQUIREMENTS FOR NICKEL AND COBALT FILLER METALS

AWS A5.8 Classification	UNS Number[b]	Chemical Composition, Weight Percent[a]
BNi-1	N99600	Ni Remainder Cr 13.0-15.0 B 2.75-3.50 Si 4.0-5.0 Fe 4.0-5.0 C 0.60-0.90 P 0.02 S 0.02 Al 0.05 Ti 0.05 Zr 0.05 Co 0.10 Se 0.005 Other Elements Total[c] 0.50
BNi-1a	N99610	Ni Remainder Cr 13.0-15.0 B 2.75-3.50 Si 4.0-5.0 Fe 4.0-5.0 C 0.06 P 0.02 S 0.02 Al 0.05 Ti 0.05 Zr 0.05 Co 0.10 Se 0.005 Other Elements Total[c] 0.50
BNi-2	N99620	Ni Remainder Cr 6.0-8.0 B 2.75-3.50 Si 4.0-5.0 Fe 2.5-3.5 C 0.06 P 0.02 S 0.02 Al 0.05 Ti 0.05 Zr 0.05 Co 0.10 Se 0.005 Other Elements Total[c] 0.50
BNi-3	N99630	Ni Remainder B 2.75-3.50 Si 4.0-5.0 Fe 0.5 C 0.06 P 0.02 S 0.02 Al 0.05 Ti 0.05 Zr 0.05 Co 0.10 Se 0.005 Other Elements Total[c] 0.50
BNi-4	N99640	Ni Remainder B 1.50-2.20 Si 3.0-4.0 Fe 1.5 C 0.06 P 0.02 S 0.02 Al 0.05 Ti 0.05 Zr 0.05 Co 0.10 Se 0.005 Other Elements Total[c] 0.50
BNi-5	N99650	Ni Remainder Cr 18.5-19.5 B 0.03 Si 9.75-10.50 C 0.06 P 0.02 S 0.02 Al 0.05 Ti 0.05 Zr 0.05 Co 0.10 Se 0.005 Other Elements Total[c] 0.50
BNi-5a	N99651	Ni Remainder Cr 18.5-19.5 B 1.0-1.5 Si 7.0-7.5 Fe 0.5 C 0.10 P 0.02 S 0.02 Al 0.05 Ti 0.05 Zr 0.05 Co 0.10 Se 0.005 Other Elements Total[c] 0.50
BNi-6	N99700	Ni Remainder C 0.06 P 10.0-12.0 S 0.02 Al 0.05 Ti 0.05 Zr 0.05 Co 0.10 Se 0.005 Other Elements Total[c] 0.50
BNi-7	N99710	Ni Remainder Cr 13.0-15.0 B 0.01 Si 0.10 Fe 0.2 C 0.06 P 9.7-10.5 S 0.02 Al 0.05 Ti 0.05 Mn 0.04 Zr 0.05 Co 0.10 Se 0.005 Other Elements Total[c] 0.50
BNi-8	N99800	Ni Remainder Si 6.0-8.0 C 0.06 P 0.02 S 0.02 Al 0.05 Ti 0.05 Mn 21.5-24.5 Cu 4.0-5.0 Zr 0.05 Co 0.10 Se 0.005 Other Elements Total[c] 0.50
BNi-9	N99612	Ni Remainder Cr 13.5-16.5 B 3.25-4.00 Fe 1.5 C 0.06 P 0.02 S 0.02 Al 0.05 Ti 0.05 Zr 0.05 Co 0.10 Se 0.005 Other Elements Total[c] 0.50
BNi-10	N99622	Ni Remainder Cr 10.0-13.0 B 2.0-3.0 Si 3.0-4.0 Fe 2.5-4.5 C 0.40-0.55 P 0.02 S 0.02 Al 0.05 Ti 0.05 Zr 0.05 W 15.0-17.0 Co 0.10 Se 0.005 Other Elements Total[c] 0.50
BNi-11	N99624	Ni Remainder Cr 9.00-11.75 B 2.2-3.1 Si 3.35-4.25 Fe 2.5-4.0 C 0.30-0.50 P 0.02 S 0.02 Al 0.05 Ti 0.05 Zr 0.05 W 11.50-12.75 Co 0.10 Se 0.005 Other Elements Total[c] 0.50
BCo-1	R39001	Ni 16.0-18.0 Cr 18.0-20.0 B 0.70-0.90 Si 7.5-8.5 Fe 1.0 C 0.35-0.45 P 0.02 S 0.02 Al 0.05 Ti 0.05 Zr 0.05 W 3.5-4.5 Co Remainder Se 0.005 Other Elements Total[c] 0.50

CHEMICAL COMPOSITION REQUIREMENTS FOR NICKEL AND COBALT FILLER METALS (Continued)

a. Single values are maximum.
b. SAE/ASTM Unified Numbering System for Metals and Alloys.
c. The filler metal shall be analyzed for those specific elements for which values are shown in this table. If the presence of other elements is indicated in the course of this work, the amount of those elements shall be determined to ensure that their total does not exceed the limit specified.

CHEMICAL COMPOSITION REQUIREMENTS FOR FILLER METALS FOR VACUUM SERVICE

AWS A5.8 Classification	UNS Number	Chemical Composition, Weight Percent[a,b,c]
Grade 1		
BVAg-0	P07017	Ag 99.95 min Cu 0.05 Zn 0.001 Cd 0.001 Pb 0.002 P 0.002 C 0.005
BVAg-6b	P07507	Ag 49.0-51.0 Cu Remainder Zn 0.001 Cd 0.001 Pb 0.002 P 0.002 C 0.005
BVAg-8	P07727	Ag 71.0-73.0 Cu Remainder Zn 0.001 Cd 0.001 Pb 0.002 P 0.002 C 0.005
BVAg-8b	P07728	Ag 70.5-72.5 Cu Remainder Ni 0.3-0.7 Zn 0.001 Cd 0.001 Pb 0.002 P 0.002 C 0.005
BVAg-18	P07607	Ag 59.0-61.0 Cu Remainder Sn 9.5-10.5 Zn 0.001 Cd 0.001 Pb 0.002 P 0.002 C 0.005
BVAg-29	P07627	Ag 60.5-62.5 Cu Remainder In 14.0-15.0 Zn 0.001 Cd 0.001 Pb 0.002 P 0.002 C 0.005
BVAg-30	P07687	Ag 67.0-69.0 Cu Remainder Pd 4.5-5.5 Zn 0.001 Cd 0.001 Pb 0.002 P 0.002 C 0.005
BVAg-31	P07587	Ag 57.0-59.0 Cu 31.0-33.0 Pd Remainder Zn 0.001 Cd 0.001 Pb 0.002 P 0.002 C 0.005
BVAg-32	P07547	Ag 53.0-55.0 Cu 20.0-22.0 Pd Remainder Zn 0.001 Cd 0.001 Pb 0.002 P 0.002 C 0.005
BVAu-2	P00807	Au 79.5-80.5 Cu Remainder Zn 0.001 Cd 0.001 Pb 0.002 P 0.002 C 0.005
BVAu-4	P00827	Au 81.5-82.5 Ni Remainder Zn 0.001 Cd 0.001 Pb 0.002 P 0.002 C 0.005
BVAu-7	P00507	Au 49.5-50.5 Ni 24.5-25.5 Co 0.06 Pd Remainder Zn 0.001 Cd 0.001 Pb 0.002 P 0.002 C 0.005
BVAu-8	P00927	Au 91.0-93.0 Pd Remainder Zn 0.001 Cd 0.001 Pb 0.002 P 0.002 C 0.005
BVPd-1	P03657	Ni 0.06 Co Remainder Pd 64.0-66.0 Zn 0.001 Cd 0.001 Pb 0.002 P 0.002 C 0.005
Grade 2		
BVAg-0	P07017	Ag 99.95 min Cu 0.05 Zn 0.002 Cd 0.002 Pb 0.002 P 0.002 C 0.005
BVAg-6b	P07507	Ag 49.0-51.0 Cu Remainder Zn 0.002 Cd 0.002 Pb 0.002 P 0.02 C 0.005
BVAg-8	P07727	Ag 71.0-73.0 Cu Remainder Zn 0.002 Cd 0.002 Pb 0.002 P 0.02 C 0.005
BVAg-8b	P07728	Ag 70.5-72.5 Cu Remainder Ni 0.3-0.7 Zn 0.002 Cd 0.002 Pb 0.002 P 0.02 C 0.005

CHEMICAL COMPOSITION REQUIREMENTS FOR FILLER METALS FOR VACUUM SERVICE (Continued)		
AWS A5.8 Classification	UNS Number	Chemical Composition, Weight Percent[a,b,c]
Grade 2 (Continued)		
BVAg-18	P07607	Ag 59.0-61.0 Cu Remainder Sn 9.5-10.5 Zn 0.002 Cd 0.002 Pb 0.002 P 0.02 C 0.005
BVAg-29	P07627	Ag 60.5-62.5 Cu Remainder In 14.0-15.0 Zn 0.002 Cd 0.002 Pb 0.002 P 0.02 C 0.005
BVAg-30	P07687	Ag 67.0-69.0 Cu Remainder Pd 4.5-5.5 Zn 0.002 Cd 0.002 Pb 0.002 P 0.02 C 0.005
BVAg-31	P07587	Ag 57.0-59.0 Cu 31.0-33.0 Pd Remainder Zn 0.002 Cd 0.002 Pb 0.002 P 0.002 C 0.005
BVAg-32	P07547	Ag 53.0-55.0 Cu 20.0-22.0 Pd Remainder Zn 0.002 Cd 0.002 Pb 0.002 P 0.002 C 0.005
BVAu-2	P00807	Au 79.5-80.5 Cu Remainder Zn 0.002 Cd 0.002 Pb 0.002 P 0.002 C 0.005
BVAu-4	P00827	Au 81.5-82.5 Ni Remainder Zn 0.002 Cd 0.002 Pb 0.002 P 0.002 C 0.005
BVAu-7	P00507	Au 49.5-50.5 Ni 24.5-25.5 Co 0.06 Pd Remainder Zn 0.002 Cd 0.002 Pb 0.002 P 0.002 C 0.005
BVAu-8	P00927	Au 91.0-93.0 Pd Remainder Zn 0.002 Cd 0.002 Pb 0.002 P 0.002 C 0.005
BVPd-1	P03657	Ni 0.06 Co Remainder Pd 64.0-66.0 Zn 0.002 Cd 0.002 Pb 0.002 P 0.002 C 0.005
BVCu-1x	C14181	Cu 99.99 min Zn 0.002 Cd 0.002 Pb 0.002 P 0.002 C 0.005

a. The filler metal shall be analyzed for those specific elements for which values are shown in this table. If the presence of other elements is indicated in the course of this work, the amount of those elements shall be determined.

b. All other elements in addition to those listed in the table above, with a vapor pressure higher than 10^{-7} torr at 932°F (500°C) (such as Mg, Sb, K, Na, Li, Ti, S, Cs, Rb, Se, Te, Sr and Ca) are limited to 0.001 percent each for Grade 1 filler metals and 0.002 percent each for Grade 2 filler metals. The total of all these high vapor pressure elements (including zinc, cadmium, and lead) is limited to 0.010 percent. The total of other elements not designated as high vapor pressure elements is limited to 0.05 percent, except for BVCu-1x, for which the total shall be 0.015 percent, max.

c. Single values are maximum, unless noted.

SOLIDUS, LIQUIDUS, AND BRAZING TEMPERATURE RANGES[a]

AWS A5.8 Classification	Solidus		Liquidus		Brazing Temperature Range	
	°F	°C	°F	°C	°F	°C
Silver						
BAg-1	1125	607	1145	618	1145-1400	618-760
BAg-1a	1160	627	1175	635	1175-1400	635-760
BAg-2	1125	607	1295	702	1295-1550	702-843
BAg-2a	1125	607	1310	710	1310-1550	710-843
BAg-3	1170	632	1270	688	1270-1500	688-816
BAg-4	1240	671	1435	779	1435-1650	779-899
BAg-5	1225	663	1370	743	1370-1550	743-843
BAg-6	1270	688	1425	774	1425-1600	774-871
BAg-7	1145	618	1205	652	1205-1400	652-760
BAg-8	1435	779	1435	779	1435-1650	779-899
BAg-8a	1410	766	1410	766	1410-1600	766-871
BAg-9	1240	671	1325	718	1325-1550	718-843
BAg-10	1275	691	1360	738	1360-1550	738-843
BAg-13	1325	718	1575	857	1575-1775	857-968
BAg-13a	1420	771	1640	893	1600-1800	871-982
BAg-18	1115	602	1325	718	1325-1550	718-843
BAg-19	1400	760	1635	891	1610-1800	877-982
BAg-20	1250	677	1410	766	1410-1600	766-871
BAg-21	1275	691	1475	802	1475-1650	802-899
BAg-22	1260	680	1290	699	1290-1525	699-830
BAg-23	1760	960	1780	970	1780-1900	970-1038
BAg-24	1220	660	1305	750	1305-1550	750-843
BAg-26	1305	705	1475	800	1475-1600	800-870
BAg-27	1125	605	1375	745	1375-1575	745-860
BAg-28	1200	649	1310	710	1310-1550	710-843
BAg-33	1125	607	1260	682	1260-1400	681-760

The Metals Blue Book

SOLIDUS, LIQUIDUS, AND BRAZING TEMPERATURE RANGES[a] (Continued)

AWS A5.8 Classification	Solidus		Liquidus		Brazing Temperature Range	
	°F	°C	°F	°C	°F	°C
Silver (Continued)						
BAg-34	1200	649	1330	721	1330-1550	721-843
BAg-35	1265	685	1390	754	1390-1545	754-841
BAg-36	1195	646	1251	677	1251-1495	677-813
BAg-37	1270	688	1435	779	1435-1625	779-885
BVAg-0	1761	961	1761	961	1761-1900	961-1038
BVAg-6b	1435	779	1602	872	1660-1800	871-982
BVAg-8	1435	779	1435	779	1435-1650	779-899
BVAg-8b	1435	779	1463	795	1470-1650	799-899
BVAg-18	1115	602	1325	718	1325-1550	718-843
BVAg-29	1155	624	1305	707	1305-1450	707-788
BVAg-30	1485	807	1490	810	1490-1700	810-927
BVAg-31	1515	824	1565	852	1565-1625	852-885
BVAg-32	1650	900	1740	950	1740-1800	950-982
Gold						
BAu-1	1815	991	1860	1016	1860-2000	1016-1093
BAu-2	1635	891	1635	891	1635-1850	891-1010
BAu-3	1785	974	1885	1029	1885-1995	1029-1091
BAu-4	1740	949	1740	949	1740-1840	949-1004
BAu-5	2075	1135	2130	1166	2130-2250	1166-1232
Bau-6	1845	1007	1915	1046	1915-2050	1046-1121
BVAu-2	1635	891	1635	891	1635-1850	891-1010
BVAu-4	1740	949	1740	949	1740-1840	949-1004
BVAu-7	2015	1102	2050	1121	2050-2110	1121-1154
BVAu-8	2190	1200	2265	1240	2265-2325	1240-1274
Palladium						
BVPd-1	2245	1230	2255	1235	2255-2285	1235-1252

SOLIDUS, LIQUIDUS, AND BRAZING TEMPERATURE RANGES[a] (Continued)

AWS A5.8 Classification	Solidus °F	Solidus °C	Liquidus °F	Liquidus °C	Brazing Temperature Range °F	Brazing Temperature Range °C
Aluminum						
BAlSi-2	1070	577	1142	617	1110-1150	599-621
BAlSi-3	970	521	1085	585	1060-1120	571-604
BAlSi-4	1070	577	1080	582	1080-1120	582-604
BAlSi-5	1070	577	1095	591	1090-1120	588-604
BAlSi-7	1038	559	1105	596	1090-1120	588-604
BAlSi-9	1044	562	1080	582	1080-1120	582-604
BAlSi-11	1038	559	1105	596	1090-1120	588-604
Copper						
Bcu-1	1981	1083	1981	1083	2000-2100	1093-1149
BCu-1a	1981	1083	1981	1083	2000-2100	1093-1149
BVCu-1X	1981	1083	1981	1083	2000-2100	1093-1149
BCu-2	1981	1083	1981	1083	2000-2100	1093-1149
RBCuZn-A	1630	888	1650	899	1670-1750	910-954
RBCuZn-B	1590	866	1620	882	1620-1800	882-982
RBCuZn-C	1590	866	1630	888	1670-1750	910-954
RBCuZn-D	1690	921	1715	935	1720-1800	938-982
BCuP-1	1310	710	1695	924	1450-1700	788-927
BCuP-2	1310	710	1460	793	1350-1550	732-843
BCuP-3	1190	643	1495	813	1325-1500	718-816
BCuP-4	1190	643	1325	718	1275-1450	691-788
BCuP-5	1190	643	1475	802	1300-1500	704-816
BCuP-6	1190	643	1450	788	1350-1500	732-816
BCuP-7	1190	643	1420	771	1300-1500	704-816
Nickel						
BNi-1	1790	977	1900	1038	1950-2200	1066-1204
BNi-1a	1790	977	1970	1077	1970-2200	1077-1204

The Metals Blue Book

SOLIDUS, LIQUIDUS, AND BRAZING TEMPERATURE RANGES[a] (Continued)

AWS A5.8 Classification	Solidus		Liquidus		Brazing Temperature Range	
	°F	°C	°F	°C	°F	°C
Nickel (Continued)						
BNi-2	1780	971	1830	999	1850-2150	1010-1177
BNi-3	1800	982	1900	1038	1850-2150	1010-1177
BNi-4	1800	982	1950	1066	1850-2150	1010-1177
BNi-5	1975	1079	2075	1135	2100-2200	1149-1204
BNi-5a	1931	1065	2111	1150	2100-2200	1149-1204
BNi-6	1610	877	1610	877	1700-2000	927-1093
BNi-7	1630	888	1630	888	1700-2000	927-1093
BNi-8	1800	982	1850	1010	1850-2000	1010-1093
BNi-9	1930	1055	1930	1055	1950-2200	1066-1204
BNi-10	1780	970	2020	1105	2100-2200	1149-1204
BNi-11	1780	970	2003	1095	2100-2200	1149-1204
Cobalt						
BCo-1	2050	1120	2100	1149	2100-2250	1149-1232
Magnesium						
BMg-1	830	443	1110	599	1120-1160	604-627

a.   Solidus and liquidus shown are for the nominal composition in each classification.

AWS A5.8 CLASSIFICATION SYSTEM

The classification method for brazing filler metals is based on chemical composition rather than on mechanical property requirements. The mechanical properties of a brazed joint depend, among other things, on the base metal and filler metal used. Therefore, a classification method based on mechanical properties would be misleading since it would only apply if the brazing filler metal were used on a given base metal with a specific joint design. If a user of brazing filler metal desires to determine the mechanical properties of a given base metal and filler metal combination, tests should be conducted using the latest edition of ANSI/AWS C3.2, *Standard Method for Evaluating the Strength of Brazed Joints*.

Brazing filler metals are standardized into seven groups of classifications as follows: silver, gold, aluminum, copper, nickel, cobalt, and magnesium filler metals. Many filler metals of these classifications are used for joining assemblies for vacuum applications, such as vacuum tubes and other electronic devices. For these critical applications, it is desirable to hold the high vapor pressure elements to a minimum, as they usually contaminate the vacuum with vaporized elements during operation of the device. Filler metals for electronic devices have been incorporated as additional "vacuum grade" classifications within this specification.

The basic groups of classifications of brazing filler metal are identified, as shown in AWS A5.8 Tables 1 through 6, by the principal element in their chemical composition. In a typical example, such as BCuP-2, the "B" is for brazing filler metal (as the "E" for electrodes and the "R" welding rods in other AWS specifications). The "RB" in RBCuZn-A, RBCuZn-C, and RBCuZn-D indicates that the filler metal is suitable as a welding rod and as a brazing filler metal. "CuP" is for copper-phosphorus, the two principal elements in this particular brazing filler metal. (Similarly, in other brazing filler metals, Si is for silicon, Ag for silver, etc., using standard chemical symbols). The numeral or letter following the chemical symbol indicates chemical composition within a group.

The vacuum grade nomenclature follows the examples above, with two exceptions. The first exception is the addition of the letter "V", yielding the generic letters "BV" for brazing filler metals for vacuum service. The second exception is the use of the grade suffix number; Grade 1 to indicate the more stringent requirements for high vapor pressure impurities, and Grade 2 to indicate less stringent requirements for high vapor pressure impurities. Vacuum grade filler metals are considered to be spatter-free. Therefore, this specification no longer lists spatter-free and nonspatter-free vacuum grades. An example of a filler metal for vacuum service is BVAg-6b, Grade 1, AWS A5.8 Table 6 lists filler metals for vacuum service.

BARE STAINLESS STEEL WELDING ELECTRODES & RODS

CHEMICAL COMPOSITION REQUIREMENTS

AWS A5.9 Classification[c,d]	UNS Number[e]	Composition, Weight Percent[a,b]
ER209	S20980	C 0.05 Cr 20.5-24.0 Ni 9.5-12.0 Mo 1.5-3.0 Mn 4.0-7.0 Si 0.90 P 0.03 S 0.03 N 0.10-0.30 Cu 0.75 Other Elements V 0.10-0.30
ER218	S21880	C 0.10 Cr 16.0-18.0 Ni 8.0-9.0 Mo 0.75 Mn 7.0-9.0 Si 3.5-4.5 P 0.03 S 0.03 N 0.08-0.18 Cu 0.75
ER219	S21980	C 0.05 Cr 19.0-21.5 Ni 5.5-7.0 Mo 0.75 Mn 8.0-10.0 Si 1.00 P 0.03 S 0.03 N 0.10-0.30 Cu 0.75
ER240	S24080	C 0.05 Cr 17.0-19.0 Ni 4.0-6.0 Mo 0.75 Mn 10.5-13.5 Si 1.00 P 0.03 S 0.03 N 0.10-0.30 Cu 0.75
ER307	S30780	C 0.04-0.14 Cr 19.5-22.0 Ni 8.0-10.7 Mo 0.5-1.5 Mn 3.3-4.75 Si 0.30-0.65 P 0.03 S 0.03 Cu 0.75
ER308	S30880	C 0.08 Cr 19.5-22.0 Ni 9.0-11.0 Mo 0.75 Mn 1.0-2.5 Si 0.30-0.65 P 0.03 S 0.03 Cu 0.75
ER308H	S30880	C 0.04-0.08 Cr 19.5-22.0 Ni 9.0-11.0 Mo 0.50 Mn 1.0-2.5 Si 0.30-0.65 P 0.03 S 0.03 Cu 0.75
ER308L	S30883	C 0.03 Cr 19.5-22.0 Ni 9.0-11.0 Mo 0.75 Mn 1.0-2.5 Si 0.30-0.65 P 0.03 S 0.03 Cu 0.75
ER308Mo	S30882	C 0.08 Cr 18.0-21.0 Ni 9.0-12.0 Mo 2.0-3.0 Mn 1.0-2.5 Si 0.30-0.65 P 0.03 S 0.03 Cu 0.75
ER308LMo	S30886	C 0.04 Cr 18.0-21.0 Ni 9.0-12.0 Mo 2.0-3.0 Mn 1.0-2.5 Si 0.30-0.65 P 0.03 S 0.03 Cu 0.75
ER308Si	S30881	C 0.08 Cr 19.5-22.0 Ni 9.0-11.0 Mo 0.75 Mn 1.0-2.5 Si 0.65-1.00 P 0.03 S 0.03 Cu 0.75
ER308LSi	S30888	C 0.03 19.5-22.0 Ni 9.0-11.0 Mo 0.75 Mn 1.0-2.5 Si 0.65-1.00 P 0.03 S 0.03 Cu 0.75
ER309	S30980	C 0.12 Cr 23.0-25.0 Ni 12.0-14.0 Mo 0.75 Mn 1.0-2.5 Si 0.30-0.65 P 0.03 S 0.03 Cu 0.75
ER309L	S30983	C 0.03 Cr 23.0-25.0 Ni 12.0-14.0 Mo 0.75 Mn 1.0-2.5 Si 0.30-0.65 P 0.03 S 0.03 Cu 0.75
ER309Mo	S30982	C 0.12 Cr 23.0-25.0 Ni 12.0-14.0 Mo 2.0-3.0 Mn 1.0-2.5 Si 0.30-0.65 P 0.03 S 0.03 Cu 0.75
ER309LMo	S30986	C 0.03 Cr 23.0-25.0 Ni 12.0-14.0 Mo 2.0-3.0 Mn 1.0-2.5 Si 0.30-0.65 P 0.03 S 0.03 Cu 0.75
ER309Si	S30981	C 0.12 Cr 23.0-25.0 Ni 12.0-14.0 Mo 0.75 Mn 1.0-2.5 Si 0.65-1.00 P 0.03 S 0.03 Cu 0.75
ER309LSi	S30988	C 0.03 Cr 23.0-25.0 Ni 12.0-14.0 Mo 0.75 Mn 1.0-2.5 Si 0.65-1.00 P 0.03 S 0.03 Cu 0.75
ER310	S31080	C 0.08-0.15 Cr 25.0-28.0 Ni 20.0-22.5 Mo 0.75 Mn 1.0-2.5 Si 0.30-0.65 P 0.03 S 0.03 Cu 0.75
ER312	S31380	C 0.15 Cr 28.0-32.0 Ni 8.0-10.5 Mo 0.75 Mn 1.0-2.5 Si 0.30-0.65 P 0.03 S 0.03 Cu 0.75
ER316	S31680	C 0.08 Cr 18.0-20.0 Ni 11.0-14.0 Mo 2.0-3.0 Mn 1.0-2.5 Si 0.30-0.65 P 0.03 S 0.03 Cu 0.75
ER316H	S31680	C 0.04-0.08 Cr 18.0-20.0 Ni 11.0-14.0 Mo 2.0-3.0 Mn 1.0-2.5 Si 0.30-0.65 P 0.03 S 0.03 Cu 0.75
ER316L	S31683	C 0.03 Cr 18.0-20.0 Ni 11.0-14.0 Mo 2.0-3.0 Mn 1.0-2.5 Si 0.30-0.65 P 0.03 S 0.03 Cu 0.75
ER316Si	S31681	C 0.08 Cr 18.0-20.0 Ni 11.0-14.0 Mo 2.0-3.0 Mn 1.0-2.5 Si 0.65-1.00 P 0.03 S 0.03 Cu 0.75
ER316LSi	S31688	C 0.03 Cr 18.0-20.0 Ni 11.0-14.0 Mo 2.0-3.0 Mn 1.0-2.5 Si 0.65-1.00 P 0.03 S 0.03 Cu 0.75
ER317	S31780	C 0.08 Cr 18.5-20.5 Ni 13.0-15.0 Mo 3.0-4.0 Mn 1.0-2.5 Si 0.30-0.65 P 0.03 S 0.03 Cu 0.75

CHEMICAL COMPOSITION REQUIREMENTS (Continued)

AWS A5.9 Classification[c,d]	UNS Number[e]	Composition, Weight Percent[a,b]
ER317L	S31783	C 0.03 Cr 18.5-20.5 Ni 13.0-15.0 Mo 3.0-4.0 Mn1.0-2.5 Si 0.30-0.65 P 0.03 S 0.03 Cu 0.75
ER318	S31980	C 0.08 Cr 18.0-20.0 Ni 11.0-14.0 Mo 2.0-3.0 Mn1.0-2.5 Si 0.30-0.65 P 0.03 S 0.03 Cu 0.75 Cb[g] 8 X C min/1.0 max
ER320	N08021	C 0.07 Cr 19.0-21.0 Ni 32.0-36.0 Mo 2.0-3.0 Mn 2.5 Si 0.60 P 0.03 S 0.03 Cu 3.0-4.0 Cb[g] 8 X C min/1.0 max
ER320LR	N08022	C 0.025 Cr 19.0-21.0 Ni 32.0-36.0 Mo 2.0-3.0 Mn 1.5 - 2.0 Si 0.15 P 0.015 S 0.02 Cu 3.0-4.0 Cb[g] 8 X C min/0.40 max
ER321	S32180	C 0.08 Cr 18.5-20.5 Ni 9.0-10.5 Mo 0.75 Mn 1.0-2.5 Si 0.30-0.65 P 0.03 S 0.03 Cu 0.75 Ti 9 X C min/1.0 max
ER330	N08331	C 0.18-0.25 Cr 15.0-17.0 Ni 34.0-37.0 Mo 0.75 Mn 1.0-2.5 Si 0.30-0.65 P 0.03 S 0.03 Cu 0.75
ER347	S34780	C 0.08 Cr 19.0-21.5 Ni 9.0-11.0 Mo 0.75 Mn 1.0-2.5 Si 0.30-0.65 P 0.03 S 0.03 Cu 0.75 Cb[g] 10 X C min/1.0 max
ER347Si	S34788	C 0.08 Cr 19.0-21.5 Ni 9.0-11.0 Mo 0.75 Mn 1.0-2.5 Si 0.65-1.00 P 0.03 S 0.03 Cu 0.75 Cb[g] 10 X C min/1.0 max
ER383	N08028	C 0.025 Cr 26.5-28.5 Ni 30.0-33.0 Mo 3.2-4.2 Mn 1.0-2.5 Si 0.50 P 0.02 S 0.03 Cu 0.70-1.5
ER385	N08904	C 0.025 Cr 19.5-21.5 Ni 24.0-26.0 Mo 4.2-5.2 Mn 1.0-2.5 Si 0.50 P 0.02 S 0.03 Cu 1.2-2.0
ER409	S40900	C 0.08 Cr 10.5-13.5 Ni 0.6 Mo 0.50 Mn 0.08 Si 0.8 P 0.03 S 0.03 Cu 0.75 Other Elements Cb[g] 10 X C min/1.5 max
ER409Cb	S40940	C 0.08 Cr 10.5-13.5 Ni 0.6 Mo 0.50 Mn 0.08 Si 1.0 P 0.04 S 0.03 Cu 0.75 Other Elements Cb[g] 10 X C min/0.75 max
ER410	S41080	C 0.12 Cr 11.5-13.5 Ni 0.6 Mo 0.75 Mn 0.6 Si 0.5 P 0.03 S 0.03 Cu 0.75
ER410NiMo	S41086	C 0.06 Cr 11.0-12.5 Ni 4.0-5.0 Mo 0.4-0.7 Mn 0.6 Si 0.5 P 0.03 S 0.03 Cu 0.75
ER420	S42080	C 0.25-0.40 Cr 12.0-14.0 Ni 0.6 Mo 0.75 Mn 0.6 Si 0.5 P 0.03 S 0.03 Cu 0.75
ER430	S43080	C 0.10 Cr 15.5-17.0 Ni 0.6 Mo 0.75 Mn 0.6 Si 0.5 P 0.03 S 0.03 Cu 0.75
ER446LMo	S44687	C 0.015 Cr 25.0-27.5 Ni ' Mo 0.75-1.50 Mn 0.4 Si 0.4 P 0.02 S 0.02 0.015 Cu '
ER502[h]	S50280	C 0.10 Cr 4.6-6.0 Ni 0.6 Mo 0.45-0.65 Mn 0.6 Si 0.5 P 0.03 S 0.03 Cu 0.75
ER505[h]	S50480	C 0.10 Cr 8.0-10.5 Ni 0.5 Mo 0.8-1.2 Mn 0.6 Si 0.5 P 0.03 S 0.03 Cu 0.75
ER630	S17480	C 0.05 Cr 16.0-16.75 Ni 4.5-5.0 Mn 0.25-0.75 Si 0.75 P 0.03 S 0.03 Cu 3.25-4.00 Cb[g] 0.15-0.30
ER19-10H	S30480	C 0.04-0.08 Cr 18.5-20.0 Ni 9.0-11.0 Mo 0.25 Mn 1.0-2.0 Si 0.30-0.65 P 0.03 S 0.03 Cu 0.75, Ti 0.5
ER16-8-2	S16880	C 0.10 Cr 14.5-16.5 Ni 7.5-9.5 Mo 1.0-2.0 Mn 1.0-2.0 Si 0.30-0.65 P 0.03 S 0.03 Cu 0.75
ER2209	S39209	C 0.03 Cr 21.5-23.5 Ni 7.5-9.5 Mo 2.5-3.5 Mn 0.50-2.0 Si 0.90 P 0.03 S 0.03 N 0.08-0.20 Cu 0.75
ER2553	S39553	C 0.04 Cr 24.0-27.0 Ni 4.5-6.5 Mo 2.9-3.9 Mn 1.5 Si 1.0 P 0.04 S 0.03 N 0.10-0.25 Cu 1.5-2.5
ER3556	R30556	C 0.05-0.15 Cr 21.0-23.0 Ni 19.0-22.5 Mo 2.5-4.0 Mn 0.50-2.00 Si 0.20-0.80 P 0.04 S 0.015 N 0.10-0.30 Co 16.0-21.0, W 2.0-3.5, Cb 0.30, Ta 0.30-1.25, Al 0.10-0.50, Zr 0.001-0.10, La 0.005-0.10, B 0.02

CHEMICAL COMPOSITION REQUIREMENTS (Continued)

a. Analysis shall be made for the elements for which specific values are shown in this table. If the presence of other elements is indicated in the course of this work, the amount of those elements shall be determined to ensure that their total, excluding iron, does not exceed 0.50 percent.

b. Single values shown are maximum percentages.

c. In the designator for composite, stranded, and strip electrodes, the "R" shall be deleted. A designator "C" shall be used for composite and stranded electrodes and a designator "Q" shall be used for strip electrodes. For example, ERXXX designates a solid wire and EQXXX designates a strip electrode of the same general analysis, and the same UNS number. However, ECXXX designates a composite metal cored or stranded electrode and may not have the same UNS number. Consult ASTM/SAE Uniform Numbering System for the proper UNS Number.

d. For special applications, electrodes and rods may be purchased with less than the specified silicon content.

e. SAE/ASTM Unified Numbering System for Metals and Alloys.

f. Nickel + copper equals 0.5 percent maximum.

g. Cb (Nb) may be reported as Cb (Nb) + Ta.

h. These classifications also will be included in the next revision of ANSI/AWS A5.28, *Specification for Low Alloy Steel Filler Metals for Gas Shielded Metal Arc Welding*. They will be deleted from ANSI/AWS A5.9 in the first revision following publication of the revised ANSI/AWS A5.28 document.

TENSILE REQUIREMENTS FOR ALL-WELD-METAL

AWS A5.9 Classification	Tensile Strength, min		% Elongation, min	Heat Treatment
	ksi	MPa		
E209-XX	100	690	15	None
E219-XX	90	620	15	None
E240-XX	100	690	15	None
E307-XX	85	590	30	None
E308-XX	80	550	35	None
E308H-XX	80	550	35	None
E308L-XX	75	520	35	None
E308Mo-XX	80	550	35	None
E308MoL-XX	75	520	35	None
E309-XX	80	550	30	None
E309L-XX	75	520	30	None
E309Cb-XX	80	550	30	None
E309Mo-XX	80	550	30	None
E309MoL-XX	75	520	30	None
E310-XX	80	550	30	None
E310H-XX	90	620	10	None
E310Cb-XX	80	550	25	None
E310Mo-XX	80	550	30	None
E312-XX	95	660	22	None
E316-XX	75	520	30	None
E316H-XX	75	520	30	None
E316L-XX	70	490	30	None
E317-XX	80	550	30	None
E317L-XX	75	520	30	None
E318-XX	80	550	25	None
E320-XX	80	550	30	None
E320LR-XX	75	520	30	None

The Metals Blue Book

TENSILE REQUIREMENTS FOR ALL-WELD-METAL

AWS A5.9 Classification	Tensile Strength, min		% Elongation, min	Heat Treatment
	ksi	MPa		
E330-XX	75	520	25	None
E330H-XX	90	620	10	None
E347-XX	75	520	30	None
E349-XX	100	690	25	None
E383-XX	75	520	30	None
E385-XX	75	520	30	None
E410-XX	75	450	20	a
E410NiMo-XX	110	760	15	c
E430-XX	65	450	20	d
E502-XX	60	420	20	b
E505-XX	60	420	20	b
E630-XX	135	930	7	e
E16-8-2-XX	80	550	35	None
E7Cr-XX	60	420	20	b
E2209-XX	100	690	20	None
E2553-XX	110	760	15	None

a. Heat to 1350 to 1400°F (730 to 760°C), hold for one hour, furnace cool at a rate of 100°F (55°C) per hour to 600°F (315°C) and air cool to ambient.

b. Heat to 1550 to 1600°F (840 to 870°C), hold for two hours, furnace cool at a rate not exceeding 100°F (55°C) per hour to 1100°F (595°C) and air cool to ambient.

c. Heat to 1100 to 1150°F (595 to 620°C), hold for one hour, and air cool to ambient.

d. Heat to 1400 to 1450°F (760 to 790°C), hold for two hours, furnace cool at a rate not exceeding 100°F (55°C) per hour to 1100°F (595°C) and air cool to ambient.

e. Heat to 1875 to 1925°F (1025 to 1050°C), hold for one hour, and air cool to ambient, and then precipitation harden at 1135 to 1165°F (610 to 630°C), hold for four hours, and air cool to ambient.

AWS A5.9 CLASSIFICATION SYSTEM

The chemical composition of the filler metal is identified by a series of numbers and, in some cases, chemical symbols, the letters L, H, and LR, or both. Chemical symbols are used to designate modifications of basic alloy types, e.g., ER308Mo. The letter "H" denotes carbon content restricted to the upper part of the range that is specified for the standard grade of the specific filler metal. The letter "L" denotes carbon content in the lower part of the range that is specified for the corresponding standard grade of filler metal. The letters "LR" denote low residuals (see AWS A5.9 appendix A8.30).

The first two designators may be "ER" for solid wires that may be used as electrodes or rods; or they may be "EC" for composite cored or stranded wires; or they may be "EQ" for strip electrodes.

The three digit number such as 308 in ER308 designates the chemical composition of the filler metal.

BARE ALUMINUM & ALUMINUM ALLOY WELDING ELECTRODES & RODS

CHEMICAL COMPOSITION REQUIREMENTS FOR ALUMINUM ELECTRODES AND RODS		
AWS A5.10 Classification	**UNS Number[c]**	**Composition, Weight Percent[a,b]**
ER1100	A91100	Al 99.0 min[f] Si d Fe d Cu 0.05-0.20 Mn 0.05 Zn 0.10 Other Elements Each 0.05[e] Other Elements Total 0.15
R1100	A91100	Al 99.0 min[f] Si d Fe d Cu 0.05-0.20 Mn 0.05 Zn 0.10 Other Elements Each 0.05[e] Other Elements Total 0.15
ER1188[g]	A91188	Al 99.88 min[f] Si 0.06 Fe 0.06 Cu 0.005 Mn 0.01 Mg 0.01 Zn 0.03 Ti 0.01 Other Elements Each 0.01[e]
R1188[g]	A91188	Al 99.88 min[f] Si 0.06 Fe 0.06 Cu 0.005 Mn 0.01 Mg 0.01 Zn 0.03 Ti 0.01 Other Elements Each 0.01[e]
ER2319[h]	A92319	Al Remainder Si 0.20 Fe 0.30 Cu 5.8-6.8 Mn 0.20-0.40 Mg 0.02 Zn 0.10 Ti 0.10-0.20 Other Elements Each 0.05[e] Other Elements Total 0.15
R2319	A92319	Al Remainder Si 0.20 Fe 0.30 Cu 5.8-6.8 Mn 0.20-0.40 Mg 0.02 Zn 0.10 Ti 0.10-0.20 Other Elements Each 0.05[e] Other Elements Total 0.15
ER4009[h]	A94009	Al Remainder Si 4.5-5.5 Fe 0.20 Cu 1.0-1.5 Mn 0.10 Mg 0.45-0.6 Zn 0.10 Ti 0.20 Other Elements Each 0.05[e] Other Elements Total 0.15
R4009	A94009	Al Remainder Si 4.5-5.5 Fe 0.20 Cu1.0-1.5 Mn 0.10 Mg 0.45-0.6 Zn 0.10 Ti 0.20 Other Elements Each 0.05[e] Other Elements Total 0.15
ER4010	A94010	Al Remainder Si 6.5-7.5 Fe 0.20 Cu0.20 Mn 0.10 Mg 0.30-0.45 Zn 0.10 Ti 0.20 Other Elements Each 0.05[e] Other Elements Total 0.15
R4010	A94010	Al Remainder Si 6.5-7.5 Fe 0.20 Cu 0.20 Mn 0.10 Mg 0.30-0.45 Zn 0.10 Ti 0.20 Other Elements Each 0.05[e] Other Elements Total 0.15
R4011[k]	A94011	Al Remainder Si 6.5-7.5 Fe 0.20 Cu 0.20 Mn 0.10 Mg 0.45-0.7 Zn 0.10 Ti 0.04-0.20 Other Elements Each 0.05[e]
ER4043	A94043	Al Remainder Si 4.5-6.0 Fe 0.8 Cu 0.30 Mn 0.05 Mg 0.05 Zn 0.10 Ti 0.20 Other Elements Each 0.05[e]
R4043	A94043	Al Remainder Si 4.5-6.0 Fe 0.8 Cu 0.30 Mn 0.05 Mg 0.05 Zn 0.10 Ti 0.20 Other Elements Each 0.05[e]
ER4047	A94047	Al Remainder Si 11.0-13.0 Fe 0.8 Cu 0.30 Mn 0.15 Mg 0.10 Zn 0.20 Other Elements Each 0.05[e] Other Elements Total 0.15
R4047	A94047	Al Remainder Si 11.0-13.0 Fe 0.8 Cu 0.30 Mn 0.15 Mg 0.10 Zn 0.20 Other Elements Each 0.05[e] Other Elements Total 0.15
ER4145	A94145	Al Remainder Si 9.3-10.7 Fe 0.8 Cu 3.3-4.7 Mn 0.15 Mg 0.15 Cr 0.15 Zn 0.20 Other Elements Each 0.05[e] Other Elements Total 0.15
R4145	A94145	Al Remainder Si 9.3-10.7 Fe 0.8 Cu 3.3-4.7 Mn 0.15 Mg 0.15 Cr 0.15 Zn 0.20 Other Elements Each 0.05[e] Other Elements Total 0.15

CHEMICAL COMPOSITION REQUIREMENTS FOR ALUMINUM ELECTRODES AND RODS (Continued)		
AWS A5.10 Classification	UNS Number[c]	Composition, Weight Percent[a,b]
ER4643	A94643	Al Remainder Si 3.6-4.6 Fe 0.8 Cu 0.10 Mn 0.05 Mg 0.10-0.30 Zn 0.10 Ti 0.15 Other Elements Each 0.05[e] Other Elements Total 0.15
R4643	A94643	Al Remainder Si 3.6-4.6 Fe 0.8 Cu 0.10 Mn 0.05 Mg 0.10-0.30 Zn 0.10 Ti 0.15 Other Elements Each 0.05[e] Other Elements Total 0.15
ER5183	A95183	Al Remainder Si 0.40 Fe 0.40 Cu 0.10 Mn 0.50-1.0 Mg 4.3-5.2 Cr 0.05-0.25 Zn 0.25 Ti 0.15 Other Elements Each 0.05[e] Other Elements Total 0.15
R5183	A95183	Al Remainder Si 0.40 Fe 0.40 Cu 0.10 Mn 0.50-1.0 Mg 4.3-5.2 Cr 0.05-0.25 Zn 0.25 Ti 0.15 Other Elements Each 0.05[e] Other Elements Total 0.15
ER5356	A95356	Al Remainder Si 0.25 Fe 0.40 Cu 0.10 Mn 0.05-0.20 Mg 4.5-5.5 Cr 0.05-0.20 Zn 0.10 Ti 0.06-0.20 Other Elements Each 0.05[e] Other Elements Total 0.15
R5356	A95356	Al Remainder Si 0.25 Fe 0.40 Cu 0.10 Mn 0.05-0.20 Mg 4.5-5.5 Cr 0.05-0.20 Zn 0.10 Ti 0.06-0.20 Other Elements Each 0.05[e] Other Elements Total 0.15
ER5554	A95554	Al Remainder Si 0.25 Fe 0.40 Cu 0.10 Mn 0.50-1.0 Mg 2.4-3.0 Cr 0.05-0.20 Zn 0.25 Ti 0.05-0.20 Other Elements Each 0.05[e] Other Elements Total 0.15
R5554	A95554	Al Remainder Si 0.25 Fe 0.40 Cu 0.10 Mn 0.50-1.0 Mg 2.4-3.0 Cr 0.05-0.20 Zn 0.25 Ti 0.05-0.20 Other Elements Each 0.05[e] Other Elements Total 0.15
ER5556	A95556	Al Remainder Si 0.25 Fe 0.40 Cu 0.10 Mn 0.50-1.0 Mg 4.7-5.5 Cr 0.05-0.20 Zn 0.25 Ti 0.05-0.20 Other Elements Each 0.05[e] Other Elements Total 0.15
R5556	A95556	Al Remainder Si 0.25 Fe 0.40 Cu 0.10 Mn 0.50-1.0 Mg 4.7-5.5 Cr 0.05-0.20 Zn 0.25 Ti 0.05-0.20 Other Elements Each 0.05[e] Other Elements Total 0.15
ER5654	A95654	Al Remainder Si[i] Fe[i] Cu 0.05 Mn 0.01 Mg 3.1-3.9 Cr 0.15-0.35 Zn 0.20 Ti 0.05-0.15 Other Elements Each 0.05[e] Other Elements Total 0.15
R5654	A95654	Al Remainder Si[i] Fe[i] Cu 0.05 Mn 0.01 Mg 3.1-3.9 Cr 0.15-0.35 Zn 0.20 Ti 0.05-0.15 Other Elements Each 0.05[e] Other Elements Total 0.15
R-206.0[j]	A02060	Al Remainder Si 0.10 Fe 0.15 Cu 4.2-5.0 Mn 0.20-0.50 Mg 0.15-0.35 Ni 0.05 Zn 0.10 Ti 0.15-0.30 Other Elements Each 0.05 Other Elements Total 0.15
R-C355.0	A33550	Al Remainder Si 4.5-5.5 Fe 0.20 Cu 1.0-1.5 Mn 0.10 Mg 0.40-0.6 Zn 0.10 Ti 0.20 Other Elements Each 0.05 Other Elements Total 0.15

The Metals Blue Book

CHEMICAL COMPOSITION REQUIREMENTS FOR ALUMINUM ELECTRODES AND RODS(Continued)

AWS A5.10 Classification	UNS Number[c]	Composition, Weight Percent[a,b]
R-A356.0	A13560	Al Remainder Si 6.5-7.5 Fe 0.20 Cu 0.20 Mn 0.10 Mg 0.25-0.45 Zn 0.10 Ti 0.20 Other Elements Each 0.05 Other Elements Total 0.15
R-357.0	A03570	Al Remainder Si 6.5-7.5 Fe 0.15 Cu 0.05 Mn 0.03 Mg 0.45-0.6 Zn 0.05 Ti 0.20 Other Elements Each 0.05 Other Elements Total 0.15
R-A357.0[k]	A13570	Al Remainder Si 6.5-7.5 Fe 0.20 Cu 0.20 Mn 0.10 Mg 0.40-0.7 Zn 0.10 Ti 0.04-0.20 Other Elements Each 0.05 Other Elements Total 0.15

a.   The filler metal shall be analyzed for those specific elements for which values are shown in this table. If the presence of other elements is indicated in the course of this work, the amount of those elements shall be determined to ensure that their total does not exceed the limits specified for "Other Elements".

b.   Single values are maximum, except where otherwise specified.

c.   SAE/ASTM Unified Numbering System for Metals and Alloys.

d.   Silicon plus iron shall not exceed 0.95 percent.

e.   Beryllium shall not exceed 0.0008 percent.

f.   The aluminum content for unalloyed aluminum is the difference between 100.00 percent and the sum of all other metallic elements present in amounts of 0.010 percent or more each, expressed to the second decimal before determining the sum.

g.   Vanadium content shall be 0.05 percent maximum. Gallium content shall be 0.03 percent maximum.

h.   Vanadium content shall be 0.05-0.15 percent. Zirconium content shall be 0.10-0.25 percent.

i.   Silicon plus iron shall not exceed 0.45 percent.

j.   Tin content shall not exceed 0.05 percent.

k.   Beryllium content shall be 0.04-0.07 percent.

GUIDE TO THE CHOICE OF FILLER METAL FOR GENERAL PURPOSE WELDING

Base Metal	201.0, 206.0, 224.0	319.0, 333.0, 354.0, 355.0, C355.0	356.0, A356.0, A357.0, 413.0, 443.0, A444.0	511.0, 512.0, 513.0, 514.0, 535.0	7004, 7005, 7039, 710.0, 712.0	6009, 6010, 6070	6005, 6061, 6063, 6101, 6151, 6201, 6351, 6951	5456	5454
1060, 1070, 1080, 1350	ER4145	ER4145	ER4043[a,b]	ER5356[c,d]	ER5356[c,d]	ER4043[a,b]	ER4043[b]	ER5356[d]	ER4043[b,d]
1100, 3003, Alc 3003	ER4145	ER4145	ER4043[a,b]	ER5356[c,d]	ER5356[c,d]	ER4043[a,b]	ER4043[b]	ER5356[d]	ER4043[b,d]
2014, 2036	ER4145[e]	ER4145[e]	ER4145	-	-	ER4145	ER4145	-	-
2219	ER2319[a]	ER4145[e]	ER4145[b,c]	ER4043	ER4043	ER4043[a,b]	ER4043[a,b]	-	ER4043[b]
3004, Alc 3004	-	ER4043[b]	ER4043[b]	ER5356[f]	ER5356[f]	ER4043[b]	ER4043[b,f]	ER5356[f]	ER5356[f]
5005, 5050	-	ER4043[b]	ER4043[b]	ER5356[f]	ER5356[f]	ER4043[b]	ER4043[b,f]	ER5356[f]	ER5356[f]
5052, 5652	-	ER4043[b]	ER4043[f]	ER5356[f]	ER5356[f]	ER4043[b]	ER5356[c,f]	ER5356[f]	ER5356[f]
5083	-	-	ER5356[c,d]	ER5356[f]	ER5183[f]	-	ER5356[f]	ER5183[d]	ER5356[d]
5086	-	-	ER5356[c,d]	ER5356[f]	ER5356[f]	-	ER5356[f]	ER5356[f]	ER5356[f]
5154, 5254	-	-	ER4043[f]	ER5356[f]	ER5356[f]	ER4043[b]	ER5356[f]	ER5356[f]	ER5356[f]
5454	-	ER4043[b]	ER4043[f]	ER5356[f]	ER5356[f]	-	ER5356[c,f]	ER5356[f]	ER5554[c,f]
5456	-	-	ER5356[c,d]	ER5356[d]	ER5556[f]	-	ER5356[f]	ER5556[d]	-
6005, 6061, 6063, 6101, 6151, 6201, 6351, 6951	ER4145	ER4145[b,c]	ER4043[b,f,g]	ER5356[f]	ER5356[c,f]	ER4043[a,b,g]	ER4043[b,f,g]	-	-
6009, 6010, 6070	ER4145	ER4145[b,c]	ER4043[a,b,g]	ER4043	ER4043	ER4043[a,b,g]	-	-	-
7004, 7005, 7039, 710.0, 712.0	-	ER4043[b]	ER4043[b,f]	ER5356[f]	ER5356[d]	-	-	-	-

GUIDE TO THE CHOICE OF FILLER METAL FOR GENERAL PURPOSE WELDING (Continued)

Base Metal	201.0, 206.0, 224.0	319.0, 333.0, 354.0, 355.0, C355.0	356.0, A356.0, A357.0, 413.0, 443.0, A444.0	511.0, 512.0, 513.0, 514.0, 535.0	7004, 7005, 7039, 710.0, 712.0	6009, 6010, 6070	6005, 6061, 6063, 6101, 6151, 6201, 6351, 6951	5456	5454
511.0, 512.0, 513.0, 514.0, 535.0	-	-	ER4043[f]	ER5356[f]	-	-	-	-	5454
356.0, A356.0, 357.0, A357.0, 413.0, 443.0, A444.0	ER4145	ER4145[b,c]	ER4043[b,h]	-	-	-	-	-	-
319.0, 333.0, 354.0, 355.0, C355.0	ER4145e	ER4145[b,c,h]	-	-	-	-	-	-	-
201.0, 206.0, 224.0	ER2319[a,h]	-	-	-	-	-	-	-	-

See footnotes below next table.

GUIDE TO THE CHOICE OF FILLER METAL FOR GENERAL PURPOSE WELDING (Continued)										
Base Metal	5154, 5254[i]	5086	5083	5052, 5652[i]	5005, 5050	3004, Alc. 3004	2219	2014, 2036	1100, 3003, Alc. 3003	1060, 1070, 1080, 1350
1060, 1070, 1080, 1350	ER5356[c,d]	ER5356[d]	ER5356[d]	ER4043[b,d]	ER1100[b,c]	ER4043[b,d]	ER4145[b,c]	ER4145	ER1100[b,c]	ER1188[b,c,h,j]
1100, 3003, Alc. 3003	ER5356[c,d]	ER5356[d]	ER5356[d]	ER4043[b,d]	ER1100[b,c]	ER4043[b,d]	ER4145[b,c]	ER4145	ER1100[b,c]	-
2014, 2036	-	-	-	-	ER4145	ER4145	ER4145[e]	ER4145[e]	-	-
2219	ER4043	-	-	ER4043[b]	ER4043[a,b]	ER4043[a,b]	ER2319[a]	-	-	-
3004, Alc. 3004	ER5356[f]	ER5356[d]	ER5356[d]	ER5356[c,f]	ER5356[c,f]	ER5356[c,f]	-	-	-	-
5005, 5050	ER5356[f]	ER5356[d]	ER5356[d]	ER5356[c,d]	ER5356[c,f]	-	-	-	-	-
5052, 5652[i]	ER5356[f]	ER5356[d]	ER5356[d]	ER5654[c,f,l]	-	-	-	-	-	-
5083	ER5356[d]	ER5356[d]	ER5183[d]	-	-	-	-	-	-	-
5086	ER5356[d]	ER5356[d]	-	-	-	-	-	-	-	-
5154, 5254[i]	ER5654[f,l]	-	-	-	-	-	-	-	-	-

a. ER4145 may be used for some applications. b. ER4047 may be used for some applications. c. ER4043 may be used for some applications. d. ER5183, ER5356, or ER5556 may be used. e. ER2319 may be used for some applications. It can supply high strength when the weldment is postweld solution heat treated and aged. f. ER5183, ER5356, ER5554, ER5556, and ER5654 may be used. In some cases, they provide: (1) improved color match after anodizing treatment, (2) highest weld ductility, and (3) higher weld strength. ER5554 is suitable for sustained elevated temperature service. g. ER4643 will provide high strength in 1/2 in. (12 mm) and thicker groove welds in 6XXX base alloys when postweld solution heat treated and aged. h. Filler metal with the same analysis as the base metal is sometimes used. The following wrought filler metals possess the same chemical composition limits as cast filler alloys: ER4009 and R4009 as R-C355.0; ER4010 and R4010 as R-A356.0; and R4011 as R-A357.0. i. Base metal alloys 5254 and 5652 are used for hydrogen peroxide service. ER5654 filler metal is used for welding both alloys for service temperatures below 150°F (66°C). j. ER1100 may be used for some applications.

Notes:
1. Service conditions such as immersion in fresh or salt water, exposure to specific chemicals, or a sustained high temperature (over 150°F or 66°C) may limit the choice of filler metals. Filler metals ER5183, ER5356, ER5556, and ER5654 are not recommended for sustained elevated temperature service.
2. Recommendations in this table apply to gas shielded arc welding processes. For oxyfuel gas welding, only ER1188, ER5356, ER1100, ER4043, ER4047, and ER4145 filler metals are ordinarily used.
3. Where no filler metal is listed, the base metal combination is not recommended for welding.

AWS A5.10 CLASSIFICATION SYSTEM

Both welding electrodes and rods are classified upon the basis of the chemical composition of the aluminum filler metal and a usability test.

The AWS classifications used in this specification are based as follows:

The Aluminum Association alloy designation nomenclature is used for the numerical portion to identify the alloy and thus its registered chemical composition.

A letter prefix designates usability of the filler metal. The letter system for identifying the filler metal classifications in this specification follows the standard pattern used in other AWS filler metal specifications. The prefix "E" indicates the filler metal is suitable for use as an electrode and the prefix "R" indicates suitability as welding rod. Since some of these filler metals are used as electrodes in gas metal arc welding, and as welding rods in oxyfuel gas, gas tungsten arc, and plasma arc welding, both letters, "ER", are used to indicate suitability as an electrode or a rod. In all cases, an electrode, that meets the tests prescribed in this specification, can be used either as an electrode or a welding rod, but the reverse is not necessarily true.

Minor changes in procedures used in the manufacture of aluminum filler metals can affect their surface quality and significantly affect the resultant weld soundness. Usability testing of the electrode is desirable on a periodic basis to assure that the product classified in this specification continues to meet the soundness requirement.

The supplier should perform the usability tests of this specification on an annual basis, as a minimum, to assure that the specified soundness and operating characteristics criteria are maintained. ANSI/AWS A5.01, Filler Metal Procurement Guidelines, should be used by a purchaser for definition of lot and frequency of testing references when purchasing aluminum filler metals.

Chapter

22

NICKEL & NICKEL ALLOY WELDING ELECTRODES FOR SHIELDED METAL ARC WELDING

CHEMICAL COMPOSITION REQUIREMENTS FOR UNDILUTED WELD METAL

AWS A5.11 Classification	UNS Number[c]	Composition, Weight Percent[a,b]
ENi-1	W82141	C 0.10 Mn 0.75 Fe 0.75 P 0.03 S 0.02 Si 1.25 Cu 0.25 Ni[d] 92.0 min. Al 1.0 Ti 1.0 to 4.0 Other Elements Total 0.50
ENi-Ci-7	W84190	C 0.15 Mn 4.00 Fe 2.5 P 0.02 S 0.015 Si 1.5 Cu Rem Ni[d] 62.0–69.0 Al 0.75 Ti 1.0 Other Elements Total 0.50
ENiCrFe-1	W86132	C 0.08 Mn 3.5 Fe 11.0 P 0.03 S 0.015 Si 0.75 Cu 0.50 Ni[d] 62.0 min. Cr 13.0 to 17.0 Cb Plus Ta 1.5 to 4.0[f] Other Elements Total 0.50
ENiCrFe-2	W86133	C 0.10 Mn 1.0 to 3.5 Fe 12.0 P 0.03 S 0.02 Si 0.75 Cu 0.50 Ni[d] 62.0 min. Co[e] 13.0 to 17.0 Cb Plus Ta 0.5 to 3.0[f] Mo 0.50 to 2.50 Other Elements Total 0.50
ENiCrFe-3	W86182	C 0.10 Mn 5.0 to 9.5 Fe 10.0 P 0.03 S 0.015 Si 1.0 Cu 0.50 Ni[d] 59.0 min. Co[e] Ti 1.0 13.0 to 17.0 Cb Plus Ta 1.0 to 2.5[f] Other Elements Total 0.50
ENiCrFe-4	W86134	C 0.20 Mn 1.0 to 3.5 Fe 12.0 P 0.03 S 0.02 Si 1.0 Cu 0.50 Ni[d] 60.0 min. 13.0 to 17.0 Cb Plus Ta 1.0 to 3.5 Mo 1.0 to 3.5 Other Elements Total 0.50
ENiMo-1	W80001	C 0.07 Mn 1.0 Fe 4.0 to 7.0 P 0.04 S 0.03 Si 1.0 Cu 0.50 Ni[d] Rem Co 2.5 Cr 1.0 Mo 26.0 to 30.0 V 0.60 W 1.0 Other Elements Total 0.50
ENiMo-3	W80004	C 0.12 Mn 1.0 Fe 4.0 to 7.0 P 0.04 S 0.03 Si 1.0 Cu 0.50 Ni[d] Rem Co 2.5 Cr 2.5 to 5.5 Mo 23.0 to 27.0 V 0.60 W 1.0 Other Elements Total 0.50
ENiMo-7	W80665	C 0.02 Mn 1.75 Fe 2.0 P 0.04 S 0.03 Si 0.2 Cu 0.50 Ni[d] Rem Co 1.0 Cr 1.0 Mo 26.0 to 30.0 W 1.0 Other Elements Total 0.50
ENiCrCoMo-1	W86117	C 0.05 to 0.15 Mn 0.30 to 2.5 Fe 5.0 P 0.03 S 0.015 Si 0.75 Cu 0.50 Ni[d] Rem Co 9.0 to 15.0 Cr 21.0 to 26.0 Cb Plus Ta 1.0 Mo 8.0 to 10.0 Other Elements Total 0.50
ENiCrMo-1	W86007	C 0.05 Mn 1.0 to 2.0 Fe 18.0 to 21.0 P 0.04 S 0.03 Si 1.0 Cu 1.5 to 2.5 Ni[d] Rem Co 2.5 Cr 21.0 to 23.5 Cb Plus Ta 1.75 to 2.50 Mo 5.5 to 7.5 W 1.0 Other Elements Total 0.50
ENiCrMo-2	W86002	C 0.05 to 0.15 Mn 1.0 Fe 17.0 to 20.0 P 0.04 S 0.03 Si 1.0 Cu 0.50 Ni[d] Rem Co 0.50 to 2.50 Cr 20.5 to 23.0 Mo 8.0 to 10.0 W 0.20 to 1.0 Other Elements Total 0.50
ENiCrMo-3	W86112	C 0.10 Mn 1.0 Fe 7.0 P 0.03 S 0.02 Si 0.75 Cu 0.50 Ni[d] 55.0 min. Co[e] Cr 20.0 to 23.0 Cb Plus Ta 3.15 to 4.15 Mo 8.0 to 10.0 Other Elements Total 0.50
ENiCrMo-4	W80276	C 0.02 Mn 1.0 Fe 4.0 to 7.0 P 0.04 S 0.03 Si 0.2 Cu 0.50 Ni[d] Rem Co 2.5 Cr 14.5 to 16.5 Mo 15.0 to 17.0 V 0.35 W 3.0 to 4.5 Other Elements Total 0.50
ENiCrMo-5	W80002	C 0.10 Mn 1.0 Fe 4.0 to 7.0 P 0.04 S 0.03 Si 1.0 Cu 0.50 Ni[d] Rem Co 2.5 Cr 14.5 to 16.5 Mo 15.0 to 17.0 V 0.35 W 3.0 to 4.5 Other Elements Total 0.50

CHEMICAL COMPOSITION REQUIREMENTS FOR UNDILUTED WELD METAL (Continued)

AWS A5.11 Classification	UNS Number[c]	Composition, Weight Percent[a,b]
ENiCrMo-6	W86620	C 0.10 Mn 2.0 to 4.0 Fe 10.0 P 0.03 S 0.02 Si 1.0 Cu 0.50 Ni[d] 55.0 min. Cr 12.0 to 17.0 Cb Plus Ta 0.5 to 2.0 Mo 5.0 to 9.0 W 1.0 to 2.0 Other Elements Total 0.50
ENiCrMo-7	W86455	C 0.015 Mn 1.5 Fe 3.0 P 0.04 S 0.03 Si 0.2 Cu 0.50 Ni[d] Rem Co 2.0 Ti 0.70 Cr 14.0 to 18.0 Mo 14.0 to 17.0 W 0.5 Other Elements Total 0.50
ENiCrMo-9	W86985	C 0.02 Mn 1.0 Fe 18.0 to 21.0 P 0.04 S 0.03 Si 1.0 Cu 1.5 to 2.5 Ni[d] Rem Co 5.0 Cr 21.0 to 23.5 Cb Plus Ta 0.5 Mo 6.0 to 8.0 W 1.5 Other Elements Total 0.50
ENiCrMo-10	W86022	C 0.02 Mn 1.0 Fe 2.0 to 6.0 P 0.03 S 0.015 Si 0.2 Cu 0.50 Ni[d] Rem Co 2.5 Cr 20.0 to 22.5 Mo 12.5 to 14.5 V 0.35 W 2.5 to 3.5 Other Elements Total 0.50
ENiCrMo-11	W86030	C 0.03 Mn 1.5 Fe 13.0 to 17.0 P 0.04 S 0.02 Si 1.0 Cu 1.0 to 2.4 Ni[d] Rem Co 5.0 Cr 28.0 to 31.5 Cb Plus Ta 0.3 to 1.5 Mo 4.0 to 6.0 W 1.5 to 4.0 Other Elements Total 0.50
ENiCrMo-12	W86040	C 0.03 Mn 2.2 Fe 5.0 P 0.03 S 0.02 Si 0.7 Cu 0.50 Ni[d] Rem Cr 20.5 to 22.5 Cb Plus Ta 1.0 to 2.8 Mo 8.8 to 10.0 Other Elements Total 0.50

a. The weld metal shall be analyzed for the specific elements for which values are shown in this table. If the presence of other elements is indicated in the course of the work, the amount of those elements shall be determined to ensure that their total does not exceed the limit specified for "Other Elements Total".

b. Single values are maximum, except where otherwise specified.

c. SAE/ASTM Unified Numbering System for Metals and Alloys.

d. Includes incidental cobalt.

e. Cobalt - 0.12 maximum, when specified.

f. Tantalum - 0.30 maximum, when specified.

ALL-WELD-METAL TENSION TEST REQUIREMENTS

AWS A5.11 Classification	Tensile Strength, min.		% Elongation[a], min.
	psi	MPa	
ENi-1	60,000	410	20
ENiCu-7	70,000	480	30
ENiCrFe-1, ENiCrFe-2, ENiCrFe-3	80,000	550	30
ENiCrFe-4	95,000	650	20

The Metals Blue Book

ALL-WELD-METAL TENSION TEST REQUIREMENTS (Continued)

AWS A5.11 Classification	Tensile Strength, min.		% Elongation[a], min.
	psi	MPa	
ENiMo-1, ENiMo-3	100,000	690	25
ENiMo-7	110,000	760	25
ENiCrCoMo-1	90,000	620	25
ENiCrMo-1	90,000	620	20
ENiCrMo-2	95,000	650	20
ENiCrMo-3	110,000	760	30
ENiCrMo-4	100,000	690	25
ENiCrMo-5	100,000	690	25
ENiCrMo-6	90,000	620	35
ENiCrMo-7	100,000	690	25
ENiCrMo-9	90,000	620	25
ENiCrMo-10	100,000	690	25
ENiCrMo-11	85,000	585	25
ENiCrMo-12	95,000	650	35

a.   The elongation shall be determined from a gage length equal to 4 times the gage diameter.

AWS A5.11 CLASSIFICATION SYSTEM

The system for identifying the electrode classifications in this specification follows the standard pattern used in other AWS filler metal specifications. The letter "E" at the beginning of the classification designation stands for electrode.

Since the electrodes are classified according to the chemical composition of the weld metal they deposit, the chemical symbol "Ni" appears right after the "E", as a means of identifying the electrodes as nickel-base alloys. The other symbols (Cr, Cu, Fe, Mo, and Co) in the designations are intended to group the electrodes according to their principal alloying elements. The individual designations are made up of these symbols and a number at the end of the designation (ENiMo-1 and ENiMo-3, for example). These numbers separate one composition from another, within a group, and are not repeated within that group.

Chapter

23

TUNGSTEN & TUNGSTEN ALLOY ELECTRODES FOR ARC WELDING & CUTTING

CHEMICAL COMPOSITION REQUIREMENTS FOR ELECTRODES[a]

AWS A5.12 Classification	UNS Number[b]	Composition, Weight Percent
EWP	R07900	W min. (difference)[c] 99.5 Other Oxides Or Elements Total 0.5
EWCe-2	R07932	W min. (difference)[c] 97.3 CeO_2 1.8-2.2 Other Oxides Or Elements Total 0.5
EWLa-1	R07941	W min. (difference)[c] 98.3 La_2O_3 0.9-1.2 Other Oxides Or Elements Total 0.5
EWTh-1	R07911	W min. (difference)[c] 98.3 ThO_2 0.8-1.2 Other Oxides Or Elements Total 0.5
EWTh-2	R07912	W min. (difference)[c] 97.3 ThO_2 1.7-2.2 Other Oxides Or Elements Total 0.5
EWZr-1	R07920	W min. (difference)[c] 99.1 ZrO_2 0.15-0.40 Other Oxides Or Elements Total 0.5
EWG[d]	-	W min. (difference)[c] 94.5 CeO_2, La_2O_3, ThO_2, ZrO_2 Not Specified Other Oxides Or Elements Total 0.5

a. The electrode shall be analyzed for the specific oxides for which values are shown in this table. If the presence of other elements or oxides is indicated in the course of this work, the amount of those elements or oxides shall be determined to ensure that their total does not exceed the limit specified for "Other Oxides Or Elements Total".

b. SAE/ASTM Unified Numbering System for Metals and Alloys.

c. Tungsten content shall be determined by determining the measured content of all specified oxides and other oxides and elements and subtracting the total from 100%.

d. Classification EWG must contain some oxide or element additive and the manufacturer must identify the type and nominal content of the oxide or element additive.

TYPICAL CURRENT RANGES FOR AWS A5.12 TUNGSTEN ELECTRODES[a]

Electrode Diameter		DCEN (DCSP) A	DCEP (DCRP) A	Alternating Current Unbalanced Wave A		Alternating Current Balanced Wave A	
in.	mm	EWX-X	EWX-X	EWP	EWX-X	EWP	EWX-X
0.010	0.30	Up to 15	na[b]	Up to 15	Up to 15	Up to 15	Up to 15
0.020	0.50	5-20	na	5-15	5-20	10-20	5-20
0.040	1.00	15-80	na	10-60	15-80	20-30	20-60
0.060	1.60	70-150	10-20	50-100	70-150	30-80	60-120
0.093	2.40	150-250	15-30	100-160	140-235	60-130	100-180
0.125	3.20	250-400	25-40	150-200	225-325	100-180	160-250
0.156	4.00	400-500	40-55	200-275	300-400	160-240	200-320
0.187	5.00	500-750	55-80	250-350	400-500	190-300	290-390
0.250	6.40	750-1000	80-125	325-450	500-630	250-400	340-525

a.   All are values based on the use of argon gas. Other current values may be employed depending on the shielding gas, type of equipment and application.

b.   na = not applicable

AWS A5.12 CLASSIFICATION SYSTEM

The system for identifying the electrode classifications in this specification follows the standard pattern used in other AWS filler metal specifications. The letter "E" at the beginning of the classification designation stands for electrode. The "W" indicates that the electrode is primarily tungsten. The "P" indicates that the electrode is essentially pure tungsten and contains no intentionally added alloying elements. The "Ce", "La", "Th", and "Zr" indicate that the electrode is alloyed with oxides of cerium, lanthanum, thorium, or zirconium, respectively. The numeral at the end of some of the classifications indicates a different chemical composition level or product within a specific group.

This specification includes electrodes classified as EWG. The "G" indicates that the electrode is of a general classification. It is "general" because not all of the particular requirements specified for each of the other classifications are specified for this classification. The intent, in establishing this classification, is to provide a means by which electrodes that differ in one respect or another (chemical composition, for example) from other classifications (meaning that the composition of the electrode - in the case of this example - does not meet the composition specified for any of the classifications in the specification) can still be classified according to the specification. The purpose is to allow a useful electrode - one that otherwise would have to await a revision of the specification - to be classified immediately, under the existing specification. This means, then, that two electrodes - each bearing the same "G" classification - may be quite different in some certain respect. To prevent the confusion that this situation could create, this specification requires the manufacturer to identify, in the label, the type and nominal content of the alloy addition made in the particular product.

Chapter
24

SOLID SURFACING
WELDING RODS & ELECTRODES

CHEMICAL REQUIREMENTS FOR BARE SURFACING WELDING RODS AND ELECTRODES[d,e]

AWS A5.13 Classification	Composition, Weight Percent
RFe5-A	C 0.7 to 1.0 Mn 0.50 W 5.0 to 7.0 Cr 3.0 to 5.0 Mo 4.0 to 6.0 Se Remainder V 1.0 to 2.5 Si 0.5 Total Other Elements 1.0
RFe5-B	C 0.5 to 0.9 Mn 0.50 W 1.0 to 2.5 Cr 3.0 to 5.0 Mo 5.0 to 9.5 Se Remainder V 0.8 to 1.3 Si 0.5 Total Other Elements 1.0
RFeCr-A1	C 3.7 to 5.0 Mn 2.0 to 6.0 Cr 27.0 to 35.0 Se Remainder Si 1.10 to 2.5 Total Other Elements 1.0
RCoCr-A	C 0.9 to 1.4 Mn 1.00 Co Remainder W 3.0 to 6.0 Ni 3.0 Cr 26.0 to 32.0 Mo 1.0 Se 3.0 Si 2.0 Total Other Elements 0.50
RCoCr-B	C 1.2 to 1.7 Mn 1.00 Co Remainder W 7.0 to 9.5 Ni 3.0 Cr 26.0 to 32.0 Mo 1.0 Se 3.0 Si 2.0 Total Other Elements 0.50
RCoCr-C	C 2.0 to 3.0 Mn 1.00 Co Remainder W 11.0 to 14.0 Ni 3.0 Cr 26.0 to 33.0 Mo 1.0 Se 3.0 Si 2.0 Total Other Elements 0.50
RCuZn-E[b]	Mn 0.30 Fe 1.50 Cu 56.0 min Al 0.01[a] Zn Remainder Si 0.04 to 0.25 Pb 0.05[a] Sn 2.00 to 3.00 Total Other Elements 0.50
ERCuSi-A[b]	Mn 1.5[c] Ni[a] Se 0.5 Cu 94.0 min Al 0.01[a] Zn 1.5[a] Si 2.8 to 4.0 Pb 0.02[a] 1.5[a] P[a] Total Other Elements 0.50
ERCuAl-A2[b]	Fe 1.5 Cu Remainder Al 9.0 to 11.0 Zn 0.02 Si 0.10 Pb 0.02[a] Total Other Elements 0.50
ERCuAl-A3	Fe 3.0 to 5.0 Cu Remainder Al 10.0 to 11.0 Zn 0.10 Si 0.10 Pb 0.02[a] Total Other Elements 0.50
RCuAl-C	Fe 3.0 to 5.0 Cu Remainder Al 12.0 to 13.0 Zn 0.02 Si 0.4 Pb 0.02 Total Other Elements 0.50
RCuAl-D	Fe 3.0 to 5.0 Cu Remainder Al 13 to 14.0 Zn 0.02 Si 0.04 Pb 0.02 Total Other Elements 0.50
RCuAl-E	Fe 3.0 to 5.0 Cu Remainder Al 14.0 to 15.0 Zn 0.02 Si 0.4 Pb 0.02 Total Other Elements 0.50
ERCuSn-A[b]	Mn[a] Ni[a] Fe[a] Cu 93.5 min Al 0.01[a] Zn[a] Si[a] Pb 0.02[a] Sn 4.0 to 6.0 P 0.10 to 0.35 Total Other Elements 0.50
RCuSn-D	Cu 88.5 min Al 0.01[a] Pb 0.05[a] Sn 9.0 to 11.0 P 0.10 to 0.30 Total Other Elements 0.50
RNiCr-A	C 0.30 to 0.60 Co 1.50 Ni Remainder Cr 8.0 to 14.0 Fe 1.25 to 3.25 V 2.00 to 3.00 percent boron Si 1.25 to 3.25 Total Other Elements 0.50
RNiCr-B	C 0.40 to 0.80 Co 1.25 Ni Remainder Cr 10.0 to 16.0 Fe 3.00 to 5.00 V 2.00 to 4.00 percent boron Si 3.00 to 5.00 Total Other Elements 0.50
RNiCr-C	C 0.50 to 1.00 Co 1.00 Ni Remainder Cr 12.0 to 18.0 Fe 3.50 to 5.50 V 2.50 to 4.50 percent boron Si 3.50 to 5.50 Total Other Elements 0.50

a.   Total other elements, including the elements marked with footnote a, shall not exceed the value specified.
b.   This AWS classification is intended to be identical with the same classification that appears in the latest edition of AWS A5.7, Specification for Copper and Copper Alloy Bare Welding Rods and Electrodes.
c.   One or more of these elements may be present within the limits specified.
d.   Analysis shall be made for the elements for which specific values are shown in this table. If, however, the presence of other elements is indicated in the course of routine analysis, further analysis shall be made to determine that the total of these other elements is not present in excess of the limits specified for "Total Other Elements" in this table.
e.   Single values shown are maximum percentages, except where otherwise specified.

CHEMICAL REQUIREMENTS FOR COVERED SURFACING ELECTRODES[d,e]

AWS A5.13 Classification	Composition, Weight Percent
EFe5-A	C 0.7 to 1.0 Mn 0.60 W 5.0 to 7.0 Cr 3.0 to 5.0 Mo 4.0 to 6.0 Fe Remainder V 1.0 to 2.5 Si 0.80 Total Other Elements 1.0
EFe5-B	C 0.5 to 0.9 Mn 0.60 W 1.0 to 2.5 Cr 3.0 to 5.0 Mo 5.0 to 9.5 Fe Remainder V 0.8 to 1.3 Si 0.80 Total Other Elements 1.0
EFe5-C	C 0.3 to 0.5 Mn 0.60 W 1.0 to 2.5 Cr 3.0 to 5.0 Mo 5.0 to 9.0 Fe Remainder V 0.8 to 1.2 Si 0.80 Total Other Elements 1.0
EFeMn-A	C 0.5 to 0.9 Mn 11.0 to 16.0 Ni 2.75 to 6.0 Cr 0.50 Fe Remainder Si 1.3 P 0.03 Total Other Elements 1.0
EFeMn-B	C 0.5 to 0.9 Mn 11.0 to 16.0 Cr 0.50 Mo 0.6 to 1.4 Fe Remainder Si 0.3 to 1.3 P 0.03 Total Other Elements 1.0
EFeCr-A1	C 3.0 to 5.0 Mn 4.0 to 8.0 Cr 26.0 to 32.0 Mo 2.0 Fe Remainder Si 1.0 to 2.5 Total Other Elements 1.0
ECoCr-A	C 0.7 to 1.4 Mn 2.0 Co Remainder W 3.0 to 6.0 Ni 3.0 Cr 25.0 to 32.0 Mo 1.0 Fe 5.0 Si 2.0 Total Other Elements 0.50
ECoCr-B	C 1.0 to 1.7 Mn 2.0 Co Remainder W 7.0 to 9.5 Ni 3.0 Cr 25.0 to 32.0 Mo 1.0 Fe 5.0 Si 2.0 Total Other Elements 0.50
ECoCr-C	C 1.75 to 3.0 Mn 2.0 Co Remainder W 11.0 to 14.0 Ni 3.0 Cr 25.0 to 33.0 Mo 1.0 Fe 5.0 Si 2.0 Total Other Elements 0.50
ECuSi[c]	Mn 1.5 Ni[b] Fe 0.50 Cu Remainder Al 0.01 Si 2.4 to 4.0 Pb 0.02[b] Sn 1.5 P[b] Total Other Elements 0.50
ECuAl-A2[c]	Mn[b] Ni[b] Fe 0.5 to 5.0 Cu Remainder Al 7.0 to 9.0 Zn[b] Si 1.0 Pb 0.02[b] Sn[b] Total Other Elements 0.60
ECuAl-B[c]	Mn[b] Ni[b] Fe 2.5 to 5.0 Cu Remainder Al 8.4 to 10.0 Zn[b] Si 1.0 Pb 0.02[b] Sn[b] Total Other Elements 0.60
ECuAl-C	Fe 3.0 to 5.0 Cu Remainder Al 12.0 to 13.0 Zn 0.02 Si 0.04 Pb 0.02 Total Other Elements 0.50
ECuAl-D	Fe 3.0 to 5.0 Cu Remainder Al 13.0 to 14.0 Zn 0.02 Si 0.04 Pb 0.02 Total Other Elements 0.50
ECuAl-E	Fe 3.0 to 5.0 Cu Remainder Al 14.0 to 15.0 Zn 0.02 Si 0.04 Pb 0.02 Total Other Elements 0.50
ECuSn-A[c]	Mn[b] Ni[b] Se 0.25 Cu Remainder Al 0.01 Zn[b] Si[b] Pb 0.02[b] Sn 4.0 to 6.0 P 0.05 to 0.35 Total Other Elements 0.50
ECuSn-C[c]	Mn[b] Ni[b] Se 0.25 Cu Remainder Al 0.01 Zn[b] Si[b] Pb 0.02[b] Sn 7.0 to 9.0 P 0.05 to 0.35 Total Other Elements 0.50
ENiCr-A	C 0.30 to 0.60 Co 1.50 Ni Remainder Cr 8.0 to 14.0 Se 1.25 to 3.25 V 2.00 to 3.00 percent boron Si 1.25 to 3.25 Total Other Elements 0.50
ENiCr-B	C 0.40 to 0.80 Co 1.25 Ni Remainder Cr 10.0 to 16.0 Se 3.00 to 5.00 V 2.00 to 4.00 percent boron Si 3.00 to 5.00 Total Other Elements 0.50
ENiCr-C	C 0.50 to 1.00 Co 1.00 Ni Remainder Cr 12.0 to 8.0 Se 3.50 to 5.50 V 2.50 to 4.50 percent boron Si 3.50 to 5.50 Total Other Elements 0.50

a.   The analysis given is for deposited weld metal.

b.   Total other elements, including the elements marked with footnote b, shall not exceed the value specified.

c.   This AWS classification is intended to be identical with the same classification that appears in the latest edition of AWS A5.6 Specification for Copper and Copper Alloy Covered Electrodes.

d.   Analysis shall be made for the elements for which specific values are shown in this table. If however, the presence of other elements is indicated in the course of routine analysis, further analysis shall be made to determine that the total of these other elements is not present in excess of the limits specified for "Total Other Elements" in this table.

e.   Single values shown are maximum percentages, except where otherwise specified.

USUAL HARDNESS OF COBALT-BASE WELD DEPOSITS[a,b] (70°F) (21°C)

Process	CoCr-A	Hardness, Rockwell C CoCr-B	CoCr-C
Oxyfuel gas welded	38 to 47	45 to 49	48 to 58
Arc welded	23 to 47	34 to 47	43 to 58

a. Lower values can be expected in single layer deposits due to dilution with the base metal.
b. See AWS A5.13 for more details.

ARC WELDED DEPOSITS[a]

Room Temperature Hardness Data On CoCr-C Arc Welded Deposits

Sample	% Carbon	Average Hardness Brinell	Average Hardness Rockwell C	Brinell 1	Brinell 2	Brinell 3	Rockwell C 1	Rockwell C 2	Rockwell C 3
B	2.28	465	47	389	497	509	41	48	53
G	1.83	391	40	328	399	444	32	43	47
H	2.12	437	49	368	455	489	44	51	53
E	2.95	448	47	381	433	528	42	46	53

a. See AWS A5.13 appendix for more details.

ARC WELDED DEPOSITS[a,b]

Room Temperature Hardness Data On CoCr-C Arc Welded Deposits

Sample	Brinell Hardness 1	Brinell Hardness 2	Brinell Hardness 3	Rockwell C Hardness 1	Rockwell C Hardness 2	Rockwell C Hardness 3
B	357 to 440	443 to 557	390 to 640	33 to 46	38 to 50	48 to 56
G	297 to 353	373 to 415	429 to 478	27 to 35	38 to 45	44 to 50
H	324 to 384	413 to 483	450 to 514	41 to 45	41 to 53	51 to 56
E	351 to 443	322 to 514	465 to 578	37 to 45	34 to 53	50 to 55

a. Range of hardness observed from 3 coupons for each sample, 3 readings for each of 3 positions on each of 3 layers
b. See AWS A5.13 appendix for more details.

INSTANTANEOUS HARDNESS VALUES[b]

Temperature		CoCr-A Rockwell C For Sample Number Given					CoCr-C Rockwell C For Sample Number Given			
°F	°C	B	G	H-a	H-b	E	B	G	HJ	E
650	345	29.6	33.3	26.5	29.8	41.0	43.7	35.7	41.2	46.1
850	455	24.3	26.8	21.5	28.5	36.6	41.0	31.9	38.0	40.6
1050	565	20.0	21.8	19.1	22.7	32.2	35.6	29.8	35.1	30.5
1200	650	15.9	19.7	16.4	21.8	25.8	29.9	24.3	29.0	27.9
1400	760	46.8[a]	45.8[a]	44.7[a]	49.8[a]	49.8[a]	53.9[a]	53.0[a]	53.4[a]	51.3[a]

a. These are Rockwell A values.
b. See AWS A5.13 appendix for more details.

HARDNESS OF WELD DEPOSITS[a]

AWS A5.13 Classification	Number of Layers	Oxyfuel Gas Weld Deposit Hardness, Rockwell C	Arc Weld Deposit Hardness of Covered and Bare Electrode, Rockwell C
NiCr-A	1	35 to 40	24 to 29
	2	35 to 40	30 to 35
NiCr-B	1	45 to 50	30 to 45
	2	45 to 50	40 to 45
NiCr-C	1	56 to 62	35 to 45
	2	56 to 62	49 to 56

a. See AWS A5.13 appendix for more details.

HARDNESS OF WELD DEPOSITS[a]

AWS A5.13 Classification	Brinell Hardness of Deposit,		Welding Process Applicable
	3000 kg load	500 kg load	
ERCuA1-A2	130-150	-	GTAW, GMAW
ECuA1-A2	115-140	-	SMAW
ERCuA1-A3	140-180	-	GTAW, SMAW
ECuA1-B	140-180	-	SMAW
ERCuA1-C	250-290	-	GTAW
ECuA1-C	180-220	-	SMAW
RCuA1-D	310-350	-	GTAW
ECuA1-D	230-270	-	SMAW
RCuA1-E	350-390	-	GTAW
ECuA1-E	280-320	-	SMAW
ECuSi	-	80-100	SMAW
ERCuSi-A	-	80-100	OFW, GMAW, GTAW
ERCuSn-A	-	70-85	GTAW, GMAW
ECuSu-A	-	70-85	SMAW
ECuSu-C	-	85-100	SMAW
RCuSn-D	-	90-110	GTAW
RCuZn-E	-	130 min	OFW

a.   See AWS A5.13 appendix for more details.

HOT HARDNESS OF WELD DEPOSITS[a]

AWS A5.13 Classification	Loading interval, min	Arc Weld Deposit Hardness, Rockwell C			Oxyfuel Gas Weld Deposit Hardness, Rockwell C		
		600°F (315°C)	800°F (430°C)	1000°F (540°C)	600°F (315°C)	800°F (430°C)	1000°F (540°C)
NiCr-A	0	30	29	24	34	33	29
	1	30	28	21	33	32	26
	2	30	28	20	33	32	25
	3	29	28	19	33	31	24
NiCr-B	0	41	39	33	46	45	42
	1	41	38	29	46	44	39
	2	41	38	28	45	43	38
	3	40	37	26	45	42	37
NiCr-C	0	49	46	39	55	52	48
	1	49	45	33	54	51	42
	2	48	45	32	54	51	41
	3	48	45	31	54	50	40

a.   See AWS A5.13 appendix for more details.

COMPRESSION PROPERTIES OF CAST COBALT-BASE ALLOYS[a,b]

	Co-Cr-A	Co-Cr-C
Yield strength (0.1 percent offset), ksi	64 to 76	85 to 110
Ultimate compression strength, ksi	150 to 230	250 to 270
Plastic deformation, percent	5 to 8	1 to 2
Brinell hardness	350 to 420	480 to 550

a.   See AWS A5.13 appendix for more details.
b.   Cast values are included because weld metal data are not available.

SURFACING FILLER METALS[a]	
New AWS A5.13 Classification	Old Designation
Fe	IA5
FeMn	IB2
FeCr	IC1
CoCr-A	IIA
CoCr-C	IIB
CuZn	IVA
CuSi	IVB
CuAl	IVC
NiCr	VB

a. See AWS A5.13 appendix for more details.

AWS A5.13 CLASSIFICATION SYSTEM

The system for identifying welding rod and electrode classifications used in this specification follows the standard pattern used in other AWS filler metal specifications. The letter "E" at the beginning of each classification indicates an electrode, and the letter "R" indicates a welding rod. The letters "ER" indicate a filler metal that may be used as either a bare electrode or rod. The letters immediately after the E, R, or ER are the chemical symbols for the principal elements in the classification. Thus, CoCr is cobalt-chromium alloy, CuZn is a copper-zinc alloy, etc. Where more than one classification is included in a basic group, the individual classifications in the group are identified by the letters A, B, C, etc. as in ECuSn-A. Further subdividing is done by using a 1, 2, etc., after the last letter, as the 2 in ECuAl-A2.

The correlation between old AWS designations and the new AWS A5.13 classifications covered by this specification is indicated in preceding table.

NICKEL & NICKEL ALLOY
BARE WELDING ELECTRODES & RODS

CHEMICAL COMPOSITION REQUIREMENTS FOR NICKEL AND NICKEL ALLOY ELECTRODES & RODS

AWS A5.14 Classification	UNS Number[c]	Composition, Weight Percent[a,b]
ERNi-1	N02061	C 0.15 Mn 1.0 Fe 1.0 P 0.03 S 0.015 Si 0.75 Cu 0.25 Ni[d] 93.0 min Al 1.5 Ti 2.0 to 3.5 Other Elements Total 0.50
ERNiCu-7	N04060	C 0.15 Mn 4.0 Fe 2.5 P 0.02 S 0.015 Si 1.25 Cu Rem Ni[d] 62.0 to 69.0 Al 1.25 Ti 1.5 to 3.0 Other Elements Total 0.50
ERNiCr-3	N06082	C 0.10 Mn 2.5 to 3.5 Fe 3.0 P 0.03 S 0.015 Si 0.50 Cu 0.50 Ni[d] 67.0 min Co[e] Ti 0.75 Cr 18.0 to 22.0 Cb plus Ta 2.0 to 3.0[f] Other Elements Total 0.50
ERNiCrFe-5	N06062	C 0.08 Mn 1.0 Fe 6.0 to 10.0 P 0.03 S 0.015 Si 0.35 Cu 0.50 Ni[d] 70.0 min Co[e] Cr 14.0 to 17.0 Cb plus Ta 1.5 to 3.0[f] Other Elements Total 0.50
ERNiCrFe-6	N07092	C 0.08 Mn 2.0 to 2.7 Fe 8.0 P 0.03 S 0.015 Si 0.35 Cu 0.50 Ni[d] 67.0 min Ti 2.5 to 3.5 Cr 14.0 to 17.0 Other Elements Total 0.50
ERNiFeCr-1	N08065	C 0.05 Mn 1.0 Fe 22.0 P 0.03 S 0.03 Si 0.50 Cu 1.50 to 3.0 Ni[d] 38.0 to 46.0 Al 0.20 Ti 0.60 to 1.2 Cr 19.5 to 23.5 Mo 2.5 to 3.5 Other Elements Total 0.50
ERNiFeCr-2[g]	N07718	C 0.08 Mn 0.35 Fe Rem P 0.015 S 0.015 Si 0.35 Cu 0.30 Ni[d] 50.0 to 55.0 Al 0.20 to 0.80 Ti 0.65 to 1.15 Cr 17.0 to 21.0 Cb plus Ta 4.75 to 5.50 Mo 2.80 to 3.30 Other Elements Total 0.50
ERNiMo-1	N10001	C 0.08 Mn 1.0 Fe 4.0 to 7.0 P 0.025 S 0.03 Si 1.0 Cu 0.50 Ni[d] Rem Co 2.5 Cr 1.0 Mo 26.0 to 30.0 V 0.20 to 0.40 W 1.0 Other Elements Total 0.50
ERNiMo-2	N10003	C 0.04 to 0.08 Mn 1.0 Fe 5.0 P 0.015 S 0.02 Si 1.0 Cu 0.50 Ni[d] Rem Co 0.20 Cr 6.0 to 8.0 Mo 15.0 to 18.0 V 0.50 W 0.50 Other Elements Total 0.50
ERNiMo-3	N10004	C 0.12 Mn 1.0 Fe 4.0 to 7.0 P 0.04 S 0.03 Si 1.0 Cu 0.50 Ni[d] Rem Co 2.5 Cr 4.0 to 6.0 Mo 23.0 to 26.0 V 0.60 W 1.0 Other Elements Total 0.50
ERNiMo-7	N10665	C 0.02 Mn 1.0 Fe 2.0 P 0.04 S 0.03 Si 0.10 Cu 0.50 Ni[d] Rem Co 1.0 Cr 1.0 Mo 26.0 to 30.0 W 1.0 Other Elements Total 0.50
ERNiCrMo-1	N06007	C 0.05 Mn 1.0 to 2.0 Fe 18.0 to 21.0 P 0.04 S 0.03 Si 1.0 Cu 1.5 to 2.5 Ni[d] Rem Co 2.5 Cr 21.0 to 23.5 Cb plus Ta 1.75 to 2.50 Mo 5.5 to 7.5 W 1.0 Other Elements Total 0.50
ERNiCrMo-2	N06002	C 0.05 to 0.15 Mn 1.0 Fe 17.0 to 20.0 P 0.04 S 0.03 Si 1.0 Cu 0.50 Ni[d] Rem Co 0.50 to 2.5 Cr 20.5 to 23.0 Mo 8.10 to 10.0 W 0.20 to 1.0 Other Elements Total 0.50
ERNiCrMo-3	N06625	C 0.10 Mn 0.50 Fe 5.0 P 0.02 S 0.015 Si 0.50 Cu 0.50 Ni 58.0 min Al 0.40 Ti 0.40 Cr 20.0 to 23.0 Cb plus Ta 3.15 to 4.15 Mo 8.10 to 10.0
ERNiCrMo-4	N10276	C 0.02 Mn 1.0 Fe 4.0 to 7.0 P. 0.04 S 0.03 Si 0.08 Cu 0.50 Ni[d] Rem Co 2.5 Cr 14.5 to 16.5 Mo 15.0 to 17.0 V 0.35 W 3.0 to 4.5 Other Elements Total 0.50

CHEMICAL COMPOSITION REQUIREMENTS FOR NICKEL AND NICKEL ALLOY ELECTRODES & RODS (Continued)		
AWS A5.14 Classification	UNS Number[c]	Composition, Weight Percent[a,b]
ERNiCrMo-7	N06455	C 0.015 Mn 1.0 Fe 3.0 P 0.04 S 0.03 Si 0.08 Cu 0.50 Ni[d] Rem Co 2.0 Ti 0.70 Cr 14.0 to 18.0 Mo 14.0 to 18.0 W 0.50 Other Elements Total 0.50
ERNiCrMo-8	N06975	C 0.03 Mn 1.0 Fe Rem P 0.03 S 0.03 Si 1.0 Cu 0.7 to 1.20 Ni[d] 47.0 to 52.0 Ti 0.70 to 1.50 Cr 23.0 to 26.0 Mo 5.0 to 7.0 Other Elements Total 0.50
ERNiCrMo-9	N06985	C 0.015 Mn 1.0 Fe 18.0 to 21.0 P 0.04 S 0.03 Si 1.0 Cu 1.5 to 2.5 Ni[d] Rem Co 5.0 Cr 21.05 to 23.5 Cb plus Ta 0.50 Mo 6.0 to 8.0 W 1.5 Other Elements Total 0.50
ERNiCrMo-10	N06022	C 0.015 Mn 0.50 Fe 2.0 to 6.0 P 0.20 S 0.010 Si 0.08 Cu 0.50 Ni Rem Co 2.5 Cr 20.0 to 22.5 Mo 12.5 to 14.5 V 0.35 W 2.5 to 4.5 Other Elements Total 0.50
ERNiCrMo-11	N06030	C 0.30 Mn 1.5 Fe 13.0 to 17.0 P 0.04 S 0.02 Si 0.80 Cu 1.0 to 2.4 Ni Rem Co 5.0 Cr 28.0 to 31.5 Cb + Ta 0.30 to 1.50 Mo 4.0 to 6.0 W 1.5 to 4.0 Other Elements Total 0.50
ERNiCrCoMo-1	N06617	C 0.05 to 0.15 Mn 1.0 Fe 3.0 P 0.03 S 0.015 Si 1.0 Cu 0.50 Ni Rem Co 10.0 to 15.0 Al 0.80 to 1.50 Ti 0.60 Cr 20.0 to 24.0 Mo 8.0 to 10.0 Other Elements Total 0.50

a. The filler metal shall be analyzed for the specific elements for which values are shown in this table. If the presence of other elements is indicated in the course of this work, the amount of those elements shall be determined to ensure that their total does not exceed the limit specified for "Other Elements Total" in this table. "Rem" stands for Remainder.

b. Single values are maximum, except where otherwise specified.

c. SAE/ASTM Unified Numbering System for Metals and Alloys.

d. Includes incidental cobalt

e. Cobalt - 0.12 maximum, when specified.

f. Tantalum - 0.30 maximum, when specified.

g. Boron is 0.006 percent maximum.

COMPARISON OF CLASSIFICATIONS

Classification in A5.14		Military Designation[b,c]	Corresponding Classification in AWS A5 11-88[c]
Present Classification	Previous Classification[a,c]		
ERNi-1	ERNi-1	EN61 & RN61	ENi-1
ERNiCu-7	ERNiCu-7	EN60 & RN60	ENiCu-7
ERNiCr-3	ERNiCr-3	EN82 & RN82	ENiCrFe-3
ERNiCrFe-5	ERNiCrFe-5	EN62 & RN62	ENiCrFe-1
ERNiCrFe-6	ERNiCrFe-6	EN6A & RN6A	-
ERNiFeCr-1	ERNiFeCr-1	-	-
ERNiFeCr-2	Not Classified	-	-
ERNiMo-1	ERNiMo-1	-	ENiMo-1
ERNiMo-2	ERNiMo-2	-	-
ERNiMo-3	ERNiMo-3	-	ENiMo-3
ERNiMo-7	ERNiMo-7	-	ENiMo-7
ERNiCrMo-1	ERNiCrMo-1	-	ENiCrMo-1
ERNiCrMo-2	ERNiCrMo-2	-	ENiCrMo-2
ERNiCrMo-3	ERNiCrMo-3	EN625 & RN625	ENiCrMo-3
ERNiCrMo-4	ERNiCrMo-4	-	ENiCrMo-4
ERNiCrMo-7	ERNiCrMo-7	-	ENiCrMo-7
ERNiCrMo-8	ERNiCrMo-8	-	-
ERNiCrMo-9	ERNiCrMo-9	-	ENiCrMo-9
ERNiCrMo-10	Not Classified	-	ENiCrMo-10
ERNiCrMo-11	Not Classified	-	ENiCrMo-11
ERNiCrCoMo-1	Not Classified	-	ENiCrCoMo-1

a.   AWS A5.14-83
b.   Mil-E-21562.
C.   Specifications are not exact duplicates. Information is supplied only for general comparison.

TYPICAL WELD METAL TENSILE STRENGTHS[a]		
AWS A5.14 Classification	psi	MPa
ERNi-1	55,000	380
ERNiCu-7	70,000	480
ERNiCr-3, ERNiCrFe-5, ERNiCrFe-6, ERNiFeCr-1	80,000	550
ERNiFeCr-2	165,000[b]	1,138[b]
ERNiMo-1, ERNiMo-2, ERNiMo-3	100,000	690
ERNiMo-7	110,000	760
ERNiCrMo-1, ERNiCrMo-8, ERNiCrMo-9, ERNiCrMo-11	85,000	590
ERNiCrCoMo-1	90,000	620
ERNiCrMo-2	95,000	660
ERNiCrMo-3	110,000	760
ERNiCrMo-4, ERNiCrMo-7, ERNiCrMo-10	100,000	690

a. Tensile strength in as-welded condition unless otherwise specified.
b. Age hardened condition: heat to 1,325°F (718°C), hold at temperature for eight hours, furnace cool to 1,150°F (620°C) at 100°F (55°C)/hour and then air cool.

AWS A5.14 CLASSIFICATION SYSTEM

The system for classifying the filler metals in this specification follows the standard pattern used in other AWS filler metal specifications. The letter "ER" at the beginning of each classification designation stands for electrode and rod, indicating that the filler metal may be used either way.

Since the filler metals are classified according to their chemical composition, the chemical symbol "Ni" appears right after the "ER" as a means of identifying the filler metals as nickel-base alloys. The other symbols (Cr, Cu, Fe, and Mo) in the designations are intended to group the filler metals according to their principal alloying elements. The individual designations are made up of these symbols and a number at the end of the designation (ERNiMo-1 and ERNiMo-2, for example). These numbers separate one composition from another within a group and are not repeated within that group.

Chapter
26

WELDING ELECTRODES & RODS FOR CAST IRON

CHEMICAL COMPOSITION REQUIREMENTS FOR UNDILUTED WELD METAL FOR SHIELDED METAL ARC

AWS A5.15 Classification[d]	UNS Number[e]	Composition, Weight Percent[a,b,c]
Shielded Metal Arc Welding Electrodes		
ENi-Cl	W82001	C 2.0 Mn 2.5 Si 4.0 S 0.03 Fe 8.0 Ni[f] 85 min. Cu[g] 2.5 Al 1.0 Other Elements Total 1.0
ENi-Cl-A	W82003	C 2.0 Mn 2.5 Si 4.0 S 0.03 Fe 8.0 Ni[f] 85 min. Cu[g] 2.5 Al 1.0-3.0 Other Elements Total 1.0
ENiFe-Cl	W82002	C 2.0 Mn 2.5 Si 4.0 S 0.03 Fe Rem Ni[f] 45-60 Cu[g] 2.5 Al 1.0 Other Elements Total 1.0
ENiFe-Cl-A	W82004	C 2.0 Mn 2.5 Si 4.0 S 0.03 Fe Rem Ni[f] 45-60 Cu[g] 2.5 Al 1.0-3.0 Other Elements Total 1.0
ENiFeMn-Cl	W82006	C 2.0 Mn 10-14 Si 1.0 S 0.03 Fe Rem Ni[f] 35-45 Cu[g] 2.5 Al 1.0 Other Elements Total 1.0
ENiCu-A	W84001	C 0.35-0.55 Mn 2.3 Si 0.75 S 0.025 Fe 3.0-6.0 Ni[f] 50-60 Cu[g] 35-45 Other Elements Total 1.0
ENiCu-B	W84002	C 0.35-0.55 Mn 2.3 Si 0.75 S 0.025 Fe 3.0-6.0 Ni[f] 60-70 Cu[g] 25-35 Other Elements Total 1.0
Flux Cored Arc Welding Electrodes		
ENiFeT3-Cl[h]	W82032	C 2.0 Mn 3.0-5.0 Si 1.0 S 0.03 Fe Rem Ni[f] 45-60 Cu[g] 2.5 Al 1.0 Other Elements Total 1.0

a. The weld metal, core wire, or the filler metal, as specified, shall be analyzed for the specific elements for which values are shown in this table. If the presence of other elements is indicated in the course of this work, the amount of those elements shall be determined to ensure that their total does not exceed the limit specified for "Other Elements Total" in this table.

b. Single values are maximum, except where otherwise specified.

c. "Rem" stands for Remainder.

d. Copper-base filler metals frequently used in the braze welding of cast irons are no longer included in this specification. For information pertaining to these materials, see AWS A5.15 appendix A7.6.

e. SAE/ASTM Unified Numbering System for Metals and Alloys.

f. Nickel plus incidental cobalt.

g. Copper plus incidental silver.

h. No shielding gas shall be used for classification ENiFeT3-Cl.

CHEMICAL COMPOSITION REQUIREMENTS FOR CORE WIRE FOR SHIELDED METAL ARC WELDING ELECTRODES		
AWS A5.15 Classification[d]	UNS Number[e]	Composition, Weight Percent[a,b,c]
Shielded Metal Arc Welding Electrodes		
ESt	K01520	C 0.15 Mn 0.60 Si 0.15 P 0.04 S 0.04 Fe Rem

a. The weld metal, core wire, or the filler metal, as specified, shall be analyzed for the specific elements for which values are shown in this table. If the presence of other elements is indicated in the course of this work, the amount of those elements shall be determined to ensure that their total does not exceed the limit specified for "Other Elements Total" in this table.

b. Single values are maximum, except where otherwise specified.

c. "Rem" stands for Remainder.

d. Copper-base filler metals frequently used in the braze welding of cast irons are no longer included in this specification. For information pertaining to these materials, see AWS A5.15 appendix A7.6.

e. SAE/ASTM Unified Numbering System for Metals and Alloys.

CHEMICAL COMPOSITION REQUIREMENTS FOR RODS AND BARE ELECTRODES		
AWS A5.15 Classification[d]	UNS Number[e]	Composition, Weight Percent[a,b,c]
Cast Iron Welding Rods for OFW		
RCI	F10090	C 3.2-3.5 Mn 0.60-0.75 Si 2.7-3.0 P 0.50-0.75 S 0.10 Fe Rem Ni[f] Trace Mo Trace
RCI-A	F10091	C 3.2-3.5 Mn 0.50-0.70 Si 2.0-2.5 P 0.20-0.40 S 0.10 Fe Rem Ni[f] 1.2-1.6 Mo 0.25-0.45
RCI-B	F10092	C 3.2-4.0 Mn 0.10-0.40 Si 3.2-3.8 P 0.05 S 0.015 Fe Rem Ni[f] 0.50 Mg 0.04-0.10 Ce 0.20
Electrodes for Gas Metal Arc Welding		
ERNi-CI	N02215	C 1.0 Mn 2.5 Si 0.75 S 0.03 Fe 4.0 Ni[f] 90 min. Cu[g] 4.0 Other Elements Total 1.0
ERNiFeMn-CI	N02216	C 0.50 Mn 10-14 Si 1.0 S 0.03 Fe Rem Ni[f] 35-45 Cu[g] 2.5 Al 1.0 Other Elements Total 1.0

a. The weld metal, core wire, or the filler metal, as specified, shall be analyzed for the specific elements for which values are shown in this table. If the presence of other elements is indicated in the course of this work, the amount of those elements shall be determined to ensure that their total does not exceed the limit specified for "Other Elements Total" in this table. b. Single values are maximum, except where otherwise specified. c. "Rem" stands for Remainder. d. Copper-base filler metals frequently used in the braze welding of cast irons are no longer included in this specification. For information pertaining to these materials, see AWS A5.15 appendix A7.6. e. SAE/ASTM Unified Numbering System for Metals and Alloys. f. Nickel plus incidental cobalt. g. Copper plus incidental silver. h. No shielding gas shall be used for classification ENiFeT3-CI.

TYPICAL MECHANICAL PROPERTIES OF UNDILUTED WELD METAL[a]

AWS A5.15 Classification	Tensile Strength		Yield Strength 0.2% Offset		% Elongation in 2 in.	Brinell Hardness
	ksi	MPa	ksi	MPa		
RCI	20-25	138-172	-	-	-	150-210
RCI-A	35-40	241-276	-	-	-	225-290
RCI-B (As-welded)	80-90	552-621	70-75	483-517	3-5	220-310
RCI-B (Annealed)	50-60	345-414	40-45	276-310	5-15	150-200
ESt	-	-	-	-	-	250-400
ENi-CI	40-65	276-448	38-60	262-414	3-6	135-218
ENi-CI-A	40-65	276-448	38-60	262-414	3-6	135-218
ENiFe-CI	58-84	400-579	43-63	296-434	6-18	165-218
ENiFe-CI-A	58-84	400-579	43-63	296-434	4-12	165-218
ENiFeMn-CI	75-95	517-655	60-70	414-483	10-18	165-210
ENiFeT3-CI	65-80	448-552	40-55	276-379	12-20	150-165
ERNiFeMn-CI	75-100	517-689	65-80	448-552	15-35	165-210

a. These are typical mechanical properties of undiluted weld metal listed in AWS A5.15 appendix Table A1.

AWS A5.15 CLASSIFICATION SYSTEM

The system for identifying welding rod and electrode classifications used in this specification follows the standard pattern used in other AWS filler metal specifications. The letter "E" at the beginning of each classification designation stands for electrode, the letters "ER" at the beginning of each classification designation stands for use as either an electrode or rod, and the letter "R" at the beginning of each classification designation stands for a welding rod. The next letters in the filler metal designation are based on the chemical composition of the filler metal or undiluted weld metal. Thus NiFe is a nickel-iron alloy, NiCu is a nickel-copper alloy, etc. Where different compositional limits in filler metals of the same alloy family result in more than one classification, the individual classifications are differentiated by the designators "A" or "B", as in ENiCu-A and ENiCu-B.

For flux cored electrodes, the designator "T" indicates a tubular electrode. The number "3" indicates that the electrode is used primarily without an external shielding gas.

Most of the classifications within this specification contain the usage designator "CI" after the hyphen which indicates that these filler metals are intended for cast iron applications. The usage designator is included to eliminate confusion with other filler metal classifications from other specifications which are designed for alloys other than cast irons. The two exceptions, ENiCu-A and ENiCu-B, preceded the introduction of the usage designator and have never had the "CI" added.

The chemical symbols have been used in all the filler metals except the cast iron and mild steel groups. Since there are no chemical symbols for cast iron and mild steel, the letters "CI" and "St" have been assigned to this group to designate cast iron and mild steel filler metals, respectively. The suffixes "A" and "B" are used to differentiate two alloys of the cast iron filler metals from other cast iron rod classifications.

Chapter
27

TITANIUM & TITANIUM ALLOY WELDING ELECTRODES & RODS

CHEMICAL COMPOSITION REQUIREMENTS FOR TITANIUM AND TITANIUM ALLOY ELECTRODES AND RODS

AWS A5.16 Classification		UNS Number[e]	Composition, Weight Percent[a,b,c,d]
1990	1970		
ERTi-1	ERTi-1	R50100	C 0.03 O 0.10 H 0.005 N 0.015 Fe 0.10
ERTi-2	ERTi-2	R50120	C 0.03 O 0.10 H 0.008 N 0.020 Fe 0.20
ERTi-3	ERTi-3	R50125	C 0.03 O 0.10-0.15 H 0.008 N 0.020 Fe 0.20
ERTi-4	ERTi-4	R50130	C 0.03 O 0.15-0.25 H 0.008 N 0.020 Fe 0.30
ERTi-5	ERTi-6Al-4V	R56400	C 0.05 O 0.18 H 0.015 N 0.030 Al 5.5-6.7 V 3.5-4.5 Fe 0.30 Other Element Amount: Y 0.005
ERTi-5ELI	ERTi-6Al-4V-1	R56402	C 0.03 O 0.10 H 0.005 N 0.012 Al 5.5-6.5 V 3.5-4.5 Fe 0.15 Other Element Amount: Y 0.005
ERTi-6	ERTi-5Al-2.5Sn	R54522	C 0.08 O 0.18 H 0.015 N 0.050 Al 4.5-5.8 Sn 2.0-3.0 Fe 0.50 Other Element Amount: Y 0.005
ERTi-6ELI	ERTi-5Al-2.5Sn-1	R54523	C 0.03 O 0.10 H 0.005 N 0.012 Al 4.5-5.8 Sn 2.0-3.0 Fe 0.20 Other Element Amount: Y 0.005
ERTi-7	ERTi-0.2Pd	R52401	C 0.03 O 0.10 H 0.008 N 0.020 Fe 0.20 Other Element Amount: Pd 0.12/0.25
ERTi-9	ERTi-3Al-2.5V	R56320	C 0.03 O 0.12 H 0.008 N 0.020 Al 2.5-3.5 V 2.0-3.0 Fe 0.25 Other Element Amount: Y 0.005
ERTi-9ELI	ERTi-3Al-2.5V-1	R56321	C 0.03 O 0.10 H 0.005 N 0.012 Al 2.5-3.5 V 2.0-3.0 Fe 0.20 Other Element Amount: Y 0.005
ERTi-12	-	R53400	C 0.03 O 0.25 H 0.008 N 0.020 Fe 0.30 Other Element Amount: Mo 0.2/0.4, Ni 0.6/0.9
ERTi-15	ERTi-6Al-2Cb-1Ta-1Mo	R56210	C 0.03 O 0.10 H 0.005 N 0.015 Al 5.5-6.5 Fe 0.15 Other Element Amount: Mo 0.5/1.5, Cb 1.5/2.5, Ta 0.5/1.5

a. Titanium constitutes the remainder of the composition.

b. Single values are maximum.

c. Analysis of the interstitial elements C, O, H and N shall be conducted on samples of filler metal taken after the filler metal has been reduced to its final diameter and all processing operations have been completed. Analysis of the other elements may be conducted on these samples or it may have been conducted on samples taken from the ingot or the rod stock from which the filler metal is made. In case of dispute, samples from the finished filler metal shall be the referee method.

d. Residual elements, total, shall not exceed 0.20 percent, with no single such element exceeding 0.05 percent. Residual elements need not be reported unless a report is specifically required by the purchaser. Residual elements are those elements (other than titanium) that are not listed in this table for the particular classification, but which are inherent in the raw material or the manufacturing practice. Residual elements can be present only in trace amounts and they cannot be elements that have been intentionally added to the product.

e. SAE/ASTM Unified Numbering System for Metals and Alloys.

SPECIFICATION CROSS INDEX[a]

AWS A5.16 Classification		Aerospace Materials Specification (AMS)	Military Specification (MIL)	ASTM/ASME Grades
1990	1970			
ERTi-1	ERTi-1	4951	MIL-R-81558	1
ERTi-2	ERTi-2	-	MIL-R-81558	2
ERTi-3	ERTi-3	-	MIL-R-81558	3
ERTi-4	ERTi-4	-	MIL-R-81558	4
ERTi-5	ERTi-6Al-4V	4954	-	5
ERTi-5ELI	ERTi-6Al-4V-1	4956	MIL-R-81558	-
ERTi-6	ERTi-5Al-2.5Sn	4953	-	6
ERTi-6ELI	ERTi-5Al-2.5Sn-1	-	MIL-R-81558	-
ERTi-7	ERTi-0.2Pd	-	-	7
ERTi-9	ERTi-3Al-2.5V	-	-	9
ERTi-9ELI	ERTi-3Al-2.5V-1	-	-	-
ERTi-12	ERTi-0.8Ni-0.3Mo	-	-	12
ERTi-15	ERTi-6Al-2Cb1Ta-1Mo	-	MIL-R-81558	-

a. Specifications are not exact duplicates. Information is supplied only for general comparison.

AWS A5.16 CLASSIFICATION SYSTEM

The system for identifying the filler metal classifications in this specification follows the standard pattern used in other AWS filler metal specifications. The letter "E" at the beginning of each classification designation stands for electrode, and the letter "R" stands for welding rod. Since these filler metals are used as electrodes in gas metal arc welding and as rods in gas tungsten arc welding, both letters are used.

The chemical symbol "Ti" appears after "R" as a means of identifying the filler metals as unalloyed titanium or a titanium-base alloy. The numeral provides a means of identifying different variations in the composition. The filler letters "ELI" designate titanium alloy filler metals with extra low content of interstitial elements (carbon, oxygen, hydrogen, and nitrogen).

Designations for individual alloys in this revision of the specification are different from those used in earlier documents. With the exception of ERTi-15, specific alloys now are identified by a number similar to the grade designation used in ASTM/ASME specifications for corresponding base metals. In the absence of a grade number in general usage for the Ti-6Al-2Cb-1Ta-1Mo alloy, the number 15 was assigned arbitrarily to designate this classification of filler metal.

The Metals Blue Book

Chapter

28

CARBON STEEL ELECTRODES & FLUXES FOR SUBMERGED ARC WELDING

CHEMICAL COMPOSITION REQUIREMENTS FOR SOLID ELECTRODES

AWS A5.17 Classification	UNS Number[c]	Composition, Weight Percent[a,b]
Low Manganese Electrodes		
EL8	K01008	C 0.10 Mn 0.25/0.60 Si 0.07 S 0.030 P 0.030 Cu[d] 0.35
EL8K	K01009	C 0.10 Mn 0.25/0.60 Si 0.10/0.25 S 0.030 P 0.030 Cu[d] 0.35
EL12	K01012	C 0.04/0.14 Mn 0.25/0.60 Si 0.10 S 0.030 P 0.030 Cu[d] 0.35
Medium Manganese Electrodes		
EM12	K01112	C 0.06/0.15 Mn 0.80/1.25 Si 0.10 S 0.030 P 0.030 Cu[d] 0.35
EM12K	K01113	C 0.05/0.15 Mn 0.80/1.25 Si 0.10/0.35 S 0.030 P 0.030 Cu[d] 0.35
EM13K	K01313	C 0.06/0.16 Mn 0.90/1.40 Si 0.35/0.75 S 0.030 P 0.030 Cu[d] 0.35
EM14K	K01314	C 0.06/0.19 Mn 0.90/1.40 (Ti 0.03/0.17) Si 0.35/0.75 S 0.025 P 0.025 Cu[d] 0.35
EM15K	K01515	C 0.10/0.20 Mn 0.80/1.25 Si 0.10/0.35 S 0.030 P 0.030 Cu[d] 0.35
High Manganese Electrodes		
EH11K	K11140	C 0.07/0.15 Mn 1.40/1.85 Si 0.80/1.15 S 0.030 P 0.030 Cu[d] 0.35
EH12K	K01213	C 0.06/0.15 Mn 1.50/2.00 Si 0.25/0.65 S 0.025 P 0.025 Cu[d] 0.35
EH14	K11585	C 0.10/0.20 Mn 1.70/2.20 Si 0.10 S 0.030 P 0.030 Cu[d] 0.35

a. The filler metal shall be analyzed for the specific elements for which values are shown in this table. If the presence of other elements is indicated in the course of this work, the amount of those elements shall be determined to ensure that their total (excluding iron) does not exceed 0.50 percent.

b. Single values are maximum.

c. SAE/ASTM Unified Numbering System for Metals and Alloys.

d. The copper limit includes any copper coating that may be applied to the electrode.

CHEMICAL COMPOSITION REQUIREMENTS FOR COMPOSITE ELECTRODE WELD METAL

AWS A17 Classification	Weight Percent[a,b,c]
EC1	C 0.15 Mn 1.80 Si 0.90 S 0.035 P 0.035 Cu 0.035

a. The filler metal shall be analyzed for the specific elements for which values are shown in this table. If the presence of other elements is indicated in the course of this work, the amount of those elements shall be determined to ensure that their total (excluding iron) does not exceed 0.50 percent.

b. Single values are maximum.

c. A low dilution area of the groove weld of AWS A5.17 Figure 3 or the fractured tension test specimen of AWS A5.17 Figure 5 may be substituted for the weld pad, and shall meet the above requirements. In case of dispute, the weld pad shall be the referee method.

TENSION TEST REQUIREMENTS

AWS A5.17 Flux	Tensile Strength		Yield Strength, min.[b]		% Elongation, min.[b]
Classification[a]	psi	MPa	psi	MPa	
F6XX-EXXX	60,000/80,000	415/550	48,000	330	22
F7XX-EXXX	70,000/95,000	480/650	58,000	400	22

a. The letter "X" used in various places in the classifications in this table stands for, respectively, the condition of heat treatment, the toughness of the weld metal, and the classification of the electrode (see AWS A5.17 Figure 1).

b. Yield strength at 0.2 percent offset and elongation in 2 in. (51 mm) gage length.

IMPACT TEST REQUIREMENTS[a]

Digit	Test Temperature		Average Energy Level, min.
	°F	°C	
Z	no impact requirements		---
0	0	-18	20 ft•lbf (27 J)
2	-20	-29	20 ft•lbf (27 J)
4	-40	-40	20 ft•lbf (27 J)
5	-50	-46	20 ft•lbf (27 J)
6	-60	-51	20 ft•lbf (27 J)
8	-80	-62	20 ft•lbf (27 J)

a. Based on the results of the impact tests of the weld metal, the manufacturer shall insert in the classification (AWS A5.17 Table 5) the appropriate digit from the table above (AWS A5.17 Table 6), as indicated in AWS A5.17 Figure 1. Weld metal from a specific flux-electrode combination that meets impact requirements at a given temperature also meets the requirements at all higher temperatures in this table (i.e. weld metal meeting the requirements for digit 5 also meets the requirements for digits 4, 2, 0, and Z). See AWS A5.17 for more details.

AWS A5.17 CLASSIFICATION SYSTEM

Classification of Electrodes

The system for identifying the electrode classifications in this specification follows the standard pattern used in other AWS filler metal specifications. The letter "E" at the beginning of each classification designation stands for electrode. The remainder of the designation indicates the chemical composition of the electrode, or, in the case of composite electrodes, of the low dilution weld metal obtained with a particular flux. See AWS A5.17 Figure 1.

The letter "L" indicates that the solid electrode is comparatively low in manganese content. The letter "M" indicates a medium manganese content, while the letter "H" indicates a comparatively high manganese content. The one or two digits following the manganese designator indicate the nominal carbon content of the electrode. The letter "K", which appears in some designations, indicates that the electrode is made from a heat of silicon-killed steel. Solid electrodes are classified only on the basis of their chemical composition, as specified in AWS A5.17 Table 1.

A composite electrode is indicated by the letter "C" after the "E", and a numerical suffix. The composition of a composite electrode is meaningless and the user is therefore referred to weld metal composition (AWS A5.17 Table 2) with a particular flux, rather than to electrode composition.

Classification of Fluxes

Fluxes are classified on the basis of the mechanical properties of the weld metal they produce with some certain classification of electrode, under the specific test conditions called in AWS A5.17 Section B.

As examples of flux classifications, consider the following:

    F6A0-EH14
    F7P6-EM12K
    F7P4-EC1

The prefix "F" designates a flux. This is followed by a single digit representing the minimum tensile strength required of the weld metal in 10,000 psi increments.

Classification of Fluxes (Continued)

When the letter "A" follows the strength designator, it indicates that the weld metal was tested (and is classified) in the as-welded condition. When the letter "P" follows the strength designator, it indicates the weld metal was tested (and is classified) after postweld heat treatment called for in the specification. The digit that follows the A or P will be a number or the letter "Z". This digit refers to the impact strength of the weld metal. Specifically, it designates the temperature at (and above) which the weld metals meets, or exceeds, the required 20 ft•lbf (27 J) Charpy V-notch impact strength (except for the letter Z, which indicates that no impact requirement is specified - see AWS A5.17 Table 6). These mechanical property designations are followed by the designation of the electrode used in classifying the flux (see AWS A5.17 Table 1). The suffix (EH14, EM14K, EC1, etc.) included after the hyphen refers to the electrode classification with which the flux will deposit weld metal that meets the specified mechanical properties when tested as called for in the specification.

It should be noted that flux of any specific trade designation may have many classifications. The number is limited only by the number of different electrode classifications and the condition of heat treatment (as-welded and postweld heat-treated) with which the flux can meet the classification requirements. The flux marking lists at least one, and may list all, classifications to which the flux conforms.

Solid electrodes having the same classification are interchangeable when used with a specific flux; composite electrodes may not be. However, the specific usability (or operating) characteristics of various fluxes of the same classification may differ in one respect or another.

Chapter

29

CARBON STEEL ELECTRODES & RODS FOR GAS SHIELDED ARC WELDING

CHEMICAL COMPOSITION REQUIREMENTS FOR SOLID ELECTRODES AND RODS

AWS A5.18 Classification[b]	UNS Number[c]	Composition, Weight Percent[a]
ER70S-2	K10726	C 0.07 Mn 0.90 to 1.40 Si 0.40 to 0.70 P 0.025 S 0.035 Ni (e) Cr (e) Mo (e) Cu[d] 0.50 Ti 0.05 to 0.15 Zr 0.02 to 0.12 Al 0.05 to 0.15
ER70S-3	K11022	C 0.07 to 0.15 Mn 0.90 to 1.40 Si 0.45 to 0.75 P 0.025 S 0.035 Ni (e) Cr (e) Mo (e) Cu[d] 0.50
ER70S-4	K11132	C 0.07 to 0.15 Mn 1.00 to 1.50 Si 0.65 to 0.85 P 0.025 S 0.035 Ni (e) Cr (e) Mo (e) Cu[d] 0.50
ER70S-5	K11357	C 0.07 to 0.19 Mn 0.90 to 1.40 Si 0.30 to 0.60 P 0.025 S 0.035 Ni (e) Cr (e) Mo (e) Cu[d] 0.50 Al 0.50 to 0.90
ER70S-6	K11140	C 0.06 to 0.15 Mn 1.40 to 1.85 Si 0.80 to 1.15 P 0.025 S 0.035 Ni (e) Cr (e) Mo (e) Cu[d] 0.50
ER70S-7	K11125	C 0.07 to 0.15 Mn 1.50 to 2.00[f] Si 0.50 to 0.80 P 0.025 S 0.035 Ni (e) Cr (e) Mo (e) Cu[d] 0.50
ER70S-G	-	Not specified[g]

a. Single values are maximum.
b. The letter "N" as a suffix to a classification indicates that the weld metal is intended for the core belt region of nuclear reactor vessels, as described in the Annex of AWS A5.18. This suffix changes the limits on the phosphorus, vanadium and copper as follows:
   P = 0.012% maximum
   V = 0.05% maximum
   Cu = 0.08% maximum
c. SAE/ASTM Unified Numbering System for Metals and Alloys.
d. Copper due to any coating on the electrode or rod plus the copper content of the filler metal itself, shall not exceed the stated 0.50% max.
e. These residual elements shall not exceed 0.50% in total.
f. In this classification, the maximum Mn may exceed 2.0%. If it does, the maximum C must be reduced 0.01% for each 0.05% increase in Mn or part thereof.
g. Chemical requirements are not specified but there shall be no intentional addition of Ni, Cr, Mo, and V. Composition shall be reported. Requirements are those agreed to by the purchaser and the supplier.

CHEMICAL COMPOSITION REQUIREMENTS FOR WELD METAL FROM COMPOSITE ELECTRODES

AWS A5.18 Classification[a]	UNS Number[b]	Shielding Gas[c]	Composition, Weight Percent[d]
Multiple-Pass Classification			
E70C-3X	W07703	75-80% Ar/Balance CO_2 or 100% CO_2	C 0.12 Mn 1.75 Si 0.90 S 0.03 P 0.03 Cu 0.50 (Ni Cr Mo V)[e]
E70C-6X	W07706	75-80% Ar/Balance CO_2 or 100% CO_2	C 0.12 Mn 1.75 Si 0.90 S 0.03 P 0.03 Cu 0.50 (Ni Cr Mo V)[e]
E70C-G(X)	-	(f)	Chemical Requirements Not Specified[g]
Single-Pass Classification			
E70C-GS(X)	-	(f)	Chemical Requirements Not Specified[h]

a. The final X shown in the classification represents a "C" or "M" which corresponds to the shielding gas with which the electrode is classified. The use of "C" designates 100% CO_2 shielding; "M" designates 75-80% Ar/balance CO_2. For E70C-G and E70C-GS, the final "C" or "M" may be omitted if these gases are not used for classification.

b. SAE/ASTM Unified Numbering System for Metals and Alloys.

c. Use of a shielding gas other than that specified will result in different weld metal composition.

d. Single values are maximums.

e. To be reported if intentionally added; the sum of Ni, Cr, Mo, and V shall not exceed 0.50%.

f. Shielding gas shall be as agreed upon between the purchaser and supplier.

g. Composition shall be reported; the requirements are those agreed to between purchaser and supplier.

h. The composition of the weld metal from this classification is not specified since electrodes of this classification are intended only for single-pass welds. Dilution, in such welds, usually is quite high.

TENSION TEST REQUIREMENTS (AS WELDED)

AWS A5.18 Classification[a]	Shielding Gas[c]	Tensile Strength, min		Yield Strength[b], min		% Elongation[b], min
		psi	MPa	psi	MPa	
ER70S-2, ER70S-3, ER70S-4,	CO_2[c]	70,000	480	58,000	400	22
ER70S-5, ER70S-6, ER70S-7						
ER70S-G	(d)	70,000	480	58,000	400	22
E70C-3X, E70C-6X	75-80% Ar/balance CO_2 or 100% CO_2	70,000	480	58,000	400	22
E70C-G(X)	(d)	70,000	480	Not specified		Not specified
E70C-GS(X)	(d)	70,000	480			

a. The final X shown in the classification represents a "C" or "M" which corresponds to the shielding gas with which the electrode is classified. The use of "C" designates 100% CO_2 shielding; "M" designates 75-80% Ar/balance CO_2. For E70C-G and E70C-GS, the final "C" or "M" may be omitted.

b. Yield strength at 0.2% offset and elongation in 2 in. (51 mm) gage length.

c. CO_2 = carbon dioxide shielding gas. The use of CO_2 for classification purposes shall not be construed to preclude the use of Ar/CO_2 or Ar/O_2 shielding gas mixtures. A filler metal tested with gas blends, such as Ar/O_2 or Ar/CO_2, may result in weld metal having higher strength and lower elongation.

d. Shielding gas shall be as agreed to between purchaser and supplier.

IMPACT TEST REQUIREMENTS (AS WELDED)	
AWS A5.18 Classification	Average Impact Strength[a,b], min.
ER70S-2	20 ft•lbf at -20°F (27 J at -29°C)
ER70S-3	20 ft•lbf at 0°F (27 J at -18°C)
ER70S-4	Not required
ER70S-5	Not required
ER70S-6	20 ft•lbf at -20°F (27 J at -29°C)
ER70S-7	20 ft•lbf at -20°F (27 J at -29°C)
ER70S-G	As agreed between supplier and purchaser
E70C-G(X)	As agreed between supplier and purchaser
E70C-3X	20 ft•lbf at 0°F (27 J at -18°C)
E70C-6X	20 ft•lbf at -20°F (27 J at -29°C)
E70C-GS(X)	Not required

a.   Both the highest and lowest of the five test values obtained shall be disregarded in computing the impact strength. Two of the remaining three values shall equal or exceed 20 ft•lbf; one of the three remaining values may be lower than 20 ft•lbf but not lower than 15 ft•lbf. The average of the three shall not be less than the 20 ft•lbf specified.

b.   For classifications with the "N" (nuclear) designation, three additional specimens shall be tested at room temperature. Two of the three shall equal, or exceed, 75 ft•lbf (102 J), and the third shall not be lower than 70 ft•lbf (95 J). Average of the three values shall equal, or exceed, 75 ft•lbf (102 J).

AWS A5.18 CLASSIFICATION SYSTEM

The system for identifying the electrode classifications in this specification follows the standard pattern used in other AWS filler metal specifications as shown in AWS A5.18 Figure A1.

The prefix "E" designates an electrode as in other specifications. The letters "ER" indicate that the filler metal may be used either as an electrode or a rod. The number 70 indicates the required minimum tensile strength as a multiple of 1,000 psi (6.9 MPa) of the weld metal in a test weld made using the electrode in accordance with the welding conditions in the specification. The letter "S" designates a solid electrode or rod. The letter "C" designates a composite electrode. The digit following the hyphen, 2, 3, 4, 5, 6, 7, G, or GS, indicates the chemical composition of the filler metal itself, in the case of solid electrodes and rods, or the impact testing temperature of the weld metal under certain test conditions, in the case of the both solid and composite stranded and metal cored electrodes. In the case of some composite stranded and metal cored electrodes, the letter "M" or "C" will follow, indicating the type of shielding gas.

The Metals Blue Book

AWS A5.18 CLASSIFICATION SYSTEM (Continued)

The addition of the letter "N" as a suffix to a classification indicates that the electrode is intended for certain very special welds in nuclear applications. These welds are found in the core belt region of the reactor vessel. This region is subject to intense neutron radiation, and it is necessary, therefore, that the phosphorus, vanadium and copper contents of the weld metal be limited in order to resist neutron radiation-induced embrittlement. It is also necessary that the weld metal has a high upper shelf energy level in order to withstand some embrittlement, yet remain serviceable over the years.

An optional supplemental diffusible hydrogen designator (H16, H18, or H4) may follow, indicating whether the electrode will meet a maximum hydrogen level of 16, 8, or 4 ml/100g of weld metal when tested as outlined in ANSI/AWS A4.3.

"G" Classification

This specification includes filler metals classified as ER70S-G, E70C-G, and E70C-GS. The "G" (multiple pass) or "GS" (single pass) indicates that the filler metal is of a "general" classification. It is general because not all of the particular requirements specified for each of the other classifications are specified for this classification. The intent in establishing these classifications is to provide a means by which filler metals that differ in one respect or another (chemical composition, for example) from all other classifications (meaning that the composition of the filler metal, in the case of the example, does not meet the composition specified for any of the classifications in the specification) can still be classified according to the specification. The purpose is to allow a useful filler metal - one that otherwise would have to await a revision of the specification - to be classified immediately under the existing specification. This means, then, that two filler metals, each bearing the same "G" classification, may be quite different in some particular respect (chemical composition, again, for example).

The point of difference (although not necessarily the amount of the difference) referred to above will be readily apparent from the use of the words "not required" and "not specified" in AWS A5.18. The use of these words is as follows:

"Not Specified" is used in those areas of the specification that refer to the results of some particular test. It indicates that the requirements for that test are *not specified* for that particular classification.

"Not Required" is used in those areas of the specification that refer to the tests that must be conducted in order to classify a filler metal. It indicates that the test is *not required* because the requirements (results) for the test have not been specified for that particular classification.

Restating the case, when a requirement is not specified, it is not necessary to conduct the corresponding test in order to classify a filler metal to that classification. When a purchaser wants the information provided by that test in order to consider a particular product of that classification for a certain application, they will have to arrange for that information with the supplier of the product. They will have to establish with that supplier just what the testing procedure and the acceptance requirements are to be, for that test. They may want to incorporate that information (via ANSI/AWS A5.01) into the purchase order.

Chapter
30

MAGNESIUM ALLOY WELDING ELECTRODES & RODS

CHEMICAL COMPOSITION REQUIREMENTS FOR MAGNESIUM ALLOY ELECTRODES AND RODS

AWS A5.19 Classification	UNS Number[c]	Composition, Weight Percent[a,b]
ER AZ61A, R AZ61A	M11611	Mg Remainder Al 5.8 to 7.2 Be 0.0002 to 0.0008 Mn 0.15 to 0.5 Zn 0.40 to 1.5 Cu 0.05 Fe 0.005 Ni 0.005 Si 0.05 Other Elements Total 0.30
ER AZ92A, R AZ92A	M11922	Mg Remainder Al 8.3 to 9.7 Be 0.0002 to 0.0008 Mn 0.15 to 0.5 Zn 1.7 to 2.3 Cu 0.05 Fe 0.005 Ni 0.005 Si 0.05 Other Elements Total 0.30
ER AZ101A, R AZ101A	M11101	Mg Remainder Al 9.5 to 10.5 Be 0.0002 to 0.0008 Mn 0.15 to 0.5 Zn 0.75 to 1.25 Cu 0.05 Fe 0.005 Ni 0.005 Si 0.05 Other Elements Total 0.30
ER EZ33A, R EZ33A	M12331	Mg Remainder Be 0.0008 Zn 2.0 to 3.1 Zr 0.45 to 1.0 Rare Earth 2.5 to 4.0 Other Elements Total 0.30

a. The filler metal shall be analyzed for the specific elements for which values are shown in this table. If the presence of other elements is indicated in the course of this work, the amount of those elements shall be determined to ensure that their total does not exceed the limits specified for "Other Elements Total". b. Single values are maximum. c. SAE/ASTM Unified Numbering System for Metals and Alloys.

GUIDE TO THE CHOICE OF FILLER METAL FOR GENERAL PURPOSE WELDING

Base Metal	Base Metal[c]											
	AM100A	AZ10A	AZ31B AZ31C	AZ61A	AZ63A	AZ80A	AZ81A	Z91C	AZ92A	EK41A	EZ33A	HK31A
	Filler Metal[a,b]											
AM100A	AZ101A AZ92A	-	-	-	-	-	-	-	-	-	-	-
AZ10A	AZ92A	AZ61A AZ92A	-	-	-	-	-	-	-	-	-	-
AZ31B AZ31C	AZ92A	AZ61A AZ92A	AZ61A AZ92A	-	-	-	-	-	-	-	-	-
AZ61A	AZ92A	AZ61A AZ92A	AZ61A AZ92A	AZ61A AZ92A	-	-	-	-	-	-	-	-
AZ63A	c	c	c	c	AZ101A AZ92A	-	-	-	-	-	-	-

GUIDE TO THE CHOICE OF FILLER METAL FOR GENERAL PURPOSE WELDING (Continued)

Base Metal	Base Metal[c]											
	AM100A	AZ10A	AZ31B AZ31C	AZ61A	AZ63A	AZ80A	AZ81A	AZ91C	AZ92A	EK41A	EZ33A	HK31A
	Filler Metal[a,b]											
AZ80A	AZ92A	AZ61A AZ92A	AZ61A AZ92A	AZ61A AZ92A	c	AZ61A AZ92A	-	-	-	-	-	-
AZ81A	AZ92A	AZ92A	AZ92A	AZ92A	c	AZ92A	AZ61A AZ92A	-	-	-	-	-
AZ91C	AZ92A	AZ92A	AZ92A	AZ92A	c	AZ92A	AZ92A	AZ101A AZ92A	-	-	-	-
AZ92A	AZ92A	AZ92A	AZ92A	AZ92A	c	AZ92A	AZ92A	AZ92A	AZ101A AZ92A	-	-	-
EK41A	AZ92A	AZ92A	AZ92A	AZ92A	c	AZ92A	AZ92A	AZ92A	AZ92A	EZ33A	-	-
EZ33A	AZ92A	AZ92A	AZ92A	AZ92A	c	AZ92A	AZ92A	AZ92A	AZ92A	EZ33A	EZ33A	-
HK31A	AZ92A	AZ92A	AZ92A	AZ92A	c	AZ92A	AZ92A	AZ92A	AZ92A	EZ33A	EZ33A	EZ33A
HM21A	AZ92A	AZ92A	AZ92A	AZ92A	c	AZ92A	AZ92A	AZ92A	AZ92A	EZ33A	EZ33A	EZ33A
HM31A	AZ92A	AZ92A	AZ92A	AZ92A	c	AZ92A	AZ92A	AZ92A	AZ92A	EZ33A	EZ33A	EZ33A
HZ32A	AZ92A	AZ92A	AZ92A	AZ92A	c	AZ92A	AZ92A	AZ92A	AZ92A	EZ33A	EZ33A	EZ33A
K1A	AZ92A	AZ92A	AZ92A	AZ92A	c	AZ92A	AZ92A	AZ92A	AZ92A	EZ33A	EZ33A	EZ33A
LA141A	d	d	EZ33A	c	c	c	c	c	c	d	d	d
M1A MG1	AZ92A	AZ61A AZ92A	AZ61A AZ92A	AZ61A AZ92A	c	AZ61A AZ92A	AZ92A	AZ92A	AZ92A	AZ92A	AZ92A	AZ92A
QE22A	d	d	AZ92A	d	c	d	d	d	d	EZ33A	EZ33A	EZ33A
ZE10A	AZ92A	AZ61A AZ92A	AZ61A AZ92A	AZ61A AZ92A	c	AZ61A AZ92A	AZ92A	AZ92A	AZ92A	EZ33A AZ92A	EZ33A	EZ33A
ZE41A	d	d	d	d	c	d	d	d	d	EZ33A	EZ33A	EZ33A
ZK21A	AZ92A	AZ61A AZ92A	AZ61A AZ92A	AZ61A AZ92A	c	AZ61A AZ92A	AZ92A	AZ92A	AZ92A	AZ92A	AZ92A	AZ92A

a. When more than one filler metal is given, they are listed in order of preference. b. The letter prefix (ER or R), designating usability of the filler metal, has been deleted, to reduce clutter in the table. c. Welding is not recommended for alloys ZH62A, ZK51A, ZK60A, ZK61A. d. No data available.

The Metals Blue Book

GUIDE TO THE CHOICE OF FILLER METAL FOR GENERAL PURPOSE WELDING

Base Metal	Base Metal										
	HM21A	HM31A	HZ32A	K1A	LA141A	M1A MG1	QE22A	ZE10A	ZE14A	ZK21A	ZH62A ZK51A ZK60A ZK61A
	Filler Metal[a,b]										
HM21A	EZ33A	-	-	-	-	-	-	-	-	-	-
HM31A	EZ33A	EZ33A	-	-	-	-	-	-	-	-	-
HZ32A	EZ33A	EZ33A	-	-	-	-	-	-	-	-	-
K1A	EZ33A	EZ33A	EZ33A	EZ33A	-	-	-	-	-	-	-
LA141A	EZ33A	d	d	d	EZ33A	-	-	-	-	-	-
M1A MG1	AZ92A	AZ92A	AZ92A	AZ92A	d	AZ61A AZ92A	-	-	-	-	-
QE22A	EZ33A	EZ33A	EZ33A	EZ33A	EZ33A	c	EZ33A	-	-	-	-
ZE10A	EZ33A AZ92A	EZ33A AZ92A	EZ33A AZ92A	EZ33A AZ92A	EZ33A	AZ61A AZ92A	EZ33A AZ92A	AZ61A AZ92A	-	-	-
ZE14A	EZ33A	EZ33A	EZ33A	EZ33A	d	d	EZ33A	d	EZ33A	-	-
ZK21A	AZ92A	AZ92A	AZ92A	AZ92A	d	AZ61A AZ92A	AZ92A	AZ61A AZ92A	AZ92A	AZ61A AZ92A	-
ZH62A ZK51A ZK60A ZK61A	c	c	c	c	c	c	c	c	c	c	EZ33A

a.   When more than one filler metal is given, they are listed in order of preference.
b.   The letter prefix (ER or R), designating usability of the filler metal, has been deleted, to reduce clutter in the table.
c.   Welding not recommended.
d.   No data available.

AWS A5.19 CLASSIFICATION SYSTEM

The welding electrodes and rods are classified upon the basis of the chemical composition. The alloys are designated by the same standard system used for base metals. That consists of a combination letter-number system composed of three parts. The first part indicates the two principal alloying elements by code letters arranged in order of decreasing percentage. The second part indicates the percentages of the two principal alloying elements in the same order as the code letters. The percentages are rounded to the nearest whole number. The third part is an assigned letter to distinguish different alloys with the same percentages of the two principal alloying elements.

A letter prefix designates usability of the filler metal. The letter system for identifying the filler metal classifications in this specification follows the standard pattern used in other AWS filler metal specifications. The prefix "E" indicates the filler metal is suitable for use as an electrode and the prefix "R" indicates suitability as welding rod. Since some of these filler metals are used as electrodes in gas metal arc welding, and as welding rods in oxyfuel gas, gas tungsten arc, and plasma arc welding, both letters, "ER", are used to indicate suitability as an electrode or a rod.

Chapter
31

CARBON STEEL ELECTRODES FOR FLUX CORED ARC WELDING

CHEMICAL COMPOSITION REQUIREMENTS FOR WELD METAL

AWS A5.20 Classification	UNS Number[c]	Composition, Weight Percent[a,b]
E7XT-1	W07601	C 0.18 Mn 1.75 Si 0.90 S 0.03 P 0.03 Cr 0.20 Ni[d] 0.50 Mo[d] 0.30 V[d] 0.08 Al[d,e] - Cu[d] 0.35
E7XT-1M	W07601	C 0.18 Mn 1.75 Si 0.90 S 0.03 P 0.03 Cr 0.20 Ni[d] 0.50 Mo[d] 0.30 V[d] 0.08 Al[d,e] - Cu[d] 0.35
E7XT-5	W07605	C 0.18 Mn 1.75 Si 0.90 S 0.03 P 0.03 Cr 0.20 Ni[d] 0.50 Mo[d] 0.30 V[d] 0.08 Al[d,e] - Cu[d] 0.35
E7XT-5M	W07605	C 0.18 Mn 1.75 Si 0.90 S 0.03 P 0.03 Cr 0.20 Ni[d] 0.50 Mo[d] 0.30 V[d] 0.08 Al[d,e] - Cu[d] 0.35
E7XT-9	W07609	C 0.18 Mn 1.75 Si 0.90 S 0.03 P 0.03 Cr 0.20 Ni[d] 0.50 Mo[d] 0.30 V[d] 0.08 Al[d,e] - Cu[d] 0.35
E7XT-9M	W07609	C 0.18 Mn 1.75 Si 0.90 S 0.03 P 0.03 Cr 0.20 Ni[d] 0.50 Mo[d] 0.30 V[d] 0.08 Al[d,e] - Cu[d] 0.35
E7XT-4	W07604	C[f] - Mn 1.75 Si 0.60 S 0.03 P 0.03 Cr 0.20 Ni[d] 0.50 Mo[d] 0.30 V[d] 0.08 Al[d,e] 1.8 Cu[d] 0.35
E7XT-6	W07606	C[f] - Mn 1.75 Si 0.60 S 0.03 P 0.03 Cr 0.20 Ni[d] 0.50 Mo[d] 0.30 V[d] 0.08 Al[d,e] 1.8 Cu[d] 0.35
E7XT-7	W07607	C[f] - Mn 1.75 Si 0.60 S 0.03 P 0.03 Cr 0.20 Ni[d] 0.50 Mo[d] 0.30 V[d] 0.08 Al[d,e] 1.8 Cu[d] 0.35
E7XT-8	W07608	C[f] - Mn 1.75 Si 0.60 S 0.03 P 0.03 Cr 0.20 Ni[d] 0.50 Mo[d] 0.30 V[d] 0.08 Al[d,e] 1.8 Cu[d] 0.35
E7XT-11	W07611	C[f] - Mn 1.75 Si 0.60 S 0.03 P 0.03 Cr 0.20 Ni[d] 0.50 Mo[d] 0.30 V[d] 0.08 Al[d,e] 1.8 Cu[d] 0.35
E7XT-G[g]	-	C[f] - Mn 1.75 Si 0.90 S 0.03 P 0.03 Cr 0.20 Ni[d] 0.50 Mo[d] 0.30 V[d] 0.08 Al[d,e] 1.8 Cu[d] 0.35
E7XT-12	W07612	C 0.15 Mn 1.60 Si 0.90 S 0.03 P 0.03 Cr 0.20 Ni[d] 0.50 Mo[d] 0.30 V[d] 0.08 Al[d,e] - Cu[d] 0.35
E7XT-12M	W07612	C 0.15 Mn 1.60 Si 0.90 S 0.03 P 0.03 Cr 0.20 Ni[d] 0.50 Mo[d] 0.30 V[d] 0.08 Al[d,e] - Cu[d] 0.35
E7XT-13	W06613	Not Specified[h]
E7XT-2	W07602	Not Specified[h]
E7XT-2M	W07602	Not Specified[h]
E7XT-3	W07603	Not Specified[h]
E7XT-10	W07610	Not Specified[h]
E7XT-13	W07613	Not Specified[h]
E7XT-14	W07614	Not Specified[h]
E7XT-GS	-	Not Specified[h]

a. The weld metal shall be analyzed for the specific elements for which values are shown in this table. b. Single values are maximums. c. SAE/ASTM Unified Numbering System for Metals and Alloys. d. The analysis of these elements shall be reported only if intentionally added. e. Applicable to self-shielding electrodes. Electrodes intended for use with gas shielding need not have significant additions of aluminum. f. The limit of this element is not specified, but the amount shall be determined and reported (see AWS A5.20 Annex A6.5). g. The total of all elements listed in this table shall not exceed 5%. h. The composition of the weld metal is not meaningful since electrodes of this classification are intended only for single pass welds. Dilution from the base metal in such welds usually is quite high (see AWS A5.20 Annex A7.2).

MECHANICAL PROPERTY REQUIREMENTS IN THE AS-WELDED CONDITION[a]

AWS A5.20 Classification[b]	Tensile Strength		Yield Strength[b]		% Elongation in. 2 in. (50 mm)[c]	Charpy V-Notch Impact[d]
	ksi	MPa	ksi	MPa		
E7XT-1, -1M[d]	70	480	58	400	22	20 ft•lbf at 0°F (27 J at -18°C)
E7XT-2, -2M[e]	70	480	Not Specified	Not Specified	Not Specified	Not Specified
E7XT-3[e]	70	480	Not Specified	Not Specified	Not Specified	Not Specified
E7XT-4	70	480	58	400	22	Not Specified
E7XT-5, 5M[d]	70	480	58	400	22	20 ft•lbf at -20°F (27 J at -29°C)
E7XT-6[d]	70	480	58	400	22	20 ft•lbf at -20°F (27 J at -29°C)
E7XT-7	70	480	58	400	22	Not Specified
E7XT-8[d]	70	480	58	400	22	20 ft•lbf at -20°F (27 J at -29°C)
E7XT-9, 9M[d]	70	480	58	400	22	20 ft•lbf at -20°F (27 J at -29°C)
E7XT-10[e]	70	480	Not Specified	Not Specified	Not Specified	Not Specified
E7XT-11[e]	70	480	58	400	20	Not Specified
E7XT-12, 12M[e]	70-90	480-620	58	400	22	20 ft•lbf at -20°F (27 J at -29°C)
E7XT-13[e]	60	415	Not Specified	Not Specified	Not Specified	Not Specified
E7XT-13[e]	70	480	Not Specified	Not Specified	Not Specified	Not Specified
E7XT-14[e]	70	480	Not Specified	Not Specified	Not Specified	Not Specified
E7XT-G	60	415	48	330	22	Not Specified
E7XT-G	70	480	58	400	22	Not Specified
E7XT-GS[e]	60	415	Not Specified	Not Specified	Not Specified	Not Specified
E7XT-GS[e]	70	480	Not Specified	Not Specified	Not Specified	Not Specified

a. Single values shown are minimum.
b. 0.2% offset.
c. In2 in. (50 mm) gage length. In 1 in. (25 mm) gage length for 0.045 in. (1.1 mm) and smaller sizes of EXXT-11 classification.
d. Electrodes with the following optional supplementary designations shall meet the lower temperature impact requirements specified on the next page:

AWS A5.20 Classification	Electrode Designation	Charpy V-Notch Imapct
E7XT-1, -1M	E7XT-1J, -1MJ	20 ft•lbf at -40°F (27 J at -40°C)
E7XT-5, 5M	E7XT-5J, -5MJ	20 ft•lbf at -40°F (27 J at -40°C)
E7XT-6	E7XT-6J	20 ft•lbf at -40°F (27 J at -40°C)
E7XT-8	E7XT-8J	20 ft•lbf at -40°F (27 J at -40°C)
E7XT-9, -9M	E7XT-9J, -9MJ	20 ft•lbf at -40°F (27 J at -40°C)
E7XT-12, -12M	E7XT-12J, -12JM	20 ft•lbf at -40°F (27 J at -40°C)

e.   These classifications are intended for single pass welding. They are not for multiple pass welding. Only tensile strength is specified and, for this reason, only transverse tension and longitudinal guided bend tests are required (see AWS A5.20 Table 3 for more required test details).

AWS A5.20 CLASSIFICATION SYSTEM

The system for identifying the electrode classifications in AWS A5.20 follows, for the most part, the standard pattern used in other AWS filler metal specifications. A generic example is as follows:

EXXT-XMJ HZ

E   Designates an electrode.

First X   This designator is either 6 or 7. It indicates the minimum tensile strength (in psi x 10 000) of the weld metal when the weld is made in the manner prescribed by AWS A5.20.

Second X   Indicates the primary welding position for which the electrode is designed:
0 - flat and horizontal positions
1 - all positions

T   This designator indicates that the electrode is a flux cored electrode.

X   This designator is some number from 1 through 14 or the letter "G" with or without an "S" following. The number refers to the usability of the electrode (see AWS A5.20 Annex A7 for more details). The "G" indicates that the external shielding, polarity, and impact properties are not specified. The "S" indicates that the electrode is suitable for a weld consisting of a single pass. Such an electrode is not suitable for a multiple-pass weld.

M   An "M" designator in this position indicates that the electrode is classified using 75-80% argon/balance CO_2 shielding gas. When the "M" designator does not appear, it signifies that either the shielding gas used for classification is CO_2 or that the product is a self-shielding product.

J   Optional supplementary designator. Designates that the electrode meets the requirements for improved toughness by meeting a requirement of 20 ft•lbf at -40°F (27 J at -40°C). Absence of the "J" indicates normal impact requirements as given in the above table.

HZ   Optional supplementary designator. Designates that the electrode meets the requirements of the diffusible hydrogen test (an optional supplemental test of the weld metal with an average value not exceeding "Z" ml of H_2 per 100g of deposited metal where "Z" is 4, 8, or 16). For more details, see AWS A5.20 Table 8 and Annex A8.2.

Note: The letter "X" as used in this example and in electrode classification designations in AWS A5.20 substitutes for specific designations indicated by this example.

Chapter

32

COMPOSITE SURFACING WELDING RODS & ELECTRODES

CHEMICAL REQUIREMENTS FOR COMPOSITE SURFACING WELDING RODS, PERCENT

AWS A5.21 Classification	Composition, Weight Percent[a, b]
RFe5-A	C 0.7 to 1.0 Mn 0.50 W 5.0 to 7.0 Cr 3.0 to 5.0 Mo 4.0 to 6.0 Fe Remainder V 1.0 to 2.5 Si 0.50 Total Other Elements 1.0
RFe5-B	C 0.5 to 0.9 Mn 0.50 W 1.0 to 2.5 Cr 3.0 to 5.0 Mo 5.0 to 9.5 Fe Remainder V 0.8 to 1.3 Si 0.50 Total Other Elements 1.0
RFeCr-A1	C 3.7 to 5.0 Mn 2.0 to 6.0 Cr 27.0 to 35.0 Fe Remainder Si 1.0 to 2.5 Total Other Elements 1.0

a.   Analysis shall be made for the elements for which specific values are shown in this table. If, however, the presence of other elements is indicated in the course of routine analysis, further analysis shall be made to determine that the total of these other elements is not present in excess of the limits specified for "Total Other Elements" in this table.

b.   Single values shown are maximum percentages, except where otherwise specified.

CHEMICAL REQUIREMENTS FOR COMPOSITE SURFACING ELECTRODES, PERCENT

AWS A5.21 Classification	Composition, Weight Percent[a, b]
EFe5-A	C 0.7 to 1.0 Mn 0.50 W 5.0 to 7.0 Cr 3.0 to 5.0 Mo 4.0 to 6.0 Fe Remainder V 1.0 to 2.5 Si 0.70 Total Other Elements 1.0
EFe5-B	C 0.5 to 0.9 Mn 0.50 W 1.0 to 2.5 Cr 3.0 to 5.0 Mo 5.0 to 9.5 Fe Remainder V 0.8 to 1.3 Si 0.70 Total Other Elements 1.0
EFe5-C	C 0.3 to 0.5 Mn 0.50 W 1.0 to 2.5 Cr 3.0 to 5.0 Mo 5.0 to 9.0 Fe Remainder V 0.8 to 1.2 Si 0.70 Total Other Elements 1.0
EFeMn-A	C 0.5 to 0.9 Mn 11.0 to 16.0 Ni 2.75 to 6.0 Cr 0.50 Fe Remainder Si 1.3 P 0.03 Total Other Elements 1.0
EFeMn-B	C 0.5 to 0.9 Mn 11.0 to 16.0 Cr 0.50 Mo 0.6 to 1.4 Fe Remainder Si 0.3 to 1.3 P 0.03 Total Other Elements 1.0
EFeCr-A1	C 3.0 to 5.0 Mn 4.0 to 8.0 Cr 26.0 to 32.0 Mo 2.0 Fe Remainder Si 1.0 to 2.5 Total Other Elements 1.0

a.   Analysis shall be made for the elements for which specific values are shown in this table. If, however, the presence of other elements is indicated in the course of routine analysis, further analysis shall be made to determine that the total of these other elements is not present in excess of the limits specified for "Total Other Elements" in this table.

b.   Single values shown are maximum percentages, except where otherwise specified.

CHEMICAL REQUIREMENTS FOR CARBON STEEL TUBES FOR TUNGSTEN-CARBIDE WELDING RODS AND ELECTRODES

Weight Percent	
	C 0.10 Mn 0.45 P 0.02 S 0.03 Si 0.01 Ni 0.30[a] Cr 0.20[a] Mo 0.30[a] V 0.08[a]

a.   Single values shown are nominal percentages, except for those marked with footnote a which indicate maximum percentages.

CHEMICAL REQUIREMENTS FOR TUNGSTEN-CARBIDE GRANULES[a, b]

Weight Percent   C 3.6 to 4.2 Co 0.3 W 94.0 min. Ni 0.3 Mo 0.60 Fe 0.5 Si 0.3

a. Single values shown are maximum percentages, except where otherwise indicated.
b. Chemical analysis shall be made on Tungsten-carbide granules that have been well cleaned to remove all flux or other surface contamination.

AWS A5.21 CLASSIFICATION SYSTEM

The system for identifying welding rod and electrode classifications used in this specification follows the standard pattern used in other AWS filler metal specifications. The letter E at the beginning of each classification indicates an electrode, and the letter R indicates a welding rod.

For the high-speed steels, austenitic manganese steels, and austenitic high chromium irons, the letters immediately after the E or R are the chemical symbols for the principal elements in the classification. Thus, FeMn is an iron-manganese steel, and FeCr is an iron-chromium alloy, etc. Where more than one classification is included in a basic group, the individual classifications in the group are identified by the letters, A, B, etc., as in EFeMn-A. Further subdividing is done by using a 1, 2, etc., after the last letter.

For the tungsten-carbide classifications, the WC immediately after the E or R indicates that the filler metal consists of a mild steel tube filled with granules of fused tungsten-carbide. The numbers following the WC indicate the mesh size limits for the tungsten-carbide granules. The number preceding the slash indicates the sieve size for the "pass" screen, and that following the slash indicates the sieve size for the "hold" screen. Where only one sieve size is shown, this indicates the size of the screen through which the granules must pass.

AWS A5.21 classifies composite surfacing filler metals. Surfacing welding rods and electrodes made from wrought core wire are covered by AWS A5.13, *Specification for Solid Surfacing Welding Rods and Electrodes.*

Chapter

33

FLUX CORED CORROSION-RESISTING CHROMIUM & CHROMIUM-NICKEL STEEL ELECTRODES

ALL-WELD-METAL CHEMICAL COMPOSITION REQUIREMENTS

AWS A.22 Classification[c]	Composition, Weight Percent[d,e]
E307T-X	C 0.13 Cr 18.0-20.5 Ni 9.0-10.5 Mo 0.5-1.5 Mn 3.3-4.75 Si 1.0 P 0.04 S 0.03 Fe Rem Cu 0.5
E308T-X	C 0.08 Cr 18.0-21.0 Ni 9.0-11.0 Mo 0.5 Mn 0.5-2.5 Si 1.0 P 0.04 S 0.03 Fe Rem Cu 0.5
E308LT-X	C[a] Cr 18.0-21.0 Ni 9.0-11.0 Mo 0.5 Mn 0.5-2.5 Si 1.0 P 0.04 S 0.03 Fe Rem Cu 0.5
E308MoT-X	C 0.08 Cr 18.0-21.0 Ni 9.0-12.0 Mo 2.0-3.0 Mn 0.5-2.5 Si 1.0 P 0.04 S 0.03 Fe Rem Cu 0.5
E308MoLT-X	C[a] Cr 18.0-21.0 Ni 9.0-12.0 Mo 2.0-3.0 Mn 0.5-2.5 Si 1.0 P 0.04 S 0.03 Fe Rem Cu 0.5
E309T-X	C 0.10 Cr 22.0-25.0 Ni 12.0-14.0 Mo 0.5 Mn 0.5-2.5 Si 1.0 P 0.04 S 0.03 Fe Rem Cu 0.5
E309CbLT-X	C[a] Cr 22.0-25.0 Ni 12.0-14.0 Mo 0.5 Cb + Ta 0.70 + 1.00 Mn 0.5-2.5 Si 1.0 P 0.04 S 0.03 Fe Rem Cu 0.5
E309LT-X	C[a] Cr 22.0-25.0 Ni 12.0-14.0 Mo 0.5 Mn 0.5-2.5 Si 1.0 P 0.04 S 0.03 Fe Rem Cu 0.5
E310T-X	C 0.20 Cr 25.0-28.0 Ni 20.0-22.5 Mo 0.5 Mn 1.0-2.5 Si 1.0 P 0.03 S 0.03 Fe Rem Cu 0.5
E312T-X	C 0.15 Cr 28.0-32.0 Ni 8.0-10.5 Mo 0.5 Mn 0.5-2.5 Si 1.0 P 0.04 S 0.03 Fe Rem Cu 0.5
E316T-X	C 0.08 Cr 17.0-20.0 Ni 11.0-14.0 Mo 2.0-3.0 Mn 0.5-2.5 Si 1.0 P 0.04 S 0.03 Fe Rem Cu 0.5
E316LT-X	C[a] Cr 17.0-20.0 Ni 11.0-14.0 Mo 2.0-3.0 Mn 0.5-2.5 Si 1.0 P 0.04 S 0.03 Fe Rem Cu 0.5
E317LT-X	C[a] Cr 18.0-21.0 Ni 12.0-14.0 Mo 3.0-4.0 Mn 0.5-2.5 Si 1.0 P 0.04 S 0.03 Fe Rem Cu 0.5
E347T-X	C 0.08 Cr 18.0-21.0 Ni 9.0-11.0 Mo 0.5 Cb + Ta 8 x C min to 1.0 max Mn 0.5-2.5 Si 1.0 P 0.04 S 0.03 Fe Rem Cu 0.5
E409T-X[b]	C 0.10 Cr 10.5-13.0 Ni 0.60 Mo 0.5 Mn 0.80 Si 1.0 P 0.04 S 0.03 Fe Rem Cu 0.5
E410T-X	C 0.12 Cr 11.0-13.5 Ni 0.60 Mo 0.5 Mn 1.2 Si 1.0 P 0.04 S 0.03 Fe Rem Cu 0.5
E410NiMoT-X	C 0.06 Cr 11.0-12.5 Ni 4.0-5.0 Mo 0.40-0.70 Mn 1.0 Si 1.0 P 0.04 S 0.03 Fe Rem Cu 0.5
E410NiTiT-X[b]	C[a] Cr 11.0-12.0 Ni 3.6-4.5 Mo 0.05 Mn 0.70 Si 0.50 P 0.03 S 0.03 Fe Rem Cu 0.5
E430T-X	C 0.10 Cr 15.0-18.0 Ni 0.60 Mo 0.5 Mn 1.2 Si 1.0 P 0.04 S 0.03 Fe Rem Cu 0.5
E502T-X	C 0.10 Cr 4.0-6.0 Ni 0.40 Mo 0.45-0.65 Mn 1.2 Si 1.0 P 0.04 S 0.03 Fe Rem Cu 0.5
E505T-X	C 0.10 Cr 8.0-10.5 Ni 0.40 Mo 0.85-1.20 Mn 1.2 Si 1.0 P 0.04 S 0.03 Fe Rem Cu 0.5
E307T-3	C 0.13 Cr 19.5-22.0 Ni 9.0-10.5 Mo 0.5-1.5 Mn 3.3-4.75 Si 1.0 P 0.04 S 0.03 Fe Rem Cu 0.5
E308T-3	C 0.08 Cr 19.5-22.0 Ni 9.0-11.0 Mo 0.5 Mn 0.5-2.5 Si 1.0 P 0.04 S 0.03 Fe Rem Cu 0.5
E308LT-3	C 0.03 Cr 19.5-22.0 Ni 9.0-11.0 Mo 0.5 Mn 0.5-2.5 Si 1.0 P 0.04 S 0.03 Fe Rem Cu 0.5
E308MoT-3	C 0.08 Cr 18.0-21.0 Ni 9.0-12.0 Mo 2.0-3.0 Mn 0.5-2.5 Si 1.0 P 0.04 S 0.03 Fe Rem Cu 0.5
E308MoLT-3	C 0.03 Cr 18.0-21.0 Ni 9.0-12.0 Mo 2.0-3.0 Mn 0.5-2.5 Si 1.0 P 0.04 S 0.03 Fe Rem Cu 0.5
E309T-3	C 0.10 Cr 23.0-25.5 Ni 12.0-14.0 Mo 0.5 Mn 0.5-2.5 Si 1.0 P 0.04 S 0.03 Fe Rem Cu 0.5
E309LT-3	C 0.03 Cr 23.0-25.5 Ni 12.0-14.0 Mo 0.5 Mn 0.5-2.5 Si 1.0 P 0.04 S 0.03 Fe Rem Cu 0.5

ALL-WELD-METAL CHEMICAL COMPOSITION REQUIREMENTS (Continued)	
AWS A.22 Classification[c]	Composition, Weight Percent[d, e]
E309CbLT-3	C 0.03 Cr 23.0-25.5 Ni 12.0-14.0 Mo 0.5 Cb + Ta 0.70-1.00 Mn 0.5-2.5 Si 1.0 P 0.04 S 0.03 Fe Rem Cu 0.5
E310T-3	C 0.20 Cr 25.0-28.0 Ni 20.0-22.5 Mo 0.5 Mn 1.0-2.5 Si 1.0 P 0.03 S 0.03 Fe Rem Cu 0.5
E312T-3	C 0.15 Cr 28.0-32.0 Ni 8.0-10.5 Mo 0.5 Mn 0.5-2.5 Si 1.0 P 0.04 S 0.03 Fe Rem Cu 0.5
E316T-3	C 0.08 Cr 18.0-20.5 Ni 11.0-14.0 Mo 2.0-3.0 Mn 0.5-2.5 Si 1.0 P 0.04 S 0.03 Fe Rem Cu 0.5
E316LT-3	C 0.03 Cr 18.0-20.5 Ni 11.0-14.0 Mo 2.0-3.0 Mn 0.5-2.5 Si 1.0 P 0.04 S 0.03 Fe Rem Cu 0.5
E317LT-3	C 0.03 Cr 18.5-21.0 Ni 13.0-15.0 Mo 3.0-4.0 Mn 0.5-2.5 Si 1.0 P 0.04 S 0.03 Fe Rem Cu 0.5
E347T-3	C 0.08 Cr 19.0-21.5 Ni 9.0-11.0 Mo 0.5 Cb + Ta 8 x C min to 1.0 max Mn 0.5-2.5 Si 1.0 P 0.04 S 0.03 Fe Rem Cu 0.5
E409T-3[b]	C 0.10 Cr 10.5-13.0 Ni 0.60 Mo 0.5 Mn 0.80 Si 1.0 P 0.04 S 0.03 Fe Rem Cu 0.5
E410T-3	C 0.12 Cr 11.0-13.5 Ni 0.60 Mo 0.5 Mn 1.0 Si 1.0 P 0.04 S 0.03 Fe Rem Cu 0.5
E410NiMoT-3	C 0.06 Cr 11.0-12.5 Ni 4.0-5.0 Mo 0.40-0.70 Mn 1.0 Si 1.0 P 0.04 S 0.03 Fe Rem Cu 0.5
E410NiTiT-3[b]	C 0.04 Cr 11.0-12.0 Ni 3.6-4.5 Mo 0.5 Mn 0.70 Si 0.50 P 0.03 S 0.03 Fe Rem Cu 0.5
E430T-3	C 0.10 Cr 15.0-18.0 Ni 0.60 Mo 0.5 Mn 1.0 Si 1.0 P 0.04 S 0.03 Fe Rem Cu 0.5
EXXXT-G	As agreed upon between supplier and purchaser.

a. The carbon content shall be 0.04% maximum when the suffix "X" is "1"; it shall be 0.03% maximum when the suffix "X" is "2".

b. Titanium - 10 x C min to 1.5% max.

c. The letter "X" as specifically presented in this table indicates a classification covering the shielding designation for both the "1" and "2" categories.

d. Analysis shall be made for the elements for which specific values are shown in this table. If, however, the presence of other elements is indicated in the course of routine analysis, further analysis shall be made to determine that the total of these other elements, except iron, is not present in excess of 0.50%.

e. Single values shown are maximum percentages.

ALL-WELD-METAL MECHANICAL PROPERTY REQUIREMENTS

AWS A5.22 Classification[a, e]	Tensile Strength, min.		% Elongation in 2 in. (50 mm), min.	Heat Treatment
	psi	MPa		
E307T-X	85,000	590	30	None
E308T-X	80,000	550	35	None
E308LT-X	75,000	515	35	None
E308MoT-X	80,000	550	35	None
E308MoLT-X	75,000	515	35	None
E309T-X	80,000	550	30	None
E309CbLT-X	75,000	515	35	None
E309LT-X	75,000	515	30	None
E310T-X	80,000	550	30	None
E312T-X	95,000	660	22	None
E316T-X	75,000	515	30	None
E316LT-X	70,000	485	30	None
E317LT-X	75,000	515	20	None
E347T-X	75,000	515	30	None
E409T-X	65,000	450	20	None
E410T-X	65,000	450	20	b
E410NiMoT-X	110,000	760	15	d
E410NiTiT-X	110,000	760	15	d
E430T-X	65,000	450	20	c
E502T-X	60,000	415	20	b
E505T-X	60,000	415	20	b
E307T-3	85,000	590	30	None
E308T-3	80,000	550	35	None
E308LT-3	75,000	515	35	None
E308MoT-3	80,000	550	35	None
E308MoLT-3	75,000	515	35	None
E309T-3	80,000	550	30	None

ALL-WELD-METAL MECHANICAL PROPERTY REQUIREMENTS (Continued)

AWS A5.22 Classification[a,e]	Tensile Strength, min.		% Elongation in 2 in. (50 mm), min.	Heat Treatment
	psi	MPa		
E309LT-3	75,000	515	30	None
E309CbLT-3	75,000	515	35	None
E310T-3	80,000	550	30	None
E312T-3	95,000	660	22	None
E316T-3	75,000	515	30	None
E316LT-3	70,000	485	30	None
E317LT-3	75,000	515	20	None
E347T-3	75,000	515	30	None
E409T-3	65,000	450	20	None
E410T-3	65,000	450	20	b
E410NiMoT-3	110,000	760	15	d
E410NiTiT-3	110,000	760	15	d
E430T-3	65,000	450	20	c

a. The letter "X" as specifically presented in this table indicates a classification covering the shielding designation for both the "1" and "2" categories. b. Specimen shall be heated to between 1550 and 1600°F (840 and 870°C) and held for two hours, furnace cooled at a rate not exceeding 100°F (55°C) per hour to 1100°F (595°C), then air cooled. c. Specimen shall be heated to between 1400 and 1450°F (760 and 790°C) and held for four hours, furnace cooled at a rate not exceeding 100°F (55°C) per hour to 1100°F (595°C), then air cooled. d. Specimen shall be heated to between 1100 and 1150°F (595 and 620°C). held for one hour, and air cooled to ambient. e. Properties as agreed upon between supplier and purchaser for EXXXT-G.

CLASSIFICATION SYSTEM AND SHIELDING MEDIUM

AWS A5.22 Designations[a] (All Classifications)	External Shielding[b] Medium	Current and Polarity
EXXT-1	CO_2	DC Electrode Positive
EXXT-2	Ar + 2% O	DC Electrode Positive
EXXT-3	None	DC Electrode Positive
EXXT-G	Not Specified	Not Specified

a. The letters "XXX" stand for the chemical composition. b. The requirement for the use of specified external shielding media for classification purposes shall not be construed to restrict the use of other media for industrial use as recommended by the producer.

The Metals Blue Book

AWS A5.22 CLASSIFICATION SYSTEM

The classification system used in this specification follows as closely as possible the standard pattern used in other AWS filler metal specifications. The inherent nature of the products being classified has, however, necessitated specific changes which more suitably classify the product.

An example of the method of classification follows;

EXXXT-X

E   Indicates an electrode.
XXX Designate classification according to composition.
T   Designates a flux cored electrode.
X   Designates the external shielding medium to be employed during welding.

In AWS A5.22, classification is on the basis of the shielding medium to be used during welding and the chemical analysis of weld deposits produced with the electrodes. The external shielding mediums recognized in this specification are carbon dioxide and argon-oxygen mixtures.

Other recognized methods of shielding are self-shielding from the core material with no externally applied gas and yet other methods not specified. The shielding designations are as follows:

EXXXT-1   Designates an electrode using carbon dioxide shielding plus a flux system.
EXXXT-2   Designates an electrode using a mixture of argon with 2 percent oxygen plus a flux oxygen.
EXXXT-3   Designates an electrode using no external shielding gas wherein shielding is provided by the flux system contained in the electrode core (self-shielding).
EXXXT-G   Indicates an electrode with unspecified method of shielding, no requirements being imposed except as agreed between purchaser and supplier. Each producer of an EXXXT-G electrode shall specify the chemical composition and mechanical property requirements for the electrode.

Electrodes classified for use with or the other of the gases required by AWS A5.22 may be operable under different shielding conditions than those tested, but no guarantee of properties is implied beyond the specific values and conditions covered by the specification.

Chapter

34

LOW ALLOY STEEL
ELECTRODES & FLUXES
FOR SUBMERGED ARC WELDING

CHEMICAL COMPOSITION REQUIREMENTS FOR SOLID ELECTRODES

AWS A5.23 Classification	UNS Number[d]	Composition, Weight Percent[a,b,c]
EL12[f]	K01012	C 0.04-0.14 Mn 0.25-0.60 Si 0.10 S 0.030 P 0.030 Cu[e] 0.35
EM12K[f]	K01113	C 0.05-0.15 Mn 0.80-1.25 Si 0.10-0.35 S 0.030 P 0.030 Cu[e] 0.35
EA1	K11222	C 0.07-0.17 Mn 0.65-1.00 Si 0.20 S 0.030 P 0.025 Mo 0.45-0.65 Cu[e] 0.35
EA2	K11223	C 0.07-0.17 Mn 0.95-1.35 Si 0.20 S 0.030 P 0.025 Mo 0.45-0.65 Cu[e] 0.35
EA3	K11423	C 0.07-0.17 Mn 1.65-2.20 Si 0.20 S 0.030 P 0.025 Mo 0.45-0.65 Cu[e] 0.35
EA3K	K21451	C 0.07-0.12 Mn 1.60-2.10 Si 0.50-0.80 S 0.025 P 0.025 Mo 0.40-0.60 Cu[e] 0.35
EA4	K11424	C 0.07-0.17 Mn 1.20-1.70 Si 0.20 S 0.030 P 0.025 Mo 0.45-0.65 Cu[e] 0.35
EB1	K11043	C 0.10 Mn 0.40-0.80 Si 0.05-0.30 S 0.025 P 0.025 Cr 0.40-0.75 Mo 0.45-0.65 Cu[e] 0.35
EB2	K11172	C 0.07-0.15 Mn 0.45-1.00 Si 0.05-0.30 S 0.030 P 0.025 Cr 1.00-1.75 Mo 0.45-0.65 Cu[e] 0.35
EB2H	K23016	C 0.28-0.33 Mn 0.45-0.65 Si 0.55-0.75 S 0.015 P 0.015 Cr 1.00-1.50 Mo 0.40-0.65 Cu[e] 0.30 V 0.20-0.30
EB3	K31115	C 0.05-0.15 Mn 0.40-0.80 Si 0.05-0.30 S 0.025 P 0.025 Cr 2.25-3.00 Mo 0.90-1.10 Cu[e] 0.35
EB5	K12187	C 0.18-0.23 Mn 0.40-0.70 Si 0.40-0.60 S 0.025 P 0.025 Cr 0.45-0.65 Mo 0.90-1.20 Cu[e] 0.30
EB6[g]	S50280	C 0.10 Mn 0.35-0.70 Si 0.05-0.50 S 0.025 P 0.025 Cr 4.50-6.50 Mo 0.45-0.70 Cu[e] 0.35
EB6H	S5018S	C 0.25-0.40 Mn 0.75-1.00 Si 0.25-0.50 S 0.030 P 0.025 Cr 4.80-6.00 Mo 0.45-0.65 Cu[e] 0.35
EB8[g]	S50480	C 0.10 Mn 0.30-0.65 Si 0.05-0.50 S 0.030 P 0.40 Cr 8.00-10.50 Mo 0.80-1.20 Cu[e] 0.35
ENi1	K11040	C 0.12 Mn 0.75-1.25 Si 0.05-0.30 S 0.020 P 0.020 Cr 0.15 Ni 0.75-1.25 Mo 0.30 Cu[e] 0.35
ENi2	K21010	C 0.12 Mn 0.75-1.25 Si 0.05-0.30 S 0.020 P 0.020 Ni 2.10-2.90 Cu[e] 0.35
ENi3	K31310	C 0.13 Mn 0.60-1.20 Si 0.05-0.30 S 0.020 P 0.020 Cr 0.15 Ni 3.10-3.80 Cu[e] 0.35
ENi4	K11485	C 0.12-0.19 Mn 0.60-1.00 Si 0.10-0.30 S 0.020 P 0.015 Ni 1.60-2.10 Mo 0.10-0.30 Cu[e] 0.35
ENi1K	K11058	C 0.12 Mn 0.80-1.40 Si 0.40-0.80 S 0.020 P 0.020 Ni 0.75-1.25 Cu[e] 0.35
EF1	K11160	C 0.07-0.15 Mn 0.90-1.70 Si 0.15-0.35 S 0.025 P 0.025 Ni 0.95-1.60 Mo 0.25-0.55 Cu[e] 0.35
EF2	K21450	C 0.10-0.18 Mn 1.70-2.40 Si 0.20 S 0.025 P 0.025 Ni 0.40-0.80 Mo 0.40-0.65 Cu[e] 0.25
EF3	K21485	C 0.10-0.18 Mn 1.70-2.40 Si 0.30 S 0.025 P 0.025 Ni 0.70-1.10 Mo 0.40-0.65 Cu[e] 0.25
EF4	K12048	C 0.16-0.23 Mn 0.60-0.90 Si 0.15-0.35 S 0.035 P 0.025 Cr 0.40-0.60 Ni 0.40-0.80 Mo 0.15-0.30 Cu[e] 0.25
EF5	K41370	C 0.10-0.17 Mn 1.70-2.20 Si 0.20 S 0.010 P 0.010 Cr 0.25-0.50 Ni 2.30-2.80 Mo 0.45-0.65 Cu[e] 0.50
EF6	K21135	C 0.07-0.15 Mn 1.45-1.90 Si 0.10-0.30 S 0.015 P 0.015 Cr 0.20-0.55 Ni 1.75-2.25 Mo 0.40-0.65 Cu[e] 0.35

CHEMICAL COMPOSITION REQUIREMENTS FOR SOLID ELECTRODES (Continued)

AWS A5.23 Classification	UNS Number[d]	Composition, Weight Percent[a,b,c]
EM2[h]	K10882	C 0.10 Mn 1.25-1.80 Si 0.20-0.60 S 0.010 P 0.010 Cr 0.30 Ni 1.40-2.10 Mo 0.25-0.55 Cu[e] 0.25 V 0.05 Al 0.10 Ti 0.10 Zr 0.10
EM3[h]	K21015	C 0.10 Mn 1.40-1.80 Si 0.20-0.60 S 0.010 P 0.010 Cr 0.55 Ni 1.90-2.60 Mo 0.25-0.65 Cu[e] 0.25 V 0.04 Al 0.10 Ti 0.10 Zr 0.10
EM4[h]	K21030	C 0.10 Mn 1.40-1.80 Si 0.20-0.60 S 0.010 P 0.010 Cr 0.60 Ni 2.00-2.80 Mo 0.30-0.65 Cu[e] 0.25 V 0.03 Al 0.10 Ti 0.10 Zr 0.10
EW	K11245	C 0.12 Mn 0.35-0.65 Si 0.20-0.35 S 0.040 P 0.030 Cr 0.50-0.80 Ni 0.40-0.80 Cu[e] 0.30-0.80
EG	-	Not specified

a. The filler metal shall be analyzed for the specific elements for which values are shown in this table. If the presence of other elements is indicated in the course of this work, the amount of those elements shall be determined to ensure that their total (excluding iron) does not exceed 0.50 percent.

b. Single values are maximum.

c. The letter "N" as a suffix to a classification indicates that the electrode is intended for welds in the core belt region of nuclear reactor vessels, as described in AWS A5.23 appendix A2.1. This suffix changes the limits on the phosphorus, vanadium, and copper, as follows:

    P = 0.012% max.   V = 0.05% max.   Cu = 0.08% max.

    "N" electrodes shall not be coated with copper or any material containing copper. The "EF5" and "EW" electrodes shall not be designated as "N" electrodes.

d. SAE/ASTM Unified Numbering System for Metals and Alloys.

e. The copper limit includes any copper coating that may be applied to the electrode.

f. The EL12 and EM12K classifications are identical to those same classifications in ANSI/AWS A5.17-89, *Specification for Carbon Steel Electrodes and Fluxes for Submerged Arc Welding*. They are included in AWS A5.23 because they are sometimes used with an alloy flux to deposit some of the weld metals classified in AWS A5.23 Table 2.

g. The EB6 and EB8 classifications are similar, but not identical, to the ER502 and ER505 classifications, respectively, in ANSI/AWS A5.9-80, *Specification for Corrosion-Resisting Chromium and Chromium-Nickel Steel Bare and Composite Metal Cored and Stranded Arc Welding Electrodes and Welding Rods*. These classifications will be dropped from the next revision of AWS A5.9.

h. The composition ranges of classifications with the "EM" prefix are intended to conform to the ranges for similar electrodes in the military specifications.

CHEMICAL COMPOSITION REQUIREMENTS FOR WELD METAL
(BOTH SOLID ELECTRODE-FLUX AND COMPOSITE ELECTRODE-FLUX COMBINATIONS)

AWS A5.23 Weld Metal Classification[e]	UNS Number[f]		Composition, Weight Percent[a,b,c,d]
	Solid	Composite	
A1	K11222	W17041	C 0.12 Mn 1.00 Si 0.80 S 0.040 P 0.030 Mo 0.40-0.65 Cu 0.35
A2	K11223	W17042	C 0.12 Mn 1.40 Si 0.80 S 0.040 P 0.030 Mo 0.40-0.65 Cu 0.35
A3	K11423	W17043	C 0.15 Mn 2.10 Si 0.80 S 0.040 P 0.030 Mo 0.40-0.65 Cu 0.35
A4	K11424	W17044	C 0.15 Mn 1.60 Si 0.80 S 0.040 P 0.030 Mo 0.40-0.65 Cu 0.35
B1	K11043	W51040	C 0.12 Mn 1.60 Si 0.80 S 0.040 P 0.030 Cr 0.40-0.65 Mo 0.40-0.65 Cu 0.35
B2	K11172	W52040	C 0.15 Mn 1.20 Si 0.80 S 0.040 P 0.030 Cr 1.00-1.50 Mo 0.40-0.65 Cu 0.35
B2H	K23016	W52240	C 0.10-0.25 Mn 1.20 Si 0.80 S 0.040 P 0.030 Cr 1.00-1.50 Mo 0.40-0.65 Cu 0.35 V 0.30
B3	K31115	W53030	C 0.15 Mn 1.20 Si 0.80 S 0.040 P 0.030 Cr 2.00-2.50 Mo 0.90-1.20 Cu 0.35
B4	--	W53346	C 0.12 Mn 1.20 Si 0.80 S 0.040 P 0.030 Cr 1.75-2.25 Mo 0.40-0.65 Cu 0.35
B5	K12187	W51348	C 0.18 Mn 1.20 Si 0.80 S 0.040 P 0.030 Cr 0.40-0.65 Mo 0.90-1.20 Cu 0.35
B6	S50280	W50240	C 0.12 Mn 1.20 Si 0.80 S 0.040 P 0.030 Cr 4.50-6.00 Mo 0.40-0.65 Cu 0.35
B6H	S50180	W50140	C 0.10-0.25 Mn 1.20 Si 0.80 S 0.040 P 0.030 Cr 4.50-6.00 Mo 0.40-0.65 Cu 0.35
B8	S50180	W50440	C 0.12 Mn 1.20 Si 0.80 S 0.040 P 0.030 Cr 8.00-10.00 Mo 0.80-1.20 Cu 0.35
Ni1[g]	K11040	W21048	C 0.12 Mn 1.60 Si 0.80 S 0.030 P 0.030 Ni 0.75-1.10 Mo 0.35 Cu 0.35 Ti 0.05[h]
Ni2[g]	K21010	W22040	C 0.12 Mn 1.60 Si 0.80 S 0.030 P 0.030 Ni 2.00-2.90 Cu 0.35
Ni3	K31310	W23040	C 0.12 Mn 1.60 Si 0.80 S 0.030 P 0.030 Ni 2.80-3.80 Cu 0.35
Ni4	K11485	W21250	C 0.14 Mn 1.60 Si 0.80 S 0.030 P 0.030 Ni 1.40-2.10 Mo 0.35 Cu 0.35
F1	K11160	W21150	C 0.12 Mn 0.70-1.50 Si 0.80 S 0.040 P 0.030 Cr 0.15 Ni 0.90-1.70 Mo 0.55 Cu 0.35
F2	K21450	W20240	C 0.17 Mn 1.25-2.25 Si 0.80 S 0.040 P 0.030 Ni 0.40-0.80 Mo 0.40-0.65 Cu 0.35
F3	K21485	W21140	C 0.17 Mn 1.25-2.25 Si 0.80 S 0.040 P 0.030 Ni 0.70-1.10 Mo 0.40-0.65 Cu 0.35
F4	K12048	W20440	C 0.17 Mn 1.60 Si 0.80 S 0.040 P 0.030 Cr 0.60 Ni 0.40-0.80 Mo 0.25 Cu 0.35 Ti 0.03[h]
F5	K41370	W22640	C 0.17 Mn 1.20-1.80 Si 0.80 S 0.030 P 0.030 Cr 0.65 Ni 2.00-2.80 Mo 0.30-0.80 Cu 0.50
F6	K21135	W21040	C 0.14 Mn 0.80-1.85 Si 0.80 S 0.030 P 0.030 Cr 0.65 Ni 1.50-2.25 Mo 0.60 Cu 0.40
M1	--	W21240	C 0.10 Mn 0.60-1.60 Si 0.80 S 0.040 P 0.030 Cr 0.15 Ni 1.25-2.00 Mo 0.35 Cu 0.30 Ti 0.03[h]
M2	--	W21340	C 0.10 Mn 0.90-1.80 Si 0.80 S 0.040 P 0.030 Cr 0.35 Ni 1.40-2.10 Mo 0.25-0.65 Cu 0.30 Ti+V+Zr 0.03

CHEMICAL COMPOSITION REQUIREMENTS FOR WELD METAL
(BOTH SOLID ELECTRODE-FLUX AND COMPOSITE ELECTRODE-FLUX COMBINATIONS) (Continued)

AWS A5.23 Weld Metal Classification[e]	UNS Number[f]		Composition, Weight Percent[a,b,c,d]
	Solid	Composite	
M3	--	W22240	C 0.10 Mn 0.90-1.80 Si 0.80 S 0.030 P 0.030 Cr 0.65 Ni 1.80-2.60 Mo 0.20-0.70 Cu 0.30 Ti+V+Zr 0.03
M4	--	W22440	C 0.10 Mn 1.30-2.25 Si 0.80 S 0.030 P 0.030 Cr 0.80 Ni 2.00-2.80 Mo 0.30-0.80 Cu 0.30 Ti+V+Zr 0.03
W	--	W21040	C 0.12 Mn 0.50-1.60 Si 0.80 S 0.040 P 0.030 Cr 0.45-0.70 Ni 0.40-0.80 Cu 0.30-0.75
G	--	--	Not Specified

a. The filler metal shall be analyzed for the specific elements for which values are shown in this table. If the presence of other elements is indicated in the course of this work, the amount of those elements shall be determined to ensure that their total (excluding iron) does not exceed 0.50 percent.

b. Single values are maximum.

c. The letter "N" as a suffix to a classification indicates that the weld metal is intended for welds in the core belt region of nuclear reactor vessels, as described in Paragraph A2.1 of the Appendix to this specification. This suffix changes the limits on the phosphorus, vanadium, and copper, as follows:
    P = 0.012% max. V = 0.05% max. Cu = 0.08% max.
    "N" electrodes shall not be coated with copper or any material containing copper. The "F5" and "W" classifications shall not be designated as "N" weld metals.

d. A low dilution area of the groove weld of AWS A5.23 Figure 3, or the reduced section of the fractured tension test specimen of AWS A5.23 Figure 5, may be substituted for the weld pad, and shall meet the above requirements. In case of dispute, the weld pad shall be the referee method.

e. The electrode designation for composite electrodes is obtained by placing an "EC" before the appropriate weld metal classification.

f. SAE/ASTM Unified Numbering System for Metals and Alloys.

g. Manganese in Ni1 and Ni2 classifications may be 1.80% maximum when carbon is restricted to 0.10% maximum.

TENSION TEST REQUIREMENTS

Electrode-Flux Combination Classification[a]	Tensile Strength			Yield Strength, min.[b]		% Elongation, min.[b]
	psi	MPa		psi	MPa	
F7XX-EXX-XX	70,000-95,000	480-660		58,000	400	22
F8XX-EXX-XX	80,000-100,000	550-690		68,000	470	20
F9XX-EXX-XX	90,000-110,000	620-760		78,000	540	17
F10XX-EXX-XX	100,000-120,000	690-830		88,000	610	16
F11XX-EXX-XX	110,000-130,000	760-900		98,000	680	15[c]
F12XX-EXX-XX	120,000-140,000	830-970		108,000	750	14[c]

a. The letter "X" used in various places in the classification in this table stands for, respectively, the condition of heat treatment, the toughness of the weld metal, the classification of the electrode, and the chemical composition of the weld metal.

b. Yield strength at 0.2 percent offset and elongation in 2 in. (51 mm) gage length.

c. Elongation may be reduced by one percentage point for F11XX-EXX-XX, F11XX-ECXX-XX, F12XX-EXX-XX, AND F12XX-ECXX-XX weld metals in the upper 25 percent of their tensile strength range.

DIFFUSIBLE HYDROGEN REQUIREMENTS

AWS A5.23 Electrode-Flux Combination Classification	Optional Supplemental Diffusible Hydrogen Designator[a,b]	Average Diffusible Hydrogen, Maximum (ml/100g Deposited Metal)[c]
All	H16	16.0
All	H8	8.0
All	H4	4.0

a. Diffusible hydrogen test is required only when specified by the purchaser and when the manufacturer puts the diffusible hydrogen designator on the label. (See also AWS A5.23 appendices A3 and A6.4).

b. This designator is added to the end of the complete electrode-flux classification designation. For example, a weathering steel electrode-flux combination meeting a maximum diffusible hydrogen of 8.0 ml/100g of deposited metal might be designated F8A2-EW-WH8.

c. Some classifications may not meet the lower average diffusible hydrogen levels (H8 and H4).

IMPACT TEST REQUIREMENTS[a,b]

Digit	Test Temperature		Average Energy Level, min.
	°F	°C	
Z	Not Specified		Not Required
0	0	-18	20 ft•lbf (27 J)
2	-20	-29	20 ft•lbf (27 J)
4	-40	-40	20 ft•lbf (27 J)
5	-50	-46	20 ft•lbf (27 J)
6	-60	-51	20 ft•lbf (27 J)
8	-80	-62	20 ft•lbf (27 J)
10	-100	-73	20 ft•lbf (27 J)
15	-150	-101	20 ft•lbf (27 J)

a. Based on the results of the impact tests of the weld metal, the manufacturer shall insert in the classification (AWS A5.23 Table 6) the appropriate digit from the table above, as indicated in AWS A5.23 Figure 1. Weld metal from a specific electrode-flux combination that meets impact requirements at a given temperature also meets the requirements at all higher temperatures in this table (i.e. weld metal meeting the requirements for digit 5 also meets the requirements for digits 4, 2, 0, and Z).

b. Weld metals with the "N" suffix shall also have a Charpy V-notch energy level of at least 75 ft•lbf (102 J) at 70°F (21°C).

AWS A5.23 CLASSIFICATION SYSTEM

The system for identifying the electrode classifications in this specification follows the standard pattern used in other AWS filler metal specifications. The letter "E" at the beginning of each classification designation stands for electrode. The remainder of the designation indicates the chemical composition of the electrode, or, in the case of composite electrodes, of the undiluted weld metal obtained with a particular flux. See AWS A5.23 appendix A2.1 for more details regarding classification of electrodes.

Fluxes are classified on the basis of the mechanical properties of the weld metal they produce with some certain classification of electrode, under the specific test conditions called for in AWS A5.23 Part B. See AWS A5.23 Appendix A2.3 for more details regarding classification of fluxes.

A generic example of electrode and flux classification follows:

FXXX-ECXXXN-XNHX

F	Indicates a submerged arc welding flux.
First X	Indicates the minimum tensile strength [in increments of 10,000 psi (69 MPa)] of weld metal with the flux and some specific classification of electrode deposited according to the welding conditions specified herein. Two digits are used for weld metal of 100,000 psi (690 MPa) tensile strength and higher (see AWS A5.23 Table 6).
Second X	Designates the condition of heat treatment in which the tests were conducted: "A" for as-welded and "P" for postweld heated treated. The PWHT is one hour at the temperature specified in AWS A5.23 Table 5.
Third X	Indicates the lowest temperature at which the impact strength of the weld metal referred to above meets or exceeds 20 ft·lbf (27 J). (See AWS A5.23 Table 7).
E	Indicates electrode.
C	Indicates that the electrode is a composite electrode. Omission of the "C" indicates that the electrode is a solid electrode.
XXX	Classification of the electrode used in producing the weld metal referred to above (see AWS A5.23 Table 1 for solid electrodes or Table 2 for composite electrodes).
N	The "N" is used only when footnote c to AWS A5.23 Tables 1 and 2 apply (see AWS A5.23 Appendix A2.1 for explanation).
X	Indicates the chemical composition of the weld metal obtained with the flux and the electrode. One or more letters or digits are used (see AWS A5.23 Table 2).
N	The "N" is used only when footnote c to AWS A5.23 Tables 1 and 2 apply (see AWS A5.23 Appendix A2.1 for explanation).
HX	Optional supplemental diffusible hydrogen designator (see AWS A5.23 Table 8).

AWS A5.23 CLASSIFICATION SYSTEM (Continued)

Two specific examples of electrode and flux classification follows:

F9P0-EB3-B3 is a complete designation for an electrode-flux combination. It refers to a flux that will produce weld metal which, in the postweld heat-treated condition, will have a tensile strength of 90,000 - 110,000 psi (620 - 760 MPa) and Charpy V-notch impact strength of at least 20 ft•lbf (27 J) at 0°F (-18°C) when made with an EB3 electrode under the conditions called for in this specification. The composition of the weld metal will be B3 (see AWS A5.23 Table 2).

F9A2-ECM1-M1 is a complete designation for a flux when the trade name of the composite electrode used in the classification tests is indicated as well [see AWS A5.23 paragraph 16.4.1 (6)]. The designation refers to a flux that will produce weld metal that, with the M1 composite electrode, in the as-welded condition, will have a tensile strength of 90,000 - 110,000 psi (620 - 760 MPa) and Charpy V-notch energy of at least 20 ft•lbf (27 J) at -20°F (-29°C) under the conditions called for in this specification. The composition of the weld meal will be M1 (see AWS A5.23 Table 2).

"G" Classification and the Use of "Not Specified" and "Not Required"

AWS A5.23 includes filler metals classified as EG or ECG. The "G" indicates that the filler metal is of a "general" classification. It is "general" because not all of the particular requirements specified for each of the other classifications are specified for this classification. The intent, in establishing this classification, is to provide a means by which filler metals that differ in one respect or another (chemical composition, for example) from all other classifications (meaning that the composition of the filler metal - in the case of the example, does not meet the composition specified for any of the classifications in the specification) can still be classified according to the specification. The purpose is to allow a useful filler metal - one that otherwise would have to await a revision of the specification - to be classified immediately, under the existing specification. This means, then, that two filler metals - each bearing the same "G" classification - may be quite different in some certain respect (chemical composition, again, for example). See AWS A5.23 Appendix A2.2 for more details regarding "G" classification.

Chapter
35

ZIRCONIUM & ZIRCONIUM ALLOY WELDING ELECTRODES & RODS

CHEMICAL COMPOSITION REQUIREMENTS FOR ZIRCONIUM AND ZIRCONIUM ELECTRODES AND RODS

AWS A5.24 Classification	UNS Number[b]	Composition, Weight Percent[a]
ERZr2	R60702	Zr + Hf 99.01 min Hf 4.5 Fe + Cr 0.20 O 0.016 H 0.005 N 0.025 C 0.05
ERZr3	R60704	Zr + Hf 97.5 min Hf 4.5 Fe + Cr 0.20 to 0.40 Sn 1.00 to 2.00 O 0.016 H 0.005 N 0.025 C 0.05
ERZr4	R60705	Zr + Hf 95.5 min Hf 4.5 Fe + Cr 0.20 O 0.016 H 0.005 N 0.025 C 0.05 Cb 2.0 to 3.0

a.   Single values are maximum, except as noted.

b.   SAE/ASTM Unified Numbering System for Metals and Alloys.

c.   Analysis of the interstitial elements C, O, H and N shall be conducted on samples of filler metal taken after the filler meal has been reduced to its final diameter and all processing operations have been completed. Analysis of the other elements may be conducted on these same samples or it may have been conducted on samples taken from the ingot or the rod stock from which the filler metal is made. In case of dispute, samples from the finished filler metal shall be the referee method.

AWS A5.24 CLASSIFICATION SYSTEM

The system of classification is similar to that used in other filler metal specifications. The letter "E" at the beginning of each designation indicates a welding electrode, and the letter "R" indicates a welding rod. Since these filler metals are used as welding electrodes in gas metal arc welding and as welding rods in gas tungsten arc welding, both letters are used.

The letters "Zr" indicate that the filler metals have a zirconium base. The digits and letters that follow provide a means for identifying the nominal composition of the filler metal.

Chapter
36

CARBON & LOW ALLOY STEEL ELECTRODES & FLUXES FOR ELECTROSLAG WELDING

CHEMICAL COMPOSITION REQUIREMENTS FOR SOLID ELECTRODES

AWS A5.25 Classification[c]	UNS Number[d]	Composition, Weight Percent[a,b]
Medium Manganese Classes		
EM5K-EW	K10726	C 0.07 Mn 0.90-1.40 P 0.025 S 0.035 Si 0.40-0.70 Cu[e] 0.35 Ti 0.05-0.15 Zr 0.02-0.12 Al 0.05-0.15 Other Elements Total 0.50
EM12-EW	K01112	C 0.06-0.15 Mn 0.80-1.25 P 0.035 S 0.035 Si 0.10 Cu[e] 0.35 Other Elements Total 0.50
EM12K-EW	K01113	C 0.05-0.15 Mn 0.80-1.25 P 0.035 S 0.035 Si 0.10-0.35 Cu[e] 0.35 Other Elements Total 0.50
EM13K-EW	K01313	C 0.07-0.19 Mn 0.90-1.40 P 0.035 S 0.035 Si 0.35-0.75 Cu[e] 0.35 Other Elements Total 0.50
EM15K-EW	K01515	C 0.10-0.20 Mn 0.80-1.25 P 0.035 S 0.035 Si 0.10-0.35 Cu[e] 0.35 Other Elements Total 0.50
High Manganese Classes		
EH14-EW	K11585	C 0.10-0.20 Mn 1.70-2.20 P 0.035 S 0.035 Si 0.10 Cu[e] 0.35 Other Elements Total 0.50
Special Classes		
EWS-EW	K11245	C 0.07-0.12 Mn 0.35-0.65 P 0.035 S 0.040 Si 0.22-0.37 Ni 0.40-0.75 Cr 0.50-0.80 Cu[e] 0.25-0.55 Other Elements Total 0.50
EH10Mo-EW	K10945	C 0.07-0.12 Mn 1.60-2.10 P 0.025 S 0.025 Si 0.50-0.80 Ni 0.15 Mo 0.40-0.60 Cu[e] 0.35 Other Elements Total 0.50
EH10K-EW	K01010	C 0.07-0.14 Mn 1.40-2.00 P 0.025 S 0.030 Si 0.15-0.30 Other Elements Total 0.50
EH11K-EW	K11140	C 0.07-0.15 Mn 1.40-1.85 P 0.025 S 0.035 Si 0.80-1.15 Cu[e] 0.35 Other Elements Total 0.50
ES-G-EW	-	Not Specified

a. The electrode shall be analyzed for the specific elements for which values are shown in this table. If the presence of other elements is indicated in the course of this work, the amount of those elements shall be determined to ensure that their total (excluding iron) does not exceed the limits specified for "Other Elements Total" in this table.

b. Single values are maximum.

c. Chemical composition requirements may be similar to those in other AWS specifications, see AWS A5.25 Appendix Table A1.

d. SAE/ASTM Unified Numbering System for Metals and Alloys.

e. The copper limit includes copper that may be applied as a coating on the electrode.

f. Composition shall be reported; the requirements are those agreed to by the purchaser and the supplier.

CHEMICAL COMPOSITION REQUIREMENTS FOR UNDILUTED WELD METAL FROM COMPOSITE METAL CORED ELECTRODES[a]

AWS A5.25 Classification	UNS Number[d]	Composition, Weight Percent[b,c]
EWT1	W06040	C 0.13 Mn 2.00 P 0.03 S 0.03 Si 0.60 Other Elements Total 0.50
EWT2	W20140	C 0.12 Mn 0.50-1.60 P 0.03 S 0.04 Si 0.25-0.80 Ni 0.40-0.80 Cr 0.40-0.70 Cu 0.25-0.75 Other Elements Total 0.50
EWT3	W22340	C 0.12 Mn 1.00-2.00 P 0.02 S 0.03 Si 0.15-0.50 Ni 1.50-2.50 Cr 0.20 Mo 0.40-0.65 V 0.05 Other Elements Total 0.50
EWT4	-	C 0.12 Mn 0.50-1.30 P 0.03 S 0.03 Si 0.30-0.80 Ni 0.40-0.80 Cr 0.45-0.70 Cu 0.30-0.75 Other Elements Total 0.50
EWTG	-	Not Specified[e]

a. The flux used shall be that with which the electrode is classified for mechanical properties (see AWS A5.25 Tables 3 and 4 for more details).

b. The weld metal shall be analyzed for the specific elements for which values are shown in this table. If the presence of other elements is indicated in the course of this work, the amount of those elements shall be determined to ensure that their total (excluding iron) does not exceed the limits specified for "Other Elements Total" in this table.

c. Single values are maximum.

d. SAE/ASTM Unified Numbering System for Metals and Alloys.

e. Composition shall be reported; the requirements are those agreed to by the purchaser and supplier.

TENSION TEST REQUIREMENTS (AS-WELDED)

AWS A5.25 Classification[a]	Tensile Strength		Yield Strength, min[b]		% Elongation[b] (min.)
	psi	MPa	psi	MPa	
FES6Z-XXX, FES60-XXX, FES62-XXX	60,000 to 80,000	(420) to (550)	36,000	(250)	24
FES7Z-XXX, FES70-XXX, FES72-XXX	70,000 to 95,000	(490) to (650)	50,000	(350)	22

a. The letters "XXX" as used in AWS Classification column in this table refer to the electrode classification used.

b. Yield strength at 0.2 percent offset and elongation in 2 in. (51 mm) gage length.

IMPACT TEST REQUIREMENTS[a] (As Welded)

AWS A5.25 Classification[b]	Average Impact Strength[c], min
FES6Z-XXX, FES7Z-XXX	Not specified
FES60-XXX, FES70-XXX	15 ft•lbf at 0°F (20J at -18°C)
FES62-XXX, FES72-XXX	15 ft•lbf at -20°F (20J at -29°C)

a. A flux-electrode combination that meets impact requirements at a given temperature also meets the requirements at all higher temperatures in this table. In this manner, FESX2-XXX may also be classified as FESX0-XXX and FESXZ-XXX and FESX0-XXX may be classified as FESXZ-XXX.

b. The letters "XXX" used in the AWS Classification column in this table refer to the electrode classification used.

c. Both the highest and lowest of the five test values obtained shall be disregarded in computing the impact strength. Two of the remaining three levels shall equal or exceed 15 ft•lbf (20J), one of the three remaining values may be lower than 15 ft•lbf but not lower than 10 ft•lbf (14J). The average of the three shall not be less than the 15 ft•lbf (20J) specified.

COMPARISON OF A5.25-91 CLASSIFICATIONS AND CLASSIFICATIONS IN OTHER AWS SPECIFICATIONS

AWS A5.25-91	Similar Classifications[a]				
	AWS A5.17-89	AWS A5.18-79	AWS A5.23-90	AWS A5.28-79	
EM5K-EW	-	ER70S-2	-	-	
EM12-EW	EM12	-	-	-	
EM12K-EW	EM12K	-	EM12K	-	
EM13K-EW	EM13K	ER70S-3	-	-	
EM15K-EW	EM15K	-	-	-	
EH14-EW	EH14	-	-	-	
EWS-EW	-	-	EW	-	
EH10Mo-EW	-	-	EA3K	ER80S-D2	
EH10K-EW	-	-	-	-	
EH11K-EW	EH11K	ER70S-6	-	-	

a. Classifications are similar, but not necessarily identical in composition:

AWS A5.17-89, Specification for Carbon Steel Electrodes and Fluxes for Submerged Arc Welding
AWS A5.18-79, Specification for Carbon Steel Filler Metals for Gas Shielded Arc Welding
AWS A5.23-90, Specification for Low Alloy Steel Electrodes and Fluxes for Submerged Arc Welding
AWS A5.28-79, Specification for Low Alloy Steel Filler Metals for Gas Shielded Arc Welding

AWS A5.25 CLASSIFICATION SYSTEM

Classification of Electrodes

The system for identifying the electrode classifications in this specification follows the standard pattern used in other AWS filler metal specifications. The letter "E" at the beginning of each classification designation stands for electrode. The remainder of the designation indicates the chemical composition of the electrode, or, in the case of composite metal cored electrodes, of the undiluted weld metal obtained with a particular flux. See AWS A5.25 Figure A1.

The letter "M" indicates that the solid electrode is of a medium manganese content, while the letter "H" would indicate a comparatively high manganese content. The one or two digits following the manganese designator indicate the nominal carbon content of the electrode. The letter "K", which appears in some designations, indicates that the electrode is made from a heat of silicon-killed steel. The designation for a solid wire is followed by the suffix "EW". Solid electrodes are classified only on the basis of their chemical composition, as specified in AWS A5.25 Table 1 of this specification. A composite electrode is indicated by the letters "WT" after the "E", and a numerical suffix. The composition of a composite electrode is meaningless, and the user is therefore referred to weld metal composition (AWS A5.25 Table 2) with a particular flux, rather than to electrode composition.

Classification of Fluxes

Fluxes are classified on the basis of the mechanical properties of the weld metal made with some certain classification of electrode, under the specific test conditions called for in this specification.

As examples of flux classifications, consider the following:

FES60-EH14-EW   FES72-EWT2

The prefix "FES" designates a flux for electroslag welding. This is followed by a single digit representing the minimum tensile strength required of the weld metal in 10,000 psi (see AWS A5.25 Table 3).

The digit that follows is a number or the letter "Z". This digit refers to the impact strength of the weld metal. Specifically, it designates the temperature at (and above) which the weld metal meets, or exceeds, the required 15 ft·lbf (20J) Charpy V-notch impact strength (except for the letter "Z", which indicates that no impact strength requirement is specified (see AWS A5.25 Table 4). These mechanical property designators are followed by the designation of the electrode used in classifying the flux (EM12-EW, EH10K-EW, EWT12, etc.) included after the first hyphen refers to the electrode classification with which the flux will produce weld metal that meets the specified mechanical properties when tested as called for in the specification.

The Metals Blue Book

AWS A5.25 CLASSIFICATION SYSTEM (Continued)

It should be noted that flux of any specific trade designation may have many classifications. The number is limited only by the number of different electrode classifications with which the flux can meet the classification requirements. The flux marking lists at least one, and may list all, classifications to which the flux conforms. Solid electrodes having the same classification are interchangeable when used with a specific flux; composite metal cored electrodes may not be. However, the specific usability (or operating) characteristics of various fluxes of the same classification may differ in one respect or another.

"G" Classification

This specification includes filler metals classified as ES-G-EW or EWTG. The letter "G" indicates that the filler metal is of a general classification. It is general because not all of the particular requirements specified for each of the other classifications are specified for this classification. The intent, in establishing this classification is to provide a means by which filler metals that differ in one respect or another (chemical composition, for example) from all other classifications (meaning that the composition of the filler metal - in the case of the example - does not meet the composition specified for any of the classifications in the specification) can still be classified according to the specification. The purpose is to allow a useful filler metal - one that otherwise would have to await a revision of the specification - to be classified immediately, under the existing specification. This means, then, that two filler metals - each bearing the same "G" classification - may be quite different in some certain respect (chemical composition, again, for example).

The point of difference (although not necessarily the amount of the difference) referred to above will be readily apparent from the use of the words "not required" and "not specified" in AWS A5.25. The use of these words is as follows:

"Not Specified" is used in those areas of the specification that refer to the results of some particular test. It indicates that the requirements for that test are not specified for that particular classification.

"Not Required" is used in those areas of the specification that refer to the test that must be conducted in order to classify a filler metal (or a welding flux). It indicates that test is not required because the requirements (results) for the test have not been specified for that particular classification. Restating the case, when a requirement is not specified, it is not necessary to conduct the corresponding test in order to classify a filler metal to that classification. When a purchaser wants the information provided by that test, in order to consider a particular product of that classification for a certain application, the purchaser will have to arrange for that information with the supplier. The purchaser and supplier also will have to establish the specific testing procedure and the acceptance requirements. These may be incorporated (via AWS A5.01, *Filler Metal Procurement Guidelines*) in the purchase order.

Chapter

37

CARBON & LOW ALLOY STEEL ELECTRODES FOR ELECTROGAS WELDING

CHEMICAL COMPOSITION REQUIREMENTS FOR SOLID ELECTRODES

AWS A5.26 Classification[c]	UNS Number[d]	Composition, Weight Percent[a,b]
EGXXS-1	K01313	C 0.07-0.19 Mn 0.90-1.40 S 0.035 P 0.025 Si 0.30-0.50 Cu[e] 0.35 Other Elements Total 0.50
EGXXS-2	K10726	C 0.07 Mn 0.90-1.40 S 0.035 P 0.025 Si 0.40-0.70 Cu[e] 0.35 Ti 0.05-0.15 Zr 0.02-0.12 Al 0.05-0.15 Other Elements Total 0.50
EGXXS-3	K11022	C 0.06-0.15 Mn 0.90-1.40 S 0.035 P 0.025 Si 0.45-0.70 Cu[e] 0.35 Other Elements Total 0.50
EGXXS-5	K11325	C 0.07-0.19 Mn 0.90-1.40 S 0.035 P 0.025 Si 0.30-0.60 Cu[e] 0.35 Al 0.50-0.90 Other Elements Total 0.50
EGXXS-6	K11140	C 0.07-0.15 Mn 1.40-1.85 S 0.035 P 0.025 Si 0.80-1.15 Cu[e] 0.35 Other Elements Total 0.50
EGXXS-D2[f]	K10945	C 0.07-0.12 Mn 1.60-2.10 S 0.035 P 0.025 Si 0.50-0.80 Ni 0.15 Mo 0.40-0.60 Cu[e] 0.35 Other Elements Total 0.50
EGXXS-G	-	Not Specified[g]

a. The filler metal shall be analyzed for the specific elements for which values are shown in this table. If the presence of other elements is indicated in the course of this work, the amount of those elements shall be determined to ensure that their total (excluding iron) does not exceed the limits specified for "Other Elements Total" in this table.

b. Single values are maximum.

c. The letters "XX" as used in the AWS classification column of this table refer respectively to "6", "7", or "8" (tensile strength of the weld metal - see AWS A5.26 Table 2) and a "Z", "0", or "2" (impact strength - see AWS A5.26 Table 3).

d. SAE/ASTM Unified Numbering System for Metals and Alloys.

e. The copper limit includes copper that may be applied as a coating on the electrode.

f. Formerly EGXXS-1B

g. Composition shall be reported; the requirements are those agreed to by the purchaser and the supplier.

TENSION TEST REQUIREMENTS (AS WELDED)

AWS A5.26 Classification[a]	Tensile Strength		Yield[b] Strength, min.		% Elongation[b], min.
	psi	MPa	psi	MPa	
EG6ZX-X, EG60X-X, EG62X-X	60,000 to 80,000	(420) to (550)	36,000	(250)	24
EG7ZX-X, EG70X-X, EG72X-X	70,000 to 95,000	(490) to (650)	50,000	(350)	22
EG8ZX-X, EG80X-X, EG82X-X	80,000 to 100,000	(550) to (690)	60,000	(420)	20

a. The letters "X-X" as they are used in the AWS Classification column in this table refer respectively to "S" or "T" (whether the electrode is solid or composite), and "1, 2, 3, 4, 5, 6, D2, Ni1, NM1, NM2, W, or G" (the designation for shielding gas and chemical composition requirements).

b. Yield strength at 0.2 percent offset and elongation in 2 in. (51 mm) gage length.

IMPACT TEST REQUIREMENTS[a] (AS WELDED)

AWS A5.26 Classification	Average Impact Strength[b], min.
EG6ZX-X, EG7ZX-X, EG8ZX-X	Not Specified
EG60X-X, EG70X-X, EG80X-X	20 ft•lbf at 0°F (27J at -18°C)
EG62X-X, EG72X-X, EG82X-X	20 ft•lbf at -20°F (27J at -29°C)

a. An electrode combination that meets impact requirements at a given temperature also meets the requirements at all higher temperatures in this table. In this manner, EGX2X-X may also be classified as EGX0X-X and EGXZX-X, and EGX0X-X may be classified as EGXZX-X.

b. Both the highest and lowest of the five test values obtained shall be disregarded in computing the impact strength. Two of the remaining three values shall equal or exceed 20 ft•lbf; one of the three remaining values may be lower than 20 ft•lbf but not lower than 15 ft•lbf. The average of the three shall not be less than the ft•lbf specified.

CHEMICAL COMPOSITION REQUIREMENTS FOR WELD METAL FROM COMPOSITE FLUX CORED AND METAL CORED ELECTRODES

AWS A5.26 Classification[c]	UNS Number[d]	Shielding Gas	Composition, Weight Percent[a,b]
EG6XT-1	W06301	None	C (e) Mn 1.7 P 0.03 S 0.03 Si 0.03 Ni 0.30 Cr 0.20 Mo 0.35 Cu 0.35 V 0.08 Other Elements Total 0.50
EG7XT-1	W07301	None	C (e) Mn 1.7 P 0.03 S 0.03 Si 0.03 Ni 0.50 Cr 0.20 Mo 0.35 Cu 0.35 V 0.08 Other Elements Total 0.50
EG6XT-2	W06302	CO_2	C (e) Mn 2.0 P 0.03 S 0.03 Si 0.03 Ni 0.90 Cr 0.20 Mo 0.35 Cu 0.35 V 0.08 Other Elements Total 0.50
EG7XT-2	W07302	CO_2	C (e) Mn 2.0 P 0.03 S 0.03 Si 0.03 Ni 0.90 Cr 0.20 Mo 0.35 Cu 0.35 V 0.08 Other Elements Total 0.50
EGXXT-Ni1[g]	W21033	CO_2	C 0.10 Mn 1.0-1.8 P 0.03 S 0.03 Si 0.50 Ni 0.70-1.10 Mo 0.30 Cu 0.35 Other Elements Total 0.50
EGXXT-NM1[h]	W22334	Ar/CO_2 or CO_2	C 0.12 Mn 1.0-2.0 P 0.02 S 0.03 Si 0.15-0.50 Ni 1.5-2.0 Cr 0.20 Mo 0.40-0.65 Cu 0.35 V 0.05 Other Elements Total 0.50
EGXXT-NM2[i]	W22333	CO_2	C 0.12 Mn 1.1-2.1 P 0.03 S 0.03 Si 0.20-0.60 Ni 1.1-2.0 Cr 0.20 Mo 0.10-0.35 Cu 0.35 V0.05 Other Elements Total 0.50
EGXXT-W[j]	W20131	CO_2	C 0.12 Mn 0.50-1.3 P 0.03 S 0.03 Si 0.30-0.80 Ni 0.40-0.80 Cr 0.45-.70 Cu 0.30-0.75 Other Elements Total 0.50
EGXXT-G	-	Not Specified[f]	

CHEMICAL COMPOSITION REQUIREMENTS FOR WELD METAL FROM COMPOSITE FLUX CORED AND METAL CORED ELECTRODES (Continued)

a. The weld metal shall be analyzed for the specific elements for which values are shown in this table. If the presence of other elements is indicated in the course of this work, the amount of those elements shall be determined to ensure that their total (excluding iron) does not exceed the limits specified for "Other Elements Total" in this table.

b. Single values are maximum.

c. The letters "XX", as used in the AWS A5.26 classification column in this table refers respectively to "6", "7", or "8" (tensile strength of the weld metal - see AWS A5.26 Table 2 for more details) and a "Z", "0", or "2" (impact strength - see AWS A5.26 Table 3 for more details). The single letter "X" as used in the AWS A5.26 classification column refers to a "Z", "0", or "2" (impact strength - see AWS A5.26 Table 3 for more details).

d. SAE/ASTM Unified Numbering System for Metals and Alloys.

e. Composition range of carbon not specified for these classifications, but the amount shall be determined and reported.

f. Composition shall be reported; the requirements are those agreed to by the purchaser and supplier.

g. Formerly EGXXT-3.

h. Formerly EGXXT-4.

i. Formerly EGXXT-6.

j. Formerly EGXXT-5.

AWS A5.26 CLASSIFICATION SYSTEM

The system for identifying the electrode classifications in this specification follows the standard pattern used in other AWS filler metal specifications. The letter "EG" at the beginning of each classification designation shows that the electrode is intended for use with the electrogas welding process.

The first digit following "EG" represents the minimum tensile strength required of the weld metal in units of 10,000 psi. The second digit (or the letter Z, when impact tests are not required), refers to the impact strength of welds in accordance with the test assembly preparation section of this specification.

The next letter, either S or T, indicates that the electrode is solid (S) or composite flux cored or metal cored (T). The designator (digits or letters) following the hyphen in the classification indicates the chemical composition (of weld metal for the composite electrodes and of the electrode itself for solid electrodes) and the type or absence of shielding gas required.

A5.26 CLASSIFICATION SYSTEM (Continued)

This specification includes filler metals classified as EGXXT-G or EGXXS-G. The last "G" indicates that the filler metal is of a general classification. It is "general" because not all of the particular requirements specified for each of the other classifications are specified for this classification. The intent in establishing this AWS classification is to provide a means by which filler metals that differ in one respect or another (chemical composition, for example) from all other classifications (meaning that the composition of the filler metal, in the case of the example, does not meet the composition specified for any of the classifications in the specification) can still be classified according to the specification. The purpose is to allow a useful filler metal - one that otherwise would have to await a revision of the specification - to be classified immediately, under the existing specification. This means, then, that two filler metals, each bearing the same "G" classification, may be quite different in some certain respect (chemical composition, again, for example).

The point of difference (although not necessarily the amount of the difference) referred to above will be readily apparent from the use of the words "not required" and "not specified" in the specification. The use of these words is as follows:

Not Specified is used in those areas of the specification that refer to the results of some particular test. It indicates that the requirements for that test are not specified for that particular classification.

Not Required is used in those areas of the specification that refer to the test that must be conducted in order to classify a filler metal. It indicates that the test is not required because the requirements (results) for the test have not been specified for that particular classification.

Restating the case, when a requirement is not specified, it is not necessary to conduct the corresponding test to classify a filler metal to that classification. When a purchaser wants the information provided by that test, in order to consider a particular product of that classification for a certain application, that information should be forwarded to the supplier of that product. The purchaser also will have to establish with the supplier the testing procedure and the acceptance requirements required for that test. This information may be incorporated (via ANSI/AWS A5.01, *Filler Metal Procurement Guidelines*) in the purchase order.

A generic example is as follows: **EGXXT-X**

EG	Designates an electrode for electrogas welding
First X	Indicates, in 10,000 psi increments, the minimum tensile strength of the weld metal produced by the electrode when tested according to this specification.
Second X	Indicates the impact strength of the weld metal produced by the electrode when tested according to this specification.
T	Indicates whether the electrode is cored (T) or solid (S)
X	Indicates the chemical composition of the weld metal produced by a composite electrode or the chemical composition of a solid electrode, and references whether shielding gas is used when welding with a composite electrode.

The Metals Blue Book

Chapter

38

LOW ALLOY STEEL FILLER METALS FOR GAS SHIELDED ARC WELDING

CHEMICAL COMPOSITION REQUIREMENTS FOR BARE SOLID ELECTRODES AND WELDING RODS[a]	
AWS A5.28 Classification[b]	**Composition, Weight Percent**[g,h]
Chromium-Molybdenum Steel Electrodes and Rods	
ER80S-B2	C 0.07-0.12 Mn 0.40-0.70 Si 0.40-0.70 P 0.025 S 0.025 Ni 0.20 Cr 1.20-1.50 Mo 0.40-0.65 Cu[c] 0.35 Total Other Elements[d] 0.50
ER80S-B2L	C 0.05 Mn 0.40-0.70 Si 0.40-0.70 P 0.025 S 0.025 Ni 0.20 Cr 1.20-1.50 Mo 0.40-0.65 Cu[c] 0.35 Total Other Elements[d] 0.50
ER90S-B3	C 0.07-0.12 Mn 0.40-0.70 Si 0.40-0.70 P 0.025 S 0.025 Ni 0.20 Cr 2.30-2.70 Mo 0.90-1.20 Cu[c] 0.35 Total Other Elements[d] 0.50
ER90S-B3L	C 0.05 Mn 0.40-0.70 Si 0.40-0.70 P 0.025 S 0.025 Ni 0.20 Cr 2.30-2.70 Mo 0.90-1.20 Cu[c] 0.35 Total Other Elements[d] 0.50
Nickel Steel Electrodes and Rods	
ER80S-Ni1	C 0.12 Mn 1.25 Si 0.40-0.80 P 0.025 S 0.025 Ni 0.80-1.10 Cr 0.15 Mo 0.35 V 0.05 Cu[c] 0.35 Total Other Elements[d] 0.50
ER80S-Ni2	C 0.12 Mn 1.25 Si 0.40-0.80 P 0.025 S 0.025 Ni 2.00-2.75 Cu[c] 0.35 Total Other Elements[d] 0.50
ER80S-Ni3	C 0.12 Mn 1.25 Si 0.40-0.80 P 0.025 S 0.025 Ni 3.00-3.75 Cu[c] 0.35 Total Other Elements[d] 0.50
Manganese-Molybdenum Steel Electrodes and Rods	
ER80S-D2[e]	C 0.07-0.12 Mn 1.60-2.10 Si 0.50-0.80 P 0.025 S 0.025 Ni 0.15 Mo 0.40-0.60 Cu[c] 0.50 Total Other Elements[d] 0.50
Other Low Alloy Steel Electrodes and Rods	
ER100S-1	C 0.08 Mn 1.25-1.80 Si 0.20-0.50 P 0.010 S 0.010 Ni 1.40-2.10 Cr 0.30 Mo 0.25-0.55 V 0.05 Ti 0.10 Zr 0.10 Al 0.10 Cu[c] 0.25 Total Other Elements[d] 0.50
ER100S-2	C 0.12 Mn 1.25-1.80 Si 0.20-0.60 P 0.010 S 0.010 Ni 0.80-1.25 Cr 0.30 Mo 0.20-0.55 Ti 0.10 Zr 0.10 Al 0.10 Cu[c] 0.35-0.65 Total Other Elements[d] 0.50
ER110S-1	C 0.09 Mn 1.40-1.80 Si 0.20-0.55 P 0.010 S 0.010 Ni 1.90-2.60 Cr 0.50 Mo 0.25-0.55 V 0.04 Ti 0.10 Zr 0.10 Al 0.10 Cu[c] 0.25 Total Other Elements[d] 0.50
ER120S-1	C 0.10 Mn 1.40-1.80 Si 0.25-0.60 P 0.010 S 0.010 Ni 2.00-2.80 Cr 0.60 Mo 0.30-0.65 V 0.03 Ti 0.10 Zr 0.10 Al 0.10 Cu[c] 0.25 Total Other Elements[d] 0.50
ERXXS-G	Subject to agreement between supplier and purchaser[f]

a. Chemical requirements for solid electrodes are based on as-manufactured composition.   b. The suffixes, B2, Ni1, etc., designate the chemical com the electrode and rod classification.   c. The maximum weight percent of copper in the rod or electrode due to any coating plus the residual copper cc steel shall comply with the stated value. d. Other elements, if intentionally added, shall be reported.   e. This composition was formerly classified E70S Specification A5.18-69. f. In order to meet the requirements of the G classification, the electrode must have as a minimum, one of either 0.50 percer percent chromium, or 0.20 percent molybdenum.   g. Single values shown are maximum.   h. Analysis shall be made for the elements for which spec shown in this table. If, however, the presence of other elements is indicated in the course of routine analysis, further analysis shall be made to detr total of these other elements, except iron, is not present in excess of the limits specified for "Total Other Elements" in this table.

CHEMICAL COMPOSITION REQUIREMENTS FOR DEPOSITED WELD METAL WITH COMPOSITE METAL CORED AND STRANDED ELECTRODES[a, g]	
AWS A5.28 Classification[b]	Composition, Weight Percent[f]
Chromium–Molybdenum	
E80C-B2L	C 0.05 Mn 0.40-1.00 Si 0.25-0.60 P 0.025 S 0.030 Ni 0.20 Cr 1.00-1.50 Mo 0.40-0.65 Cu[c] 0.35 Total Other Elements[d] 0.50
E80C-B2	C 0.07-0.12 Mn 0.40-1.00 Si 0.25-0.60 P 0.025 S 0.030 Ni 0.20 Cr 1.00-1.50 Mo 0.40-0.65 Cu[c] 0.35 Total Other Elements[d] 0.50
E90C-B3L	C 0.05 Mn 0.40-1.00 Si 0.25-0.60 P 0.025 S 0.030 Ni 0.20 Cr 2.00-2.50 Mo 0.90-1.20 Cu[c] 0.35 Total Other Elements[d] 0.50
E90C-B3	C 0.07-0.12 Mn 0.40-1.00 Si 0.25-0.60 P 0.025 S 0.030 Ni 0.20 Cr 2.00-2.50 Mo 2.00-2.50 Cu[c] 0.35 Total Other Elements[d] 0.50
Nickel Steel	
E80C-Ni1	C 0.12 Mn 1.25 Si 0.60 P 0.025 S 0.030 Ni 0.80-1.10 Mo 0.65 V 0.05 Cu[c] 0.35 Total Other Elements[d] 0.50
E80C-Ni2	C 0.12 Mn 1.25 Si 0.60 P 0.025 S 0.030 Ni 2.00-2.75 Cu[c] 0.35 Total Other Elements[d] 0.50
E80C-Ni3	C 0.12 Mn 1.25 Si 0.60 P 0.025 S 0.030 Ni 3.00-3.75 Cu[c] 0.35 Total Other Elements[d] 0.50
Other Low Alloy	
EXXC-G	Subject to agreement between supplier and purchaser[e]

a. Chemical requirements for composite electrodes are based on analysis of their weld metal in the as-welded condition and using the shielding gas specified in AWS A5.28 Table 4 (also see AWS A5.28 paragraph 3.3).

b. The suffixes B2, Ni1, etc., designate the chemical composition of the electrode classification.

c. The maximum weight percent of copper in the electrode due to any coating plus the residual copper content in the steel shall comply with the stated value.

d. Other elements, if intentionally added, shall be reported.

e. In order to meet the requirements of the G classification, the electrode must have as a minimum, one if either 0.50 percent nickel, 0.30 percent chromium, or 0.20 percent molybdenum.

f. Single values shown are maximum.

g. Composite electrodes are not recommended for gas tungsten arc welding (GTAW) or plasma arc welding (PAW).

TENSILE STRENGTH, YIELD STRENGTH, AND ELONGATION REQUIREMENTS FOR ALL-WELD-METAL TENSION TEST[d]

AWS A5.28 Classification	Shielding Gas	Current and Polarity[e]	Tensile Strength, min.		Yield Strength at 0.2% Offset, min.		% Elongation in 2. in. (50 mm), min	Condition
			ksi	MPa	ksi	MPa		
ER80S-B2, ER80S-B2L, E80C-B2, E80C-B2L	Ar plus 1-5% O_2	dcep	80	550	68	470	19	PWHT[a]
ER90S-B3, ER90S-B3L, E90C-B3, E90C-B3L	Ar plus 1-5% O_2	dcep	90	620	78	540	17	PWHT[a]
ER80S-Ni1	Ar plus 1-5% O_2	dcep	80	550	68	470	24	As-welded
ER80S-Ni2, ER80S-Ni3	Ar plus 1-5% O_2	dcep	80	550	68	470	24	PWHT[a]
E80C-Ni1	Ar plus 1-5% O_2	dcep	80	550	68	470	24	As-welded
E80C-Ni2, E80C-Ni3	Ar plus 1-5% O_2	dcep	80	550	68	470	24	PWHT[a]
ER80S-D2	CO_2	dcep	80	550	68	470	17	As-welded
ER100S-1, ER100S-2	Ar plus 2% O_2	dcep	100	690	88 to 102	610 to 700	16	As-welded
ER110S-1	Ar plus 2% O_2	dcep	110	760	95 to 107	660 to 740	15	As-welded
ER120S-1	Ar plus 2% O_2	dcep	120	830	102 to 122	730 to 840	14	As-welded
ERXXS-G, EXXC-G	(c)	dcep	(b)		(c)		(c)	As-welded

a.   Postweld heat treated in accordance with AWS A5.28 Table 12.

b.   Tensile strength shall be consistent with the level placed after the "ER" or "E" prefix; e.g. ER90S-G shall have 90,000 psi minimum ultimate tensile strength.

c.   Subject to agreement between supplier and purchaser.

d.   The values stated in U.S. customary units are to be regarded as the standard. The SI units are given as equivalent values to the US customary units. The published sizes and dimensions in the two systems are not identical and for this reason conversion from a published size or dimension in one system will not always coincide with the published size or dimension in the other. Suitable conversions, encompassing published sizes of both, can be made, however, if appropriate tolerances are applied in each case.

e.   dcep means direct current electrode positive.

IMPACT PROPERTIES[e]

AWS A5.28 Classification	Minimum Required Impact Properties
ER80S-B2, ER80S-B2L, ER90S-B3, ER90S-B3L	Not required
ER80S-Ni1	20 ft•lbf at -50°F[a,b] (27 J at -46°C)[a,b]
ER80S-Ni2	20 ft•lbf at -80°F[a,b] (27 J at -62°C)[a,c]
ER80S-Ni3	20 ft•lbf at -100°F[a,b] (27 J at -73°C)[a,c]
ER80S-D2	20 ft•lbf at -20°F[a,b] (27 J at -29°C)[a,b]
ER100S-1, ER100S-2, ER110S-1, ER120S-1	50 ft•lbf at -60°F[d] (68 J at -51°C)[d]
E80C-B2L, E80C-B2, E90C-B3L, E90C-B3	Not required
E80C-Ni1	20 ft•lbf at -50°F[a,b] (27 J at -46°C)[a,b]
E80C-Ni2	20 ft•lbf at -80°F[a,c] (27 J at -62°C)[a,c]
E80C-Ni3	20 ft•lbf at -100°F[a,c] (27 J at -73°C)[a,c]
ERXXS-G, EXXC-G	As agreed between supplier and purchaser

a. The lowest and the highest values obtained shall be disregarded for this test. Two of the three remaining values shall be greater than the specified 20 ft•lbf (27 J) energy level; one of the three may be lower but shall not be less than 15 ft•lbf (20 J). The computed average value of the three values shall be equal to or greater than the 20 ft•lbf (27 J) energy level.

b. As-welded impact properties.

c. Required impact properties after postweld heat treatment. (See AWS A5.28 Table 12 for more details).

d. Impact properties for the ER100S-1, ER100S-2, ER110S-1, and ER120S shall be obtained at test temperature specified in the table, ± 3°F (1.7°C). The lowest and the highest impact values thus obtained shall be disregarded for this test. Two of the three remaining values shall be greater than the specified 50 ft•lbf (68 J) energy level; one of the three may be lower but shall not be less than 40 ft•lbf (54 J); the computed average values of these three shall be greater than the specified 50 ft•lbf (68 J) level.

e. The values stated in U.S. customary units are to be regarded as the standard. The SI units are given as equivalent values to the US customary units. The published sizes and dimensions in the two systems are not identical and for this reason conversion from a published size or dimension in one system will not always coincide with the published size or dimension in the other. Suitable conversions, encompassing published sizes of both, can be made, however, if appropriate tolerances are applied in each case.

AWS A5.28 Classification System

The classification system used in this specification follows as closely as possible the standard pattern used in other AWS filler metal specifications. The inherent nature of the products being classified have, however, necessitated specific changes which more ably classify the product.

As an example, consider ER80S-B2 and E80C-B2. The prefix E designates an electrode, as in other specifications. The letters ER at the beginning of a classification indicate that the filler metal may be used as an electrode or welding rod. The number 80 indicates the required minimum tensile strength in multiples of 1000 psi (6.9 MPa) of the weld metal in a test weld made using the electrode in accordance with specified welding conditions. Three digits are used for weld metal of 100,000 psi (690 MPa) tensile strength and higher. The letter S designates a bare solid electrode or rod. The letter C designates a composite metal cored or stranded electrode. The suffix B2 indicates a particular classification based on as-manufactured chemical composition.

(Optional) At the option and expense of the purchaser, acceptance may be based on the results of any or all of the tests required by this specification made on GTAW test assembly described in AWS A5.28 Figure A1 with tension specimen as described in AWS A5.28 Figure A2 (and the impact specimen described in AWS A5.28 Figure 7). Composite electrodes are not recommended for GTAW or PAW.

Chapter
39

LOW ALLOY STEEL ELECTRODES FOR FLUX CORED ARC WELDING

CHEMICAL REQUIREMENTS FOR DEPOSITED WELD METAL[a]

AWS A5.29 Classification	Composition, Weight Percent
Carbon-Molybdenum Steel Electrodes	
E70T5-A1, E80T1-A1, E81T1-A1	C 0.12 Mn 1.25 P 0.03 S 0.03 Si 0.80 Mo 0.40/0.65
Chromium-Molybdenum Steel Electrodes	
E81T1-B1	C 0.12 Mn 1.25 P 0.03 S 0.03 Si 0.80 Cr 0.40/0.65 Mo 0.40/0.65
E80T5-B2L	C 0.05 Mn 1.25 P 0.03 S 0.03 Si 0.80 Cr 1.00/1.50 Mo 0.40/0.65
E80T1-B2, E81T1-B2, E80T5-B2	C 0.12 Mn 1.25 P 0.03 S 0.03 Si 0.80 Cr 1.00/1.50 Mo 0.40/0.65
E80T1-B2H	C 0.10/0.15 Mn 1.25 P 0.03 S 0.03 Si 0.80 Cr 1.00/1.50 Mo 0.40/0.65
E90T1-B3L	C 0.05 Mn 1.25 P 0.03 S 0.03 Si 0.80 Cr 2.00/2.50 Mo 0.90/1.20
E90T1-B3, E91T1-B3, E90T5-B3, E100T1-B3	C 0.12 Mn 1.25 P 0.03 S 0.03 Si 0.80 Cr 2.00/2.50 Mo 0.90/1.20
E90T1-B3H	C 0.10/0.15 Mn 1.25 P 0.03 S 0.03 Si 0.80 Cr 2.00/2.50 Mo 0.90/1.20
Nickel-Steel Electrodes	
E71T8-Ni1, E80T1-Ni1, E81T1-Ni1, E80T5-Ni1	C 0.12 Mn 1.50 P 0.03 S 0.03 Si 0.80 Ni 0.80/1.10 Cr 0.15 Mo 0.35 V 0.05 Al[b] 1.8
E71T8-Ni2, E80T1-Ni2, E81T1-Ni2, E80T5-Ni2, E90T1-Ni2, E91T1-N2	C 0.12 Mn 1.50 P 0.03 S 0.03 Si 0.80 Ni 1.75/2.75 Al[b] 1.8
E80T5-Ni3, E90T5-Ni3	C 0.12 Mn 1.50 P 0.03 S 0.03 Si 0.80 Ni 2.75/3.75
Manganese-Molybdenum Steel Electrodes	
E91T1-D1	C 0.12 Mn 1.25/2.00 P 0.03 S 0.03 Si 0.80 Mo 0.25/0.55
E90T5-D2, E100T5-D2	C 0.15 Mn 1.65/2.25 P 0.03 S 0.03 Si 0.80 Mo 0.25 /0.55
E90T1-D3	C 0.12 Mn 1.00/1.75 P 0.03 S 0.03 Si 0.80 Mo 0.40/0.65
All Other Low Alloy Steel Electrodes	
E80T5-K1	C 0.15 Mn 0.80/1.40 P 0.03 S 0.03 Si 0.80 Ni 0.80/1.10 Cr 0.15 Mo 0.20/0.65 V 0.05
E70T4-K2, E71T8-K2, E80T1-K2, E90T1-K2, E91T1-K2, E80T5-K2, E90T5-K2	C 0.15 Mn 0.50/1.75 P 0.03 S 0.03 Si 0.80 Ni 1.00/2.00 Cr 0.15 Mo 0.35 V 0.05 Al[b] 1.8

CHEMICAL REQUIREMENTS FOR DEPOSITED WELD METAL[a] (Continued)

AWS A5.29 Classification	Composition, Weight Percent
All Other Low Alloy Steel Electrodes (Continued)	
E100T1-K3, E110T1-K3, E100T5-K3, E110T5-K3	C 0.15 Mn 0.75/2.25 P 0.03 S 0.03 Si 0.80 Ni 1.25/2.60 Cr 0.15 Mo 0.25/0.65 V 0.05
E110T5-K4, E111T1-K4, E120T5-K4	C 0.15 Mn 1.20/2.25 P 0.03 S 0.03 Si 0.80 Ni 1.75/2.60 Cr 0.20/0.60 Mo 0.30/0.65 V 0.05
E120T1-K5	C 0.10/0.25 Mn 0.60/1.60 P 0.03 S 0.03 Si 0.80 Ni 0.75/2.00 Cr 0.20/0.70 Mo 0.15/0.55 V 0.05
E61T8-K6, E71T8-K6	C 0.15 Mn 0.50/1.50 P 0.03 S 0.03 Si 0.80 Ni 0.40/1.10 Cr 0.15 M0 0.15 V 0.05 Al[b] 1.8
E101T1-K7	C 0.15 Mn 1.00/1.75 P 0.03 S 0.03 Si 0.80 Ni 2.00/2.75
EXXXTX-G	Mn 1.00 min[c] P 0.03 S 0.03 Si 0.80 min[c] Ni 0.50 min[c] Cr 0.30 min[c] Mo 0.20 min[c] V 0.10 min[c] Al[b] 1.8
E80T1-W	C 0.12 Mn 0.50/1.30 P 0.03 S 0.03 Si 0.35/0.80 Ni 0.40/0.80 Cr 0.45/0.70 Cu 0.30/0.75

a. Single values are maximum unless otherwise noted.
b. For self-shielded electrodes only.
c. In order to meet the alloy requirements of the G group, the weld deposit need have the minimum, as specified in the table, of only one of the elements listed.

MECHANICAL PROPERTY REQUIREMENTS[a]

AWS A5.29 Classification	Tensile Strength Range		Yield Strength at 0.2% Offset, min.		% Elongation in 2 in. (50 mm), min.
	ksi	MPa	ksi	MPa	
E6XTX-X	60 to 80	410 to 550	50	340	22
E7XTX-X	70 to 90	490 to 620	58	400	20
E8XTX-X	80 to 100	550 to 690	68	470	19
E9XTX-X	90 to 110	620 to 760	78	540	17
E10XTX-X	100 to 120	690 to 830	88	610	16
E11XTX-X	110 to 130	760 to 900	98	680	15
E12XTX-X	120 to 140	830 to 970	108	750	14
EXXXTX-G	Properties as agreed upon between supplier and purchaser				

a. Properties of electrodes that use external gas shielding (EXXT1-X and EXXT5-X) vary with gas mixtures. Electrodes classified under this specification should not be used with gases other than those listed in AWS A5.29 Table 9 without first consulting the manufacturer.

The Metals Blue Book

IMPACT REQUIREMENTS

AWS A5.29 Classification	Condition[a]	Impact Strength
E80T1-A1	PWHT	Not required
E81T1-A1	PWHT	Not required
E70T5-A1	PWHT	20 ft•lbf at -20°F (27 J at -30°C)
E81T1-B1	PWHT	Not required
E81T1-B2	PWHT	Not required
E80T1-B2	PWHT	Not required
E80T5-B2	PWHT	Not required
E80T1-B2H	PWHT	Not required
E80T5-B2L	PWHT	Not required
E90T1-B3	PWHT	Not required
E91T1-B3	PWHT	Not required
E90T5-B3	PWHT	Not required
E100T1-B3	PWHT	Not required
E90T1-B3L	PWHT	Not required
E90T1-B3H	PWHT	Not required
E71T8-Ni1	AW	20 ft•lbf at -20°F (27 J at -30°C)
E80T1-Ni1	AW	20 ft•lbf at -20°F (27 J at -30°C)
E81T1-Ni1	AW	20 ft•lbf at -20°F (27 J at -30°C)
E80T5-Ni1	PWHT	20 ft•lbf at -60°F (27 J at -51°C)
E71T8-Ni2	AW	20 ft•lbf at -20°F (27 J at -30°C)
E80T1-Ni2	AW	20 ft•lbf at -40°F (27 J at -40°C)
E81T1-Ni2	AW	20 ft•lbf at -40°F (27 J at -40°C)
E80T5-Ni2[b]	PWHT	20 ft•lbf at -75°F (27 J at -60°C)
E90T1-Ni2	AW	20 ft•lbf at -40°F (27 J at -40°C)
E91T1-Ni2	AW	20 ft•lbf at -40°F (27 J at -40°C)
E80T5-Ni3[b]	PWHT	20 ft•lbf at -100°F (27 J at -73°C)
E90T5-Ni3[b]	PWHT	20 ft•lbf at -100°F (27 J at -73°C)
E91T1-D1	AW	20 ft•lbf at -40°F (27 J at -40°C)

IMPACT REQUIREMENTS (Continued)

AWS A5.29 Classification	Condition[a]	Impact Strength
E90T5-D2	PWHT	20 ft•lbf at -60°F (27J at -51°C)
E100T5-D2	PWHT	20 ft•lbf at -40°F (27J at -40°C)
E90T1-D3	AW	20 ft•lbf at -20°F (27J at -30°C)
E80T5-K1	AW	20 ft•lbf at -40°F (27J at -40°C)
E70T4-K2	AW	20 ft•lbf at 0°F (27J at -18°C)
E71T8-K2	AW	20 ft•lbf at -20°F (27J at -30°C)
E80T1-K2	AW	20 ft•lbf at -20°F (27J at -30°C)
E90T1-K2	AW	20 ft•lbf at 0°F (27J at -18°C)
E91T1-K2	AW	20 ft•lbf at 0°F (27J at -18°C)
E80T5-K2	AW	20 ft•lbf at -20°F (27J at -30°C)
E90T5-K2	AW	20 ft•lbf at -60°F (27J at -51°C)
E100T1-K3	AW	20 ft•lbf at 0°F (27J at -18°C)
E110T1-K3	AW	20 ft•lbf at 0°F (27J at -18°C)
E100T5-K3	AW	20 ft•lbf at -60°F (27J at -51°C)
E110T5-K3	AW	20 ft•lbf at -60°F (27J at -51°C)
E110T5-K4	AW	20 ft•lbf at -60°F (27J at -51°C)
E111T1-K4	AW	20 ft•lbf at -60°F (27J at -51°C)
E120T5-K4	AW	20 ft•lbf at -60°F (27J at -51°C)
E120T1-K5	AW	Not required
E61T8-K6	AW	20 ft•lbf at -20°F (27J at -30°C)
E71T8-K6	AW	20 ft•lbf at -20°F (27J at -30°C)
E101T1-K7	AW	20 ft•lbf at -60°F (27J at -51°C)
E80T1-W	AW	20 ft•lbf at -20°F (27J at -30°C)
EXXXTX-G	Properties as agreed upon between supplier and purchaser

a.   AW = As welded; PWHT = Postweld heat treated in accordance with AWS A5.29 paragraph 3.6.
b.   PWHT temperatures in excess of 1150°F (620°C) will decrease the impact value.

SHIELDING AND POLARITY

AWS A5.29 Classification[a]	External Shielding Medium[b]	Current and Polarity
EXXT1-X (multiple pass)	CO_2	DC, electrode positive
EXXT4-X (multiple pass)	None	DC, electrode positive
EXXT5-X (multiple pass)	CO_2	DC, electrode positive
EXXT8-X (multiple pass)	None	DC, electrode positive
EXXTX-G (multiple pass)	Not Specified	Not Specified

a. The "X" indicates the tensile strength, usability, and chemical composition designations (see also AWS A5.29 paragraph A2.2).

b. Some T1 and T5 electrodes may be classified with other than CO_2 shielding gas upon agreement between supplier and purchaser. See footnote of AWS A5.29 Table 3 and Appendix paragraph A5.5.1.

AWS A5.29 CLASSIFICATION SYSTEM

The classification system used in this specification follows as closely as possible the standard pattern used in other AWS filler metal specifications. The inherent nature of the products being classified has, however, necessitated specific changes that more suitably classify the product.

A generic example of the method of classification of electrodes is shown below.

EXXTX-X

E   Designates an electrode.

First X   Indicates the minimum tensile strength of the deposited weld metal in a test weld made with the electrode and in accordance with specified welding conditions.

Second X   Indicates the primary welding position for which the electrode is designed:

0 - flat and horizontal positions

1 - all positions

T   Indicates a flux cored electrode.

Third X   Indicates usability and performance capabilities

Fourth X   Designates the chemical composition of the deposited weld metal (see AWS A5.29 Table 1). Specific chemical compositions are not always identified with specific mechanical properties in the specification. A supplier is required by the specification to include the mechanical properties appropriate for a particular electrode in classification of that electrode. Thus, for example, a complete designation is E80T5-Ni3; EXXT5-Ni3 is not a complete classification.

AWS A5.29 CLASSIFICATION SYSTEM (Continued)

Note: The letter "X" as used in this example and in electrode classification designations in AWS A5.29 substitutes for specific designations indicated by this example.

Some products may be designed for the flat and horizontal positions regardless of size. Others may be designed for out-of-position welding in the smaller sizes and flat and horizontal positions in the larger sizes. The mandatory section of this specification allows dual classification for the primary weld positions on the latter types.

Electrode Classification

Electrodes of the T1 group are classified with CO_2 shielding gas by this specification. However, gas mixtures of argon-CO_2 may be used where recommended by the manufacturer to improve usability, especially for out-of-position applications. These electrodes are designed for single- and multiple-pass welding. The larger diameters (usually 5/64 in. [2.0 mm] and larger) are used for welding in the flat position and for horizontal fillets. The smaller diameters (usually 1/16 in. [1.6 mm] and smaller) are used for welding in all positions. The T1 electrodes are characterized by a spray transfer, low spatter loss, flat to slightly convex bead configuration, and a moderate volume of slag, which completely covers the weld bead. Most electrodes in this group have a rutile base slag.

Chapter
40

CONSUMABLE INSERTS

CHEMICAL COMPOSITION PERCENT - MILD STEEL[a]

	AWS A5.30 Classification	Composition, Weight Percent
Group A	INMs1	C 0.07 Mn 0.90 to 1.40 P 0.025 S 0.035 Si 0.40 to 0.70 Al 0.05[b] to 0.15 Zr 0.02[b] to 0.12 Ti 0.05[b] to 0.15
	INMs2	C 0.06 to 0.15 Mn 0.90 to 1.40 P 0.025 S 0.035 Si 0.45 to 0.70
	INMs3	C 0.07 to 0.15 Mn 1.40 to 1.85 P 0.025 S 0.035 Si 0.80 to 1.15

a.　Single values are maximum percentages, except where otherwise specified.

b.　Al + Zr + Ti = 0.15 minimum.

CHEMICAL COMPOSITION - CHROMIUM MOLYBDENUM STEELS[a]

	AWS A5.30 Classification	Composition, Weight Percent
Group B	IN502	C 0.10 Mn 0.40 to 0.75 P 0.025 S 0.025 Si 0.25 to 0.50 Al 0.15 Cr 4.5 to 6.0 Mo 0.45 to 0.65
	IN515	C 0.12 Mn 0.40 to 0.70 P 0.025 S 0.025 Si 0.40 to 0.70 Al 0.15 Cr 1.20 to 1.50 Mo 0.40 to 0.65 Ni 0.20 Cu 0.35 Total Other Elements 0.50
	IN521	C 0.12 Mn 0.40 to 0.70 P 0.025 S 0.025 Si 0.40 to 0.70 Al 0.15 Cr 2.30 to 2.70 Mo 0.90 to 1.20 Ni 0.20 Cu 0.35 Total Other Elements 0.50

a.　Singles values are maximum percentages, except where otherwise specified.

CHEMICAL COMPOSITION - AUSTENITIC CHROMIUM NICKEL STEELS[a]

AWS A5.30 Classification	Composition, Weight Percent
Group C	
IN308[b,d]	C 0.08 Cr 19.5 to 22.0 Ni 9.0 to 11.0 Mn 1.0 to 2.5 Si 0.25 to 0.60 P 0.03 S 0.03
IN308L[b,d]	C 0.03 Cr 19.5 to 22.0 Ni 9.0 to 11.0 Mn 1.0 to 2.5 Si 0.25 to 0.60 P 0.03 S 0.03
IN310	C 0.08 to 0.15 Cr 25.0 to 28.0 Ni 20.0 to 22.5 Mn 1.0 to 2.5 Si 0.25 to 0.60 P 0.03 S 0.03
IN312[b,d]	C 0.08 to 0.15 Cr 28.0 to 32.0 Ni 8.0 to 10.5 Mn 1.0 to 2.5 Si 0.25 to 0.60 P 0.03 S 0.03
IN316	C 0.08 Cr 18.0 to 20.0 Ni 11.0 to 14.0 Mo 2.0 to 3.0 Mn 1.0 to 2.5 Si 0.25 to 0.60 P 0.03 S 0.03
IN316L	C 0.03 Cr 18.0 to 20.0 Ni 11.0 to 14.0 Mo 2.0 to 3.0 Mn 1.0 to 2.5 Si 0.25 to 0.60 P 0.03 S 0.03
IN348[b,d]	C 0.08 Cr 19.0 to 21.5 Ni 9.0 to 11.0 Cb + Ta 10 X C min[c] to 1.0 max Mn 1.0 to 2.5 Si 0.25 to 0.60 P 0.03 S 0.03

a. Single values shown are maximum percentages, except where otherwise specified. b. Delta ferrite may be specified upon agreement between supplier and purchaser. c. Tantalum content shall not exceed 0.10 percent. d. The cobalt content shall not exceed 0.10 percent.

CHEMICAL COMPOSITION - NICKEL ALLOYS (INCLUDES COPPER-NICKEL)[a,b]

AWS A5.30 Classification	Composition, Weight Percent
Group E	
IN60	C 0.15 Mn 4.00 Fe 2.50 S 0.015 P 0.020 Si 1.25 Cu Balance Ni + Co 62.0 to 69.0 Al 1.25 Ti 1.5 to 3.0 Other Elements Total 0.50
IN61	C 0.15 Mn 1.00 Fe 1.00 S 0.015 P 0.030 Si 0.75 Cu 0.25 Ni + Co 93 min Al 1.50 Ti 2.0 to 3.5 Other Elements Total 0.50
IN62	C 0.08 Mn 1.00 Fe 6.0 to 10.0 S 0.015 P 0.030 Si 0.35 Cu 0.50 Ni + Co 70 min[c] Cr 14.0 to 17.0 Cb + Ta 1.5 to 3.0[d] Other Elements Total 0.50
IN67	Mn 1.00 Fe 0.40 to 0.75 S 0.01 P 0.020 Si 0.15 Cu Balance Ni + Co 29.0 to 32.0 Ti 0.2 to 0.5 Other Elements Total 0.50
IN6A	C 0.08 Mn 2.00 to 2.75 Fe 8.00 S 0.015 P 0.030 Si 0.35 Cu 0.50 Ni + Co 67.0 min[c] Ti 2.5 to 3.5 Cr 14.0 to 17.0
IN82	C 0.10 Mn 2.50 to 3.50 Fe 3.00 S 0.015 P 0.030 Si 0.50 Cu 0.50 Ni + Co 67.0 min[c] Ti 0.75 Cr 18.0 to 22.0 Cb + Ta 2.0 to 3.0[d] Other Elements Total 0.50

a. Analysis shall be made for the elements for which specific values are shown. If, however, the presence of other elements is indicated in the course of routine analysis, further analysis shall be made to determine that the total of these other elements is not present in excess of the limits specified for "Other Elements Total". b. Single values shown are maximum percentages, except where otherwise specified. c. Cobalt shall not exceed 0.10 percent. d. Tantalum shall not exceed 0.30 percent.

The Metals Blue Book

CLASS 3, STYLE D INSERTS - DIMENSIONS

| Nominal Diameter | | Pipe Dimensions | | | | Ring Diameters[b] | | | |
| | | Schedule Number[a] | ID[b] | | Ring OD for Nominal Pipe Dia. | | Ring ID for Nominal Pipe Dia. | | |
in.	mm		in.	mm	in.	mm	in.	mm
2	51	10S	2.157	54.78	2.43	61.7	2.06	52.3
		40	2.067	52.50	2.34	59.4	1.97	50.0
		80	1.939	49.25	2.22	56.4	1.85	47.0
2-1/2	64	10S	2.635	66.93	2.91	73.9	2.54	64.5
		40	2.469	62.71	2.75	69.9	2.38	60.4
		80	2.323	59.00	2.60	66.0	2.23	56.6
3	76	10S	3.260	82.80	3.54	88.9	3.17	80.5
		40	3.068	77.93	3.35	85.1	2.98	75.7
		80	2.900	73.66	3.18	80.8	2.81	71.4
3-1/2	89	10S	3.760	95.50	4.04	102.6	3.67	93.2
		40	3.548	90.12	3.82	97.0	3.45	87.6
		80	3.364	85.45	3.64	92.5	3.27	83.1
4	102	10S	4.260	108.20	4.54	115.3	4.17	105.9
		40	4.026	102.26	4.30	109.2	3.93	99.8
		80	3.826	97.18	4.10	104.1	3.73	94.7
5	127	5S	5.345	135.76	5.62	142.7	5.25	133.4
		10S	5.295	134.79	5.57	141.5	5.20	132.1
		40	5.047	128.19	5.32	135.1	4.95	125.7
		80	4.813	122.25	5.09	129.3	4.72	119.9
6	152	5S	6.407	162.74	6.68	169.7	6.31	160.3
		10S	6.357	161.47	6.63	168.4	6.26	159.0
		40	6.065	154.05	6.34	161.0	5.97	151.6
		80	5.761	146.33	6.04	153.4	5.67	144.0
8	203	5S	8.407	213.54	8.68	220.4	8.31	211.1
		10S	8.329	211.56	8.61	218.7	8.24	209.3
		40	7.981	202.72	8.26	209.8	7.89	200.4
		80	7.625	193.68	7.90	200.7	7.53	191.3

CLASS 3, STYLE D INSERTS - DIMENSIONS (Continued)

| Pipe Dimensions | | | | | Ring Diameters[b] | | | |
| Nominal Diameter | | Schedule | ID[b] | | Ring OD for Nominal Pipe Dia. | | Ring ID for Nominal Pipe Dia. | |
in.	mm	Number[a]	in.	mm	in.	mm	in.	mm
10	254	5S	10.482	266.24	10.76	273.3	10.39	263.9
		10S	10.420	264.67	10.70	271.8	10.33	262.4
		40	10.020	254.51	10.30	261.6	9.93	252.2
		80S	9.750	247.65	10.03	254.8	9.66	245.4
		80	9.564	242.93	9.84	249.9	9.47	240.5
12	305	5S	12.420	315.47	12.70	322.6	12.33	313.2
		10S	12.390	314.71	12.67	321.8	12.30	312.4
		40S	12.000	304.80	12.28	311.9	11.91	302.5
		40	11.938	303.22	12.22	310.4	11.85	301.0
		80S	11.750	298.45	12.03	305.6	11.66	296.2
		80	11.376	288.95	11.65	295.9	11.28	286.5
14	356	10	13.500	342.90	13.78	350.0	13.41	340.6
		40	13.126	333.40	13.40	340.4	13.03	331.0
		80	12.500	317.50	12.78	324.6	12.41	315.2
16	406	10	15.500	393.70	15.78	400.8	15.41	391.4
		40	15.000	381.00	15.28	388.1	14.91	378.7
		80	14.314	363.58	14.59	370.6	14.22	361.2

a.   Schedule number followed by "S" applied to Group C types.

b.   Class 3, style D insert rings are normally furnished to match nominal internal diameters of the pipe. Special ring sizes (non-standard) shall be obtainable when specified by contract or purchase order. The dimensions of such rings are determined as follows:

Ring ID (inside diameter) = pipe ID - 0.094 in. (2.38 mm)

Ring OD (outside diameter) = ring ID + 0.375 in. (9.53 mm)

Where special applications demand, rings have 1/8 in. (3.2 mm) instead of the 3/16 in. (4.8 mm) dimension shown on AWS A5.30 Figure 1, the narrower width (style E) shall be made available as shown in AWS A5.30 Table 8. To determine the required ID and OD dimensions for the narrower type, substitute 0.250 in. (6.35 mm) for 0.375 in. (9.53 mm) in the above equations.

CLASS 3, STYLE E INSERTS - DIMENSIONS

Nominal Diameter		Pipe Dimensions	ID[b]		Ring Diameters[b]			
		Schedule			Ring OD for Nominal Pipe Dia.		Ring ID for Nominal Pipe Dia.	
in.	mm	Number[a]	in.	mm	in.	mm	in.	mm
1/4	6.4	10S	0.410	10.41	0.57	14.5	0.32	8.1
		40	0.364	9.25	0.52	13.2	0.27	6.8
		80	0.302	7.67	0.46	11.7	0.21	5.3
3/8	9.5	10S	0.545	13.84	0.70	17.8	0.45	11.4
		40	0.493	12.52	0.65	16.5	0.40	10.1
		80	0.423	10.74	0.58	14.7	0.33	8.4
1/2	12.7	5S	0.710	18.03	0.87	22.1	0.62	15.7
		10S	0.674	17.12	0.83	21.1	0.58	14.7
		40	0.622	15.80	0.78	19.8	0.53	13.5
		80	0.546	13.87	0.70	17.8	0.45	11.4
3/4	19.1	5S	0.920	23.37	1.08	27.4	0.83	21.1
		10S	0.884	22.45	1.04	26.4	0.79	20.1
		40	0.824	20.93	0.98	24.9	0.73	18.5
		80	0.742	18.85	0.90	22.9	0.65	16.5
1	25.0	5S	1.185	30.10	1.34	34.0	1.09	27.7
		10S	1.097	27.86	1.25	31.8	1.00	25.4
		40	1.049	26.64	1.21	30.7	0.96	24.4
		80	0.957	24.31	1.11	28.2	0.86	21.8
1-1/4	32.0	5S	1.530	38.86	1.69	42.9	1.44	36.6
		10S	1.442	36.63	1.60	40.6	1.35	34.3
		40	1.380	35.05	1.54	39.1	1.29	32.8
		80	1.278	32.46	1.43	36.3	1.18	30.0
1-1/2	38.0	5S	1.770	44.96	1.93	49.0	1.68	42.7
		10S	1.682	42.72	1.84	46.7	1.59	40.4
		40	1.610	40.89	1.77	45.0	1.52	38.6
		80	1.500	38.10	1.65	41.9	1.41	35.8

CLASS 3, STYLE E INSERTS - DIMENSIONS (Continued)

					Ring Diameters[b]			
Pipe Dimensions					Ring OD for Nominal Pipe Dia.		Ring ID for Nominal Pipe Dia.	
Nominal Diameter		Schedule	ID[b]					
in.	mm	Number[a]	in.	mm	in.	mm	in.	mm
2	51	5S	2.245	57.02	2.40	61.0	2.15	54.6
		10S	2.157	54.79	2.30	58.7	2.06	52.3
		40	2.067	52.50	2.22	56.4	1.97	50.0
		80	1.939	49.25	2.10	53.3	1.85	47.0
2-1/2	64	5S	2.709	68.81	2.87	72.9	2.62	66.5
3	76	5S	3.334	84.68	3.49	88.6	3.24	82.3
3-1/2	89	5S	3.834	97.38	3.99	101.3	3.74	95.0
4	102	5S	4.334	110.08	4.49	114.0	4.24	107.7

a.   Schedule number followed by "S" applied to Group C types.

b.   Class 3, style D insert rings are normally furnished to match nominal internal diameters of the pipe. Special ring sizes (non-standard) shall be obtainable when specified by contract or purchase order. The dimensions of such rings are determined as follows:

Ring ID (inside diameter) = pipe ID - 0.094 in. (2.38 mm)

Ring OD (outside diameter) = ring ID + 0.375 in. (9.53 mm)

Where special applications demand, rings have 1/8 in. (3.2 mm) instead of the 3/16 in. (4.8 mm) dimension shown on AWS A5.30 Figure 1, the narrower width (style E) shall be made available as shown in AWS A5.30 Table 8. To determine the required ID and OD dimensions for the narrower type, substitute 0.250 in. (6.35 mm) for 0.375 in. (9.53 mm) in the above equations.

AWS A5.30 CLASSIFICATION SYSTEM

The classification system used in this specification follows as closely as possible the standard pattern used in other AWS filler metal specifications. The inherent nature of the products being classified have, however, necessitated specific changes that more ably classify the product. As an example, consider IN308. The prefix IN designates a consumable insert. The numbers 308 designate the chemical composition.

The solid products are classified on the basis of their chemical composition. However, their cross-sectional configurations are another consideration that must be selected and specified when ordering.

Chapter
41

FLUXES FOR BRAZING & BRAZE WELDING

CLASSIFICATION OF BRAZING FLUXES WITH BRAZING OR BRAZE WELDING FILLER MATERIALS[a,b]

AWS A5.31 Classification[c]	Form	Filler Metal Type	Activity Temperature Range	
			°F	°C
FB1-A	Powder	BAlSi	1080-1140	580-615
FB1-B	Powder	BAlSi	1040-1140	560-615
FB1-C	Powder	BAlSi	1000-1140	540-615
FB2-A	Powder	BMg	900-1150	480-620
FB3-A	Paste	BAg and BCuP	1050-1600	565-870
FB3-C	Paste	BAg and BCuP	1050-1700	565-925
FB3-D	Paste	BAg, BCu, BNi, BAu & RBCuZn	1400-2200	760-1205
FB3-E	Liquid	BAg and BCuP	1050-1600	565-870
FB3-F	Powder	BAg and BCuP	1200-1600	650-870
FB3-G	Slurry	BAg and BCuP	1050-1600	565-870
FB3-H	Slurry	BAg	1050-1700	565-925
FB3-I	Slurry	BAg, BCu, BNi, BAu & RBCuZn	1400-2200	760-1205
FB3-J	Powder	BAg, BCu, BNi, BAu & RBCuZn	1400-2200	760-1205
FB3-K	Liquid	BAg and RBCuZn	1400-2200	760-1205
FB4-A	Paste	BAg and BCuP	1100-1600	595-870

a. The selection of a flux designation for a specific type of work may be based on the form, the filler metal type, and the classification above, but the information here is generally not adequate for flux selection. Refer to AWS A5.31 Appendix A6 and the latest issue of the AWS Brazing Handbook for further assistance.

b. Difference between paste flux and slurry flux based on AWS A5.31 paragraphs 11.2 and 11.3, as follows:
   - Paste flux shall have a water content of 15 to 35 percent.
   - Slurry flux shall have a water content of 30 to 60 percent.

c. Flux 3B in the Brazing Manual, 3rd Edition, 1976 has been discontinued. Type 3B has been divided into types FB3-C and FB3-D.

Chapter 11

CONDITIONS FOR FLUXING ACTION, FLOW AND LIFE TESTS

AWS A5.31 Flux Classification	AWS A5.31 Filler Metal Classification	Base Metal		Test Temperature[a]		Flow Area	
		Common Name	UNS Number[b]	°F	°C	Sq. in.	Sq. mm
FB1-A	BAlSi-4	3003 aluminum	A93003	1135	613	0.2	129
FB1-B	BAlSi-4	3003 aluminum	A93003	1135	613	0.2	129
FB1-C	BAlSi-4	3003 aluminum	A93003	1135	613	NR[3]	NR[3]
FB2-A	BMg-1	AZ31B magnesium	M11311	1130	610	NR[3]	NR[3]
FB3-A	BAg-7	1008 carbon steel	G10080	1300	705	0.25	161
FB3-C	BAg-24	304 stainless steel	S30400	1400	760	0.25	161
FB3-D	RBCuZn-D, BNi-2	304 stainless steel	S30400	1850	1010	0.25	161
		304 stainless steel	S30400	1900	1040	0.25	161
FB3-E	BAg-7	304 stainless steel	S30400	1300	705	NR[3]	NR[3]
FB3-F	BAg-7	1008 carbon steel	G10080	1300	705	0.25	161
FB3-G	BAg-7	1008 carbon steel	G10080	1300	705	0.25	161
FB3-H	BAg-24	1008 carbon steel	G10080	1400	760	0.25	161
FB3-I	RBCuZn-D, BNi-2	304 stainless steel	S30400	1850	1010	0.25	161
		304 stainless steel	S30400	1900	1040	0.25	161
FB3-J	RBCuZn-D, BNi-2	304 stainless steel	S30400	1850	1010	0.25	161
		304 stainless steel	S30400	1900	1040	0.25	161
FB3-K	RBCuZn-D	1008 carbon steel	G10080	1750	955	NR[c]	NR[c]
FB4-A	BAg-6	C613 aluminum bronze	C61300	1525	830	0.25	161

a. Temperature tolerances shall be ± 15°F (± 8°C).   b. SAE/ASTM Unified Numbering System for Metals and Alloys.   c. NR - No flow requirement. Wetting of base metal by the filler metal is all that is required.

AWS A5.31 CLASSIFICATION SYSTEM

The system for identifying the flux classifications in this specification is based on three factors: applicable base metal, applicable filler metal, and activity temperature range. The letters FB at the beginning of each classification designation stands for "Flux for Brazing or Braze Welding". The third character is a number that stands for a group of applicable base metals. The fourth character, a letter, designates a change in form and attendant composition within the broader base metal classification.

The Metals Blue Book

Chapter
42

INTERNATIONAL CROSS REFERENCES CARBON & ALLOY STEEL WELDING FILLER METAL STANDARDS

INTERNATIONAL CROSS REFERENCES - CARBON & ALLOY STEEL WELDING FILLER METAL STANDARDS[a]

USA	Germany		Canada	France
AWS	W.-Nr.	DIN	CSA	AFNOR NF A 81
A5.1 E 6010	---	---	W 48.1 E41010	---
A5.1 E 6011	---	---	W 48.1 E41011	---
A5.1 E 6012	---	1913 E 43 21 R 3	W 48.1 E41012	309 E 43 2/1 R 22
A5.1 E 6012; E 6013	---	1913 E 51 22 R (C) 3	W 48.1 E41012; E41013	309 E 51 2/2 R 12
A5.1 E 6013	---	1913 E 43 43 RR (B) 7	W 48.1 E41013	309 E 43 4/3 RR 22
A5.1 E 6013	---	1913 E 51 21 RR (C) 6	W 48.1 E41013	309 E 51 2/1 RR 12
A5.1 E 6013	---	1913 E 51 22 RR 6	W 48.1 E41013	309 E 51 2/2 RR 22
A5.1 E 6013	---	1913 E 51 22 RR (C) 6	W 48.1 E41013	309 E 51 2/2 RR 12
A5.1 E 6013	---	1913 E 51 32 RR 8	W 48.1 E41013	309 E 51 3/2 RR 22
A5.1 E 6013	---	1913 E 51 32 RR (B) 8	W 48.1 E41013	309 E 51 4/3 RR 22
A5.1 E 6013	---	1913 E 51 43 AR 7	W 48.1 E41013	309 E 51 4/3 AR 22
A5.1 E 6022	---	---	W 48.1 E41022	---
A5.1 E 6027	---	---	W 48.1 E41027	---
A5.1 E 7014	---	1913 (E 51 32 RR (C) 11 120)	W 48.1 E48014	309 E 51 3/2 RR 120 12
A5.1 E 7015	---	---	W 48.1 E48015	---
A5.1 E 7016	---	1913 E 51 54 B (R) 10	W 48.1 E48016	309 E 51 5/4 B 26 H
A5.1 E 7016-1	---	---	---	---
A5.1 E 7018	---	1913 E 51 53 B 10	W 48.1 E48018	309 E 51 5/3 B 110 20
A5.1 E 7018-1	---	1913 E 51 55 B 10	W 48.1 E48018-1	309 E 51 5/5 B 120 20 (H)
A5.1 E 7024	---	1913 E 51 32 RR 11 160	W 48.1 E48024	309 E 51 3/2 RR 160 42
A5.1 E 7024	---	1913 E 51 32 RR 11 180	W 48.1 E48024	309 E 51 3/2 RR 180 42
A5.1 E 7024	---	1913 E 51 32 RR 11 240	W 48.1 E48024	309 E 51 3/2 RR 240 42
A5.1 E 7024-1	---	---	W 48.1 E48024-1	---
A5.1 E 7027	---	---	W 48.1 E48027	---
A5.1 E 7028	---	---	W 48.1 E48028	---
A5.1 E 7024	---	1913 E 51 32 RR 11 200	W 48.1 E48024	309 E 51 3/2 RR 200 42
A5.1 E 7048	---	---	W 48.1 E48048	---

INTERNATIONAL CROSS REFERENCES - CARBON & ALLOY STEEL WELDING FILLER METAL STANDARDS[a] (Continued)				
USA	Germany		Canada	France
AWS	W.-Nr.	DIN	CSA	AFNOR NF A 81
A5.2 R 60	---	8554 G II 10	---	---
A5.2 R 65	---	8554 G II 10	---	---
A5.5 E 7010 -A1	---	---	W 48.3 E48010-A1	---
A5.5 E 7011 -A1	---	---	W 48.3 E48011-A1	---
A5.5 E 7015 -A1	---	---	W 48.3 E48015-A1	---
A5.5 E 7016 -A1	---	---	W 48.3 E48016-A1	---
A5.5 E 7018 -A1	---	---	W 48.3 E48018-A1	---
A5.5 E 7020 -A1	---	---	W 48.3 E48020-A1	---
A5.5 E 7027 -A1	---	---	W 48.3 E48027-A1	---
A5.5 E 7018-G	---	8529 EY 46 65 Mn B H 5 20 +	W 48.3 E48018-G	340 EY 46 62 Mo B 120 20 BH
A5.5 E 8013-A1	---	8575 E Mo R 22 (E Ti Mo 22)		345 EC Mo R 22
A5.5 E 8018-A1	---	8575 E Mo B 20 + (E Kb Mo 20 +)	---	345 EC Mo B 20
A5.5 E 8018-A1	---	8529 EY 50 54 Mo B H 5 20 +		---
A5.5 E 8016-B1	---	---	W 48.3 E55016-B1	---
A5.5 E 8018-B1	---	---	W 48.3 E55018-B1	---
A5.5 E 8015-B2L	---	---	W 48.3 E55015-B2L	---
A5.5 E 8016-B2	---	---	W 48.3 E55016-B2	---
A5.5 E 8018-B2	---	8575 E1 CrMo B 20 + (E Kb CrMo 1 20+)	W 48.3 E55018-B2	345 EC 1 CrMo B 20
A5.5 E 8018-B2L	---	---	W 48.3 E62018-B2L	---
A5.5 E 9015-B3L	---	---	W 48.3 E62015-B3L	---
A5.5 E 9015-B3	---	---	W 48.3 E62015-B3	---
A5.5 E 9016-B3	---	---	W 48.3 E62016-B3	---
A5.5 E 9018-B3	---	8575 E2 CrMo B 20 + (E Kb CrMo 2 20 +)	W 48.3 E62018-B3	345 EC 2 CrMo B 20
A5.5 E 9018-B3L	---	---	W 48.3 E62018-B3L	---
A5.5 E 8015-B4L	---	---	W 48.3 E55015-B4L	---
A5.5 E 8016-B5	---	---	W 48.3 E55016-B5	---
A5.5 E 8016-C1	---	---	W 48.3 E55016-C1	---

INTERNATIONAL CROSS REFERENCES - CARBON & ALLOY STEEL WELDING FILLER METAL STANDARDS[a] (Continued)

USA		Germany		Canada	France
AWS	W.-Nr.	DIN		CSA	AFNOR NF A 81
A5.5 E 8018-C1	---	8529 EY 46 87 2 Ni B H 5 20 +		W 48.3 E55018-C1	340 EY 46 2 Ni B 110 20 BH
A5.5 E 7015-C1L	---	---		W 48.3 E48015-C1L	---
A5.5 E 7016-C1L	---			W 48.3 E48016-C1L	---
A5.5 E 7018-C1L	---			W 48.3 E48018-C1L	---
A5.5 E 8016-C3	---			W 48.3 E55016-C3	---
A5.5 E 8018-C3	---			W 48.3 E55018-C3	---
A5.5 E 8018-NM	---			W 48.3 E55018-NM	---
A5.5 E 8018-D1	---	8529 EY 50 65 MnMo B H 5 20 +		---	340 EY 50 2 MnMo B 110 20 BH
A5.5 E 9015-D1	---	---		W 48.3 62015-D1	---
A5.5 E 9018-D1	---	---		W 48.3 62018-D1	---
A5.5 E 8016-D3	---	---		W 48.3 E55016-D3	---
A5.5 E 8016-C2	---	---		W 48.3 E55016-C2	---
A5.5 E 8018-C2	---	---		W 48.3 E55018-C2	---
A5.5 E 7016-C2L	---	---		W 48.3 E48016-C2L	---
A5.5 E 7018-C2L	---	---		W 48.3 E48018-C2L	---
A5.5 E 8018-D3	---	---		W 48.3 E55018-D3	---
A5.5 E 10015-D2	---	---		W 48.3 E69015-D2	---
A5.5 E 7015-C2L	---	---		W 48.3 E48015-C2L	---
A5.5 E 10016-D2	---	---		W 48.3 E69016-D2	---
A5.5 E 10018-D2	---	---		W 48.3 E69018-D2	---
A5.5 E 8018-G	---	8529 EY 50 54 Mn B H 5 26		W 48.3 E55018-G	340 EY 50 Mn B 120 20
A5.5 E 8018-G	---	8529 EY 50 76 1 NiCu B H 5 20 +		W 48.3 E55018-G	---
A5.5 E 9018-G	---	8529 EY 55 54 Mn 8 H 10 20 +		W 48.3 E62018-G	340 EY 55 2 Mn B 120 20
A5.5 E 9018-M	---	8529 EY 55 76 2 NiMo B H 5 20 +		W 48.3 E62018-M	340 EY 55 1.5 NiMo B 110 20 BH
A5.5 E 10018-G	---	8529 EY 62 65 Mn 1 NiCr B H 5 20 +		W 48.3 E69018-G	---
A5.5 E 11018-M	---	8529 EY 62 76 2 NiMo B H 5 20 +		W 48.3 E76018-M	---

INTERNATIONAL CROSS REFERENCES - CARBON & ALLOY STEEL WELDING FILLER METAL STANDARDS[a] (Continued)

USA AWS	Germany W.-Nr.	Germany DIN	Canada CSA	France AFNOR NF A 81
A5.5 E 11018-M	---	8529 EY 69 76 Mn 2 NiCrMo B H 5 20 +	W 48.3 E76018-M	340 EY 69 2 Mn 2 NiCrMo B 110 20 BH
A5.5 E 12018-M	---	---	W 48.3 E83018-M	---
A5.5 E 8018-W	---	---	W 48.3 E55018-W	---
A5.18 ER 70S-2	---	---	W 48.4 ER480S-2	---
A5.18 ER 70S-3	1.5112	80 SG 1	W 48.4 ER480S-3	---
A5.18 ER 70S-4	---	---	W 48.4 ER480S-4	---
A5.18 ER 70S-5	---	---	W 48.4 ER480S-5	---
A5.18 ER 70S-6	1.5130	8559 SG 3	W 48.4 ER480S-6	---
A5.20 E 70T-3	---	---	W 48.5 E4800T-3	---
A5.20 E 70T-4	---	---	W 48.5 E4800T-4	---
A5.20 E 70T-5	---	8559 SG B1 C 47 54	W 48.5 E4800T-5B	---
A5.20 E 70T-6	---	---	W 48.5 E4800T-6	---
A5.20 E 70T-7	---	---	W 48.5 E4800T-7	---
A5.20 E 70T-8	---	---	W 48.5 E4800T-8	---
A5.20 E 70T-10	---	---	W 48.5 E4800T-10	---
A5.18 ER 70S-7	---	---	W 48.4 ER480S-7	---
A5.18 ER 70S-G	1.5125	8559 SG 2	W 48.4 ER480S-G	---
A5.20 E 70T-1	---	---	W 48.5 E4800T-1	---
A5.20 E 70T-2	---	8559 SG R 1	W 48.5 E4800T-2	---
A5.20 E 70T-11	---	---	W 48.5 E4800T-11	---
A5.20 E 70T-G	---	8559 SG B1	W 48.5 E4800T-G	---
A5.20 E 70T-GS	---	---	W 48.5 E4800T-GS	---
A5.28 ER 80 S-G	1.7339	8575 SG CrMo 1	---	---
A5.28 ER 80 S-G	1.5424	8575 SG Mo	---	---
A5.28 ER 90 S-G	1.7384	8575 SG CrMo 2	---	---

INTERNATIONAL CROSS REFERENCES - CARBON & ALLOY STEEL WELDING FILLER METAL STANDARDS[a] (Continued)

a.   It is not practical to directly correlate the various welding filler metal designations from country to country, let alone comparing several countries and their filler metal designations; from the view that chemical composition may be similar, but not identical, and that manufacturing technologies and testing procedures may differ greatly.   Consequently, the cross references made in this table are, at best, only listed as a guide to assist in finding comparable filler metal designations, rather than equivalent filler metal designations.

Chapter
43

INTERNATIONAL CROSS REFERENCES STAINLESS STEEL WELDING FILLER METAL STANDARDS

INTERNATIONAL CROSS REFERENCES - STAINLESS STEEL WELDING FILLER METAL STANDARDS[a]

USA	Germany		Great Britain	France
AWS	W.-Nr.	DIN	B.S.	AFNOR NF A 81
A5.11 E NiCrMo-4	2.4887	1736 EL-NiMo 15 Cr 15 W	---	---
5.11 E NiCrFe-3	2.4620	1736 S-NiCr 16 Fe Mn	---	---
A5.11 E NiCrMo-3	2.4621	1736 S-NiCr 20 Mo 9 Nb	---	---
A5.14 ER NiCr-3	2.4806	1736 S-NiCr 20 Nb	2901 Part 5 NA 35	---
A5.14 E (R) NiCrMo-3	2.4831	1736 S-NiCr 21 Mo 9 Nb	---	---
A5.11 E NiCu-7	2.4366	1736 S-NiCu 30 Mn	---	---
A5.14 ER NiCu-7	2.4377	1736 S-NiCu 30 MnTi	2901 Part 5 NA 33	---
A5.11 E Ni-1	2.4156	1736 S-NiTi 3	---	---
A5.14 ER Ni-1	2.4155	1736 S-NiTi 4	2901 Part 5 NA 32	---
A5.4 E 410 NiMo	1.4351	8556 E 13 4 B 20 +	---	---
A5.4 E 316-15	---	8556 E 17 11 B 20 +	---	---
A5.4 E 430-15	1.4502	8556 E 17 B 20 +	2926 17	343 EZ 17 B 20
---	1.4370	8556 E 18 8 Mn 6 B 20 +	---	343 EZ 1.8 Mn B 20
---	1.4370	8556 E 18 8 Mn 6 MPR 36 160	---	343 EZ 1.8 Mn 160 36
---	1.4370	8556 E 18 8 Mn 6 R R 26	---	343 EZ 1.8 Mn R 26
A5.4 E 308-15	1.4948	8556 E 18 11 B 20 +	---	---
A5.4 E 330-15	---	8556 E 18 36 Nb B 20 +	---	---
A5.4 E 308-15	1.4302	8556 E 19 9 B 20 +	2926 19.9	343 EZ 19.9 B 20
A5.4 E 347-15	1.4551	8556 E 19 9 Nb B 20 +	2926 19.9 Nb	343 EZ 19.9 Nb B 20
A5.4 E 347-16	1.4551	8556 E 19 9 Nb R 16	2926 19.9 Nb	343 EZ 19.9 Nb R 16
A5.4 E 347-16	1.4551	8556 E 19 9 Nb R 26	2926 19.9 Nb	343 EZ 19.9 Nb R 26
A5.4 E 347-16		8556 E 19 9 Nb R 36 160	2926 19.9 Nb	343 EZ 19.9 Nb R 160 36
A5.4 E 308 L-15	1.4316	8556 E 19 9 nC B 20 +	2926 19.9 L	343 EZ 19.9 L B 20
A5.4 E 308 L-16	1.4316	8556 E 19 9 nC R 16	2926 19.9 L	343 EZ 19.9 L R 16
A5.4 E 308 L-16	1.4316	8556 E 19 9 nC R 26	2926 19.9 L	343 EZ 19.9 L R 26
A5.4 E 308 L-16	1.4316	8556 E 19 9 nC R 36 160	2926 19.9 L	343 EZ 19.9 L R 160 36
A5.4 E 308-16	1.4302	8556 E 19 9 R 26	2926 19.9	343 EZ 19.9 R 26

INTERNATIONAL CROSS REFERENCES - STAINLESS STEEL WELDING FILLER METAL STANDARDS[a] (Continued)

USA AWS	Germany W.-Nr.	Germany DIN	Great Britain B.S.	France AFNOR NF A 81
A5.4 E 316 L-15	---	8556 E 19 10 3 nC B 20 +	---	---
A5.4 E 316-16	---	8556 E 19 12 2 R 26	---	---
A5.4 E 316-15	1.4403	8556 E 19 12 3 B 20 +	2926 19.12.3	343 EZ 19.12.3 B 20
A5.4 E 318-15	1.4576	8556 E 19 12 3 Nb B 20 +	2926 19.12.3 Nb	343 EZ 19.12.3 Nb B 20
A5.4 E 318-16	1.4576	8556 E 19 12 3 Nb R 16	2926 19.12.3 Nb	343 EZ 19.12.3 Nb R 16
A5.4 E 318-16	1.4576	8556 E 19 12 3 Nb R 26	2926 19.12.3 Nb	343 EZ 19.12.3 Nb R 26
A5.4 E 318-16	1.4576	8556 E 19 12 3 Nb R 36 160	2926 19.12.3 Nb	343 EZ 19.12.3 Nb R 160 36
A5.4 E 316 L-15	1.4430	8556 E 19 12 3 nC B 20 +	2926 19.12.3 L	343 EZ 19.12.3 L B 20
A5.4 E 316 L-16	1.4430	8556 E 19 12 3 nC R 16	2926 19.12.3 L	343 EZ 19.12.3 LR 16
A5.4 E 316 L-16	1.4430	8556 E 19 12 3 nC R 26	2926 19.12.3 L	343 EZ 19.12.3 LR 26
A5.4 E 316 L-16	1.4460	8556 E 19 12 3 nC R 36 160	2926 19.12.3 L	343 EZ 19.12.3 LR 160 36
A5.4 E 316-16	1.4403	8556 E 19 12 3 R 26	2926 19.12.3	343 EZ 19.12.3 R 26
---	1.4539	8556 E 20 25 5 Cu nC B 20 +	---	343 EZ 20.25.5 L Cu 8 20
A5.4 E 309-15	1.4829	8556 E 22 12 B 20 +	2926 23.12 B	343 EZ 22.12 B 20
A5.4 E 309-16	1.4829	8556 E 22 12 R 26	2926 23.12 R	343 EZ 22.12 R 26
A5.4 E 309 Mo-15	---	8556 E 22 14 3 nC B 20 +	---	343 EZ 23.12.2 R 20
A5.4 E 309 Mo-16	---	8556 E 22 14 3 nC R 26	---	343 EZ 23.12.2 R 26
A5.4 E 309 Cb-16	1.4556	8556 E 23 12 Nb R 26	2926 23.12 Nb	343 EZ 23.12 Nb R 26
A5.4 E 309 L-16	1.4332	8556 E 23 12 nC R 26	---	343 EZ 23.12 LR 26
A5.4 E 310-15	1.4842	8556 E 25 20 B 20 +	2926 25.20 B	343 EZ 25.20 B 20
A5.4 E 310 H-15	1.4846	8556 E 25 20 nC B 20 +	2926 25.20 H B	343 EZ 25.20 C B 20
A5.4 E 310-16	1.4842	8556 E 25 20 R 26	2926 25.20 R	343 EZ 25.20 R 25
A5.4 E 312-16	1.4337	8556 E 29 9 R 26	---	343 EZ 29.9 R 26
A5.9 ER 308 L Si	1.4316	8556 SG X 2 CrNi 19 9	---	---
A5.9 ER 309 L	1.4332	8556 SG X 2 CrNi 24 12	---	---
A5.9 ER 316 L Si	1.4430	8556 SG X 2 CrNiMo 19 12	---	---
A5.9 ER 317 L	1.4429	8556 SG X 2 CrNiMoN 18 14	---	---

INTERNATIONAL CROSS REFERENCES - STAINLESS STEEL WELDING FILLER METAL STANDARDS[a] (Continued)				
USA	Germany		Great Britain	France
AWS	W.-Nr.	DIN	B.S.	AFNOR NF A 81
A5.9 ER 308 Si	1.4302	8556 SG X 5 CrNi 19 9	---	---
A5.9 ER 316 Si	1.4403	8556 SG X 5 CrNiMo 19 12	---	---
A5.9 ER 347 Si	1.4551	8556 SG X 5 CrNiNb 19 9	---	---
A5.9 ER 318	1.4576	8556 SG X 5 CrNiMoNb 19 12	---	---
A5.9 ER 308	1.4948	8556 SG X 6 CrNi 18 11	---	---
A5.9 ER 430	1.4502	8556 SG X 8 CrTi 18	---	---
A5.9 ER 312	1.4337	8556 SG X 10 CrNi 30 9	---	---
A5.9 ER 309 Si	1.4829	8556 SG X 12 CrNi 22 12	2901 Part 2 309 S 94	---
A5.9 ER 310	1.4842	8556 SG X 12 CrNi 25 20	2901 Part 2 310 S 94	---
A5.9 ER 310-H	1.4846	8556 SG X 40 CrM 25 21	2901 Part 2 310 S 98	---
A5.22 E 307 T-2	1.4370	8556 Typ 18 8 Mn 6	---	---
A5.22 E 347 T-2	1.4551	8556 Typ 19 9 Nb	---	---
A5.22 E 308 LT-2	1.4316	8556 Typ 19 9 nC	---	---
A5.22 E 316 T-2	1.4576	8556 Typ 19 12 3 Nb	---	---
A5.22 E 316 LT-2	1.4430	8556 Typ 19 12 3 nC	---	---
A5.22 E 309 LT-2	1.4332	8556 Typ 24 12 nC	---	---

a. It is not practical to directly correlate the various welding filler metal designations from country to country, let alone comparing several countries and their filler metal designations; from the view that chemical composition may be similar, but not identical, and that manufacturing technologies and testing procedures may differ greatly. Consequently, the cross references made in this table are, at best, only listed as a guide to assist in finding comparable filler metal designations, rather than equivalent filler metal designations.

Chapter
44

LIST OF INTERNATIONAL FILLER METAL STANDARDS

AMERICAN AWS - FILLER METAL SPECIFICATIONS

AWS	Title
A5.01	Filler Metal Procurement Guidelines
A5.1	Specification for Carbon Steel Electrodes for Shielded Metal Arc Welding
A5.2	Specification for Carbon and Low Alloy Steel Rods for Oxyfuel Gas Welding
A5.3	Specification for Aluminum and Aluminum Alloy Electrodes for Shielded Metal Arc Welding
A5.4	Specification for Stainless Steel Electrodes for Shielded Metal Arc Welding
A5.5	Specification for Low Alloy Steel Covered Arc Welding Electrodes
A5.6	Specification for Covered Copper and Copper Alloy Arc Welding Electrodes
A5.7	Specification for Copper and Copper Alloy Bare Welding Rods and Electrodes
A5.8	Specification for Filler Metals for Brazing and Braze Welding
A5.9	Specification for Bare Stainless Steel Welding Electrodes and Rods
A5.10	Specification for Bare Aluminum and Aluminum Alloy Welding Electrodes and Rods
A5.11	Specification for Nickel and Nickel Alloy Welding Electrodes for Shielded Metal Arc Welding
A5.12	Specification for Tungsten and Tungsten Alloy Electrodes for Arc Welding and Cutting
A5.13	Specification for Solid Surfacing Welding Rods and Electrodes
A5.14	Specification for Nickel and Nickel Alloy Bare Welding Electrodes and Rods
A5.15	Specification for Welding Electrodes and Rods for Cast Iron
A5.16	Specification for Titanium and Titanium Alloy Welding Electrodes and Rods
A5.17	Specification for Carbon Steel Electrodes and Fluxes for Submerged Arc Welding
A5.18	Specification for Carbon Steel Electrodes and Rods for Gas Shielded Arc Welding
A5.19	Specification for Magnesium Alloy Welding Electrodes and Rods
A5.20	Specification for Carbon Steel Electrodes for Flux Cored Arc Welding
A5.21	Specification for Composite Surfacing Welding Rods and Electrodes
A5.22	Specification for Flux Cored Corrosion-Resisting Chromium and Chromium-Nickel Steel Electrodes
A5.23	Specification for Low Alloy Steel Electrodes and Fluxes for Submerged Arc Welding
A5.24	Specification for Zirconium and Zirconium Alloy Welding Electrodes and Rods
A5.25	Specification for Carbon and Low Alloy Steel Electrodes and Fluxes for Electroslag Welding
A5.26	Specification for Carbon and Low Alloy Steel Electrodes for Electrogas Welding
A5.28	Specification for Low Alloy Steel Filler Metals for Gas Shielded Arc Welding

AMERICAN AWS - FILLER METAL SPECIFICATIONS (Continued)

AWS	Title
A5.29	Specification for Low Alloy Steel Electrodes for Flux Cored Arc Welding
A5.30	Specification for Consumable Inserts
A5.31	Specification for Fluxes for Brazing and Braze Welding

AWS - American Welding Society

CANADIAN CSA - FILLER METAL SPECIFICATIONS

CSA	Title
W48.1	Carbon Steel Covered Electrodes for Shielded Metal Arc Welding
W48.2	Chromium and Chromium- Nickel Steel Covered Electrodes for Shielded Metal Arc Welding
W48.3	Low Alloy Steel Covered Electrodes for Shielded Metal Arc Welding
W48.4	Solid Mild Steel Filler Metals for Gas Shielded Arc Welding
W48.5	Carbon Steel Electrodes for Flux- and Metal- Cored Arc Welding
W48.6	Bare Mild Steel Electrodes and Fluxes for Submerged-Arc Welding

CSA - Canadian Standards Association

BRITISH BSI - FILLER METAL SPECIFICATIONS

BS	Title
639	Covered Carbon and Carbon Manganese Steel Electrodes for Manual Metal-Arc Welding
1453	Filler Materials for Gas Welding
1845	Filler Metals for Brazing
2493	Low Alloy Steel Electrodes for Manual Metal-Arc Welding
2901: Part 1	Filler Rods and Wires for Gas-Shielded Arc Welding Part 1: Ferritic Steels
2901: Part 2	Filler Rods and Wires for Gas-Shielded Arc Welding Part 2: Stainless Steels
2901: Part 3	Filler Rods and Wires for Gas-Shielded Arc Welding Part 3: Specification for Copper and Copper Alloys
2901: Part 4	Filler Rods and Wires for Gas-Shielded Arc Welding Part 4: Specification for Aluminium and Aluminium Alloys and Magnesium Alloys
2901: Part 5	Filler Rods and Wires for Gas-Shielded Arc Welding Part 5: Specification for Nickel and Nickel Alloys
2926	Chromium and Chromium- Nickel Steel Electrodes for Manual Metal-Arc Welding
4165	Electrode Wires and Fluxes for the Submerged Arc Welding of Carbon Steel and Medium-Tensile Steel
4300/13: 5554	Welding Wire: Wrought Aluminium and Aluminium Alloys For General Engineering Purposes
4577	Materials for Resistance Welding Electrodes and Ancillary Equipment
5465	Electrodes and Fluxes for the Submerged Arc Welding of Austenitic Stainless Steel
6678	(Withdrawn) 1986 Tungsten Electrodes for Inert Gas Shielded Arc Welding and for Plasma Cutting and Welding Superseded by BS EN 26848
6787	Determination of Ferrite Number in Austenitic Weld Metal Deposited by Covered Cr - Ni Steel Electrodes
7084	Carbon and Carbon - Manganese Steel Tubular Cored Welding Electrodes
7084	Carbon and Carbon -Manganese Steel Tubular Cored Welding
29692	Metal-Arc Welding with Covered Electrode, Gas-Shielded Metal-Arc Welding and Gas Welding - Joint Preparations for Steel (ISO 9692)
EN 20544	Sizes for Filler Metals for Manual Welding (ISO 544)
EN 22401	Covered Electrodes - Determination of the Efficiency, Metal Recovery and Deposition Coefficient (ISO 2401)
EN 26847	Covered Electrodes for Manual Metal Arc Welding - Deposition of a Weld Metal Pad for Chemical Analysis (ISO 6847)
EN 26848	Tungsten Electrodes for Inert Gas Shielded Arc Welding and for Plasma Cutting and Welding (ISO 6848)

BSI - British Standards Institution; BS - British Standard

GERMAN DIN - FILLER METAL SPECIFICATIONS

DIN	Title
1732 Part 1	Filler Metals for Welding Aluminium and Aluminium Alloys; Composition, Application and Technical Delivery Conditions
1732 Part 2	Welding Filler Metals for Aluminium; Testing by Welded Joints
1732 Part 3	Filler Metals for Welding Aluminium and Aluminium Alloys; Sample and Specimen Preparation, Mechanical Properties of All-Weld Metal
1733 Part 1	Filler Metals for Welding Copper and Copper Alloys; Composition, Application and Technical Delivery Conditions
1736 Part 1	Welding Filler Metals for Nickel and Nickel Alloys; Composition, Application and Technical Delivery Conditions
1736 Part 2	Welding Filler Metals for Nickel and Nickel Alloys; Sample, Test Pieces, Mechanical Properties
1737 Part 1	Filler Metals for Welding Titanium and Titanium-Palladium Alloys; Chemical Composition, Technical Delivery Conditions
1737 Part 2	Filler Metals for Welding Titanium and Titanium-Palladium Alloys; Sample and Specimen Preparation, Mechanical Properties of All-Weld Metal
1913 Part 1	Covered Electrodes for the Joint Welding of Unalloyed and Low Alloy Steel; Classification, Designation, Technical Delivery Conditions
8513 Part 1	Brazing and Braze Weld Filler Metals; Copper Base Brazing Alloys; Composition, Use, Technical Conditions of Delivery
8513 Part 2	Brazing and Braze Weld Filler Metals; Silver-Bearing Brazing Alloys with Less Than 20% by Wt. Silver; Composition, Use, Technical Conditions of Delivery
8513 Part 3	Brazing Filler Metals; Silver Brazing Filler Metals Containing Not Less Than 20% of Silver; Composition, Application, Technical Delivery Conditions
8513 Part 4	Brazing and Braze Weld Filler Metals; Aluminium-Base Brazing Alloys; Composition; Use; Technical Conditions of Delivery
8513 Part 5	Brazing Filler Metals; Nickel Base Filler Metals for High Temperature Brazing; Application, Composition, Technical Delivery Conditions
8529 Part 1	Covered Electrodes for Joint Welding of High Tensile Fine-Grained Structural Steels; Basic Covered Electrodes; Classification, Designation, Technical Delivery Conditions
8554 Part 1	Unalloyed and Low Alloy Filler Rods for Gas Welding; Designation, Technical Delivery Condition
8555 Part 1	Filler Metals Used for Surfacing; Filler Wires, Filler Rods, Wire Electrodes, Covered Electrodes; Designation; Technical Delivery Conditions
8556 Part 1	Filler Metals for Welding Stainless and Heat-Resisting Steels; Designation, Technical Delivery Conditions
8556 Part 2	Filler Metals for Welding Stainless and Heat Resisting Steels; Testing of Covered Rod Electrodes; Weld Metal Specimen
8557 Part 1	Filler Metals for Submerged Arc Welding; Joint Welding of Unalloyed and Alloy Steels; Designation; Technical Delivery Conditions
8559 Part 1	Filler Metals for Gas-Shielded Arc Welding; Wires Electrodes, Filler Wires, Solid Rods and Solid Wires for Gas-Shielded Arc Welding of Unalloyed and Alloyed Steels
8566 Part 2	Filler Metals for Thermal Spraying; Solid Wires for Arc Spraying; Technical Delivery Conditions
8566 Part 3	Metal Filler Wire and Rod and Plastics Filler Materials for Thermal Spraying; Technical Delivery Conditions
8571	Filler Metals and Auxiliary Metals for Metal Welding; Concepts; Classification
8573 Part 1	Filler Metals for Welding Unalloyed and Low Alloy Cast Iron Materials; Designation; Technical Delivery Conditions

GERMAN DIN - FILLER METAL SPECIFICATIONS (Continued)

DIN	Title
8575 Part 1	Filler Metals for Arc Welding of Creep-Resisting Steels; Classification, Designation, Technical Delivery Conditions
17145	Round Wire Rod for Welding Filler Metals; Technical Conditions of Delivery
32525 Part 1	Testing Filler Metals by Means of Weld Metal Specimens; Arc Welded Test Pieces; Specimens for Mechanical-Technological Tests
32525 Part 2	Testing Filler Metals by Means of Weld Metal Specimens; Test Pieces for Determining Chemical Composition with Low Heat Input
32525 Part 4	Testing of Filler Metals by Means of Weld Metal Specimens; Test Piece for Determining the Hardness for Surfacing
50129	Testing of Metallic Materials; Testing of Welding Filler Metals for Liability to Cracking
EN 20544	Filler Materials for Manual Welding; Size Requirements (ISO 544-1989)
EN 26848	Tungsten Electrodes for Inert Gas Shielded Arc Welding and for Plasma Cutting and Welding; Codification (ISO 6848- 1984)

DIN - Deutsches Institut für Normung e.V.

JAPANESE JIS - FILLER METAL SPECIFICATIONS

JIS	Title
C 9304	Shapes and Dimensions of Spot Welding Electrodes
C 9304	Shapes and Dimensions of Spot Welding Electrodes
G 3503	Wire Rods for Core Wire of Covered Electrode
G 3523	Core Wires for Covered Electrode
Z 3181	Method of Test for Fillet Weld of Covered Electrode
Z 3182	Method of Deposition Rate Measurement for Covered Electrodes
Z 3191	Method of Spreading Test for Brazing
Z 3211	Covered Electrodes for Mild Steel
Z 3212	Covered Electrodes for High Tensile Strength Steel
Z 3214	Covered Electrodes for Atmospheric Corrosion Resisting Steel
Z 3214	Covered Electrodes for Atmospheric Corrosion Resisting Steel
Z 3221	Stainless Steel Covered Electrodes
Z 3223	Molybdenum Steel and Chromium Molybdenum Steel Covered Electrodes
Z 3223	Molybdenum Steel and Chromium Molybdenum Steel Covered Electrodes
Z 3224	Nickel and Nickel-Alloy Covered Electrodes

JAPANESE JIS - FILLER METAL SPECIFICATIONS (Continued)

JIS	Title
Z 3225	Covered Electrodes for 9% Nickel Steel
Z 3231	Copper and Copper Alloy Covered Electrodes
Z 3232	Aluminium and Aluminium Alloy Welding Rods and Wires
Z 3233	Tungsten Electrodes for Inert Gas Shielded Arc Welding
Z 3234	Copper Alloys for Resistance Welding Electrode
Z 3241	Covered Electrodes for Low Temperature Service Steel
Z 3241	Covered Electrodes for Low Temperature Service Steel
Z 3251	Covered Electrodes for Hardfacing
Z 3252	Covered Electrodes for Cast Iron
Z 3261	Silver Brazing Filler Metals
Z 3262	Copper and Copper - Zinc Brazing Filler Metals
Z 3263	Aluminium Alloy Brazing Filler Metals and Brazing Sheets
Z 3264	Copper Phosphorus Brazing Filler Metals
Z 3265	Nickel Brazing Filler Metals
Z 3266	Gold Brazing Filler Metals
Z 3267	Palladium Brazing Filler Metals
Z 3268	Vacuum Grade Precious Brazing Filler Metals
Z 3312	MAG Welding Solid Wires for Mild Steel and High Strength Steel
Z 3312	MAG Welding Solid Wires for Mild Steel and High Strength Steel
Z 3313	Flux Cored Wires for Gas Shielded and Self-Shielded Metal Arc Welding of Mild Steel, High Strength Steel and Low Temperature Service Steel
Z 3313	Arc Welding Flux Cored Wires for Mild Steel and High Strength Steel
Z 3315	Solid Wires for CO_2 Gas Shielded Arc Welding of Atmospheric Corrosion Resisting Steel
Z 3315	Solid Wires for CO_2 Gas Shielded Arc Welding of Atmospheric Corrosion Resisting Steel
Z 3316	TIG Welding Rods and Wires for Mild Steel and Low Alloy Steel
Z 3317	MAG Welding Solid Wires for Molybdenum Steel and Chromium Molybdenum Steel
Z 3318	MAG Welding Flux Cored Wires for Molybdenum Steel and Chromium Molybdenum Steel
Z 3319	Flux Cored Wires for Electrogas Arc Welding
Z 3320	Flux Cored Wires for CO_2 Gas Shielded Arc Welding of Atmospheric Corrosion Resisting Steel

JAPANESE JIS - FILLER METAL SPECIFICATIONS (Continued)

JIS	Title
Z 3320	Flux Cored Wires for CO_2 Gas Shielded Arc Welding of Atmospheric Corrosion Resisting Steel
Z 3322	Materials for Stainless Steel Overlay Welding with Strip Electrode
Z 3323	Stainless Steel Flux Cored Wires
Z 3324	Stainless Steel Solid Wires and Fluxes for Submerged Arc Welding
Z 3326	Arc Welding Flux Cored Wires for Hardfacing
Z 3331	Titanium and Titanium Alloy Rods and Wires for Inert Gas Shielded Arc Welding
Z 3332	Filler Rods and Wires for TIG Welding of 9% Nickel Steel
Z 3333	Submerged Arc Welding Wires and Fluxes for 9% Nickel Steel
Z 3334	Nickel and Nickel Alloy Filler Rods and Wires for Arc Welding
Z 3341	Copper and Copper Alloy Rods and Wires for Inert Gas Shielded Arc Welding
Z 3351	Submerged Arc Welding Wires for Carbon Steel and Low Alloy Steel
Z 3900	Methods for Sampling of Precious Brazing Filler Metals
Z 3901	Methods for Chemical Analysis of Silver Brazing Filler Metals
Z 3902	Methods for Chemical Analysis of Brass Brazing Filler Metals
Z 3903	Methods for Chemical Analysis of Copper Phosphorus Brazing Filler Metals
Z 3904	Methods for Chemical Analysis of Gold Brazing Filler Metals
Z 3905	Methods for Chemical Analysis of Nickel Brazing Filler Metals
Z 3905	Methods for Chemical Analysis of Nickel Brazing Filler Metals
Z 3906	Methods for Chemical Analysis of Palladium Brazing Filler Metals
Z 3930	Method of Measuring Total Amount of Weld Fumes Generated by Covered Electrode

JIS - Japanese Industrial Standards

AUSTRALIAN SAA - FILLER METAL SPECIFICATIONS

SAA	Title
AS 1167.1	Welding and Brazing - Filler Metals - Part 1: Filler Metal for Brazing and Braze Welding
AS 1167.2	Welding and Brazing - Filler Metals - Part 2: Filler Metal for Welding
AS 1553.1	Covered Electrodes for Welding - Part 1: Low Carbon Steel Electrodes for Manual Metal-Arc Welding of Carbon and Carbon-Manganese Steels
AS 1553.2	Covered Electrodes for Welding - Part 2: Low and Intermediate Alloy Steel Electrodes for Manual Metal-Arc Welding of Carbon Steels and Low and Intermediate Alloy Steels
AS 1553.3	Covered Electrodes for Welding - Part 3: Corrosion-Resisting Chromium and Chromium- Nickel Steel Electrodes
AS 1858.1	Electrodes and Fluxes for Submerged-Arc Welding - Part 1: Carbon Steels and Carbon-Manganese Steels
AS 1858.2	Electrodes and Fluxes for Submerged-Arc Welding - Part 2: Low and Intermediate Alloy Steels
AS 2203.1	Cored Electrodes for Arc- Welding - Part 1: Ferritic Steel Electrodes
AS 2576	Welding Consumables for Build-Up and Wear Resistance - Classification System
AS 2717.1	Welding - Electrodes - Gas Metal Arc - Part 1: Ferritic Steel Electrodes
AS 2717.2	Welding - Electrodes - Gas Metal Arc - Part 2: Aluminium and Aluminium Alloy
AS 2717.3	Welding - Electrodes - Gas Metal Arc - Part 3: Corrosion-Resisting Chromium and Chromium- Nickel Steel Electrodes
AS 2826	Manual Metal-Arc Welding Electrode Holders

SAA - Standards Association of Australia;   AS - Australian Standard

EUROPEAN CEN - FILLER METAL SPECIFICATIONS

CEN	Title
EN 20544	Filler Materials for Manual Welding - Size Requirements
EN 22401	Covered Electrodes - Determination of the Efficiency, Metal Recovery and Deposition Coefficient (ISO 2401)
EN 26847	Covered Electrodes for Manual Metal Arc Welding - Deposition of a Weld Metal Pad for Chemical Analysis (ISO 6847)
EN 26848	Tungsten Electrodes for Inert Gas Shielded Arc Welding and for Plasma Cutting and Welding - Codification
EN 29692	Metal-Arc Welding with Covered Electrode, Gas- Shielded Metal-Arc Welding and Gas Welding - Joint Preparations for Steel (ISO 9692)

CEN - Comité Européen de Normalisation (European Committee for Standardization)

INTERNATIONAL ISO - FILLER METAL SPECIFICATIONS

ISO	Title
544	Filler Materials for Manual Welding - Size Requirements Second Edition; (Supersedes 545, 546 and 547)
R615	Methods for Determining the Mechanical Properties of the Weld Metal Deposited by Electrodes 3.15 mm or More in Diameter First Edition (Recommendation)
698	Filler Rods for Braze Welding - Determination of Conventional Bond Strength on Steel, Cast Iron and Other Metals First Edition
864	Arc Welding - Solid and Tubular Cored Wires Which Deposit Carbon and Carbon Manganese Steel - Dimensions of Wires, Spools, Rims and Coils Second Edition
1071	Covered Electrodes for Manual Arc Welding of Cast Iron - Symbolization First Edition
2401	Covered Electrodes - Determination of the Efficiency, Metal Recovery and Deposition Coefficient First Edition (CEN EN 22401)
2560	Covered Electrodes for Manual Arc Welding of Mild Steel and Low Alloy Steel - Code of Symbols for Identification First Edition
3580	Covered Electrodes for Manual Arc Welding of Creep-Resisting Steels - Code of Symbols for Identification First Edition
3581	Covered Electrodes for Manual Arc Welding of Stainless and Other Similar High Alloy Steels - Code of Symbols for Identification First Edition
3677	Filler Metals for Soft Soldering, Brazing and Braze Welding - Designation Second Edition
3690	Welding - Determination of Hydrogen in Deposited Weld Metal Arising from the Use of Covered Electrodes for Welding Unalloyed and Low Alloy Steels First Edition
5182	Welding - Materials for Resistance Welding Electrodes and Ancillary Equipment Second Edition
5184	Straight Resistance Spot Welding Electrodes First Edition
5187	Welding and Allied Processes - Assemblies Made with Soft Solders and Brazing Filler Metals - Mechanical Test Methods First Edition
6847	Covered Electrodes for Manual Metal Arc Welding - Deposition of a Weld Metal Pad for Chemical Analysis First Edition (CEN EN 26847)
6848	Tungsten Electrodes for Inert Gas Shielded Arc Welding, and for Plasma Cutting and Welding - Codification First Edition
8249	Welding - Determination of Ferrite Number in Austenitic Weld Metal Deposited by Covered Cr-Ni Steel Electrodes First Edition
9692	Metal-Arc Welding with Covered Electrode, Gas- Shielded Metal-Arc Welding and Gas Welding - Joint Preparations for Steel First Edition (CEN EN 29692)
10446	Welding - All-Weld Metal Test Assembly for the Classification of Corrosion-Resisting Chromium and Chromium- Nickel Steel Covered Arc Welding Electrodes First Edition

ISO - International Organization for Standardization

GLOSSARY OF WELDING TERMS

The terms and definitions listed in this glossary are consistent with the International Institute of Welding (IIW) terms and essentially agree with the American Welding Society (AWS) definitions and terms.

The first term listed in each group starting on the next page is English (**bold type**), the second term is French (***bold, italic type***) and the third term is Spanish (***bold, italic and underlined type***). Definitions for the English and French terms are provided where possible, however, Spanish term definitions are not provided at this time. The terms/definitions in the following list are arranged in alphabetical order by the English term.

These terms are also listed in Spanish and French alphabetical order following the term/definition section (see pages 364 and 376, respectively).

Translation of technical terms into various languages is an extremely difficult and complex undertaking and while this information was compiled with great effort and review, the publishers do not warrant its suitability and assume no liability or responsibility of any kind in connection with this information. We encourage all readers to inform *CASTI* Publishing Inc. of technical or language inaccuracies, and/or to offer suggestions for improving this glossary of terms. Your co-operation and assistance in this matter are greatly appreciated.

A

actual throat thickness in a butt weld. The perpendicular distance between two lines, each parallel to a line joining the outer toes, one being a tangent at the weld face and the other being through the furthermost point of fusion penetration.
épaisseur totale de la soudure (dans une soudure bout à bout). Distance comprise entre deux droites parallèles au plan joignant les lignes de raccordement, l'une de ces droites étant tangente à la surface du cordon et l'autre au point extrême de la pénétration à la racine.
espesor total de la soldadura (en una unión a tope)

actual throat thickness of a fillet weld. The perpendicular distance between two lines, each parallel to a line joining the outer toes, one being a tangent at the weld face and the other being through the furthermost point of fusion penetration.
épaisseur totale de la soudure (dans une soudure d'angle). Distance comprise entre deux droites parallèles au plan joignant les lignes de raccordement, l'une de ces droites étant tangente à la surface du cordon et l'autre au point extrême de la pénétration à la racine.
espesor total de soldadura (en una unión de ángulo)

age hardening. Hardening by aging, usually by rapid cooling or cold working.
endurcissement par vieillissement. Accroissement de dureté apparaissant après refroidissement rapide (vieillissement après trempe) ou après déformation (vieillissement après écrouissage).
endurecimiento por envejecimiento

age hardening crack. A crack due to hardening by aging, usually after rapid cooling or cold working.
fissure due au vieillissement. Fissure due à un durcissement par vieillissement, généralement après refroidissement rapide ou écrouissage.
grieta debida al envejecimiento

aging. A change in properties of a metal or alloy that generally occurs slowly at room temperature but more rapidly at higher temperatures.
vieillissement. Modification des propriétés d'un métal ou d'un alliage. Le phénomène, généralement lent à température ambiante, s'accélère lorsque la température croît.
envejecimiento

air hardening.   Heating a steel containing sufficient carbon and alloying elements above the transformation range and cooling in air or other gaseous medium.
trempe à l'air.   Trempe obtenue par chauffage d'un acier contenant suffisamment de carbone et d'éléments d'alliage, à une température supérieure au domaine de transformation, suivi d'un refroidissement à l'air ou dans un autre milieu gazeux.
temple al aire

angular misalignment.   Misalignment between two welded pieces such that their surface planes are not parallel (or at the intended angle).
déformation angulaire.   Non respect de l'angle prévu entre deux pièces soudées.
deformación angular

annealing.   Heating to and holding at a suitable temperature and then cooling at a suitable rate, for such purposes as reducing hardness, improving machinability, facilitating cold working, producing a desired microstructure, or obtaining desired mechanical physical or other properties.
recuit.   Traitement thermique comportant un chauffage et un maintien à une température convenable, suivi d'un refroidissement approprié. Ce traitement a pour objet d'abaisser la dureté, d'augmenter l'aptitude à l'usinage, de faciliter le travail à froid, de produire la microstructure souhaitée ou d'obtenir certaines caractéristiques mécaniques ou propriétés physiques ou autres.
recocido

arm protection; sleeve; welding sleeve.
manchette; manchette de soudeur.
manguito de soldador

arc pressure welding.   Pressure welding processes where the workpieces are heated on the faying surfaces by a short duration arc and the weld is completed by the application of a force. It is preferred not to use a filler material.
soudage par pression à l'arc; soudage à l'arc par pression.   Procédé de soudage par pression dans lequel les extrémités des pièces sont chauffées par un arc de courte durée et la soudure réalisée par application d'un effort. Il est préférable de ne pas utiliser de produit d'apport.
soldeo por arco con percusión

arc strike; stray flash.   Local damage to the surface of the parent metal adjacent to the weld resulting from accidental arcing or striking the arc outside the weld groove. (Not to be confused with flash; arc eye).
coup d'arc (sur la pièce).   Altération locale et superficielle du métal de base résultant d'un amorçage accidentel de l'arc au voisinage de la soudure. (Ne pas confondre avec coup d'arc [irritation de l'oeil]).
corte de arco (sobre la pieza)

arc welding. Fusion welding processes in which the heat for welding is obtained from an electric arc or arcs. Note: The arc and molten pool may be protected from the atmosphere by a gas, a slag or a flux.
soudage à l'arc. Procédés de soudage par fusion dans lesquels la chaleur nécessaire au soudage est fournie par un ou plusieurs arc électriques. Note: L'arc et le bain de fusion peuvent être isolés de l'atmosphère ambiante par un gaz, un laitier ou un flux.
soldeo por arco

as-welded. The condition of weld metal, welded joints and weldments after welding, prior to any subsequent thermal, mechanical or chemical treatment.
brut de soudage. État du métal fondu, des joints soudés et des ensembles soudés, après soudage et avant tout traitement thermique, mécanique ou chimique ultérieur.
bruto de soldeo

automatic welding. Welding operation in which all variables are controlled automatically including the loading and unloading operation.
soudage automatique. Opération de soudage dans laquelle tous les paramètres de soudage sont commandés automatiquement y compris le chargement et le déchargement des pièces.
soldeo automático

B

back-gouged weld
soudure gougée à la racine
soldadura acanalada en la raíz

back-step welding. A welding sequence in which short lengths of run are deposited in a direction opposite to the general progress of welding the joint; the short lengths eventually produce a continuous or intermittent weld.
soudage à pas de pèlerin. Séquence de soudage dans laquelle des cordons de faible longueur sont déposés dans un sens opposé au sens général d'avancement du soudage de l'assemblage. En fin d'opération, ces cordons courts forment soit une soudure continue, soit une soudure discontinue.
soldeo a paseo de peregrino

backing ring. Backing in the form of a ring used in the welding of pipes.
bague support; anneau support. Support en forme de bague utilisé pour le soudage des tubes.
anillo soporte

backing weld; support run; backing run. (This weld is made before the main weld is laid in the groove.)
cordon support envers. (Cordon déposé avant exécution de la soudure à proprement parler dans le joint.)
cordón soporte de raíz

bad restart. A local surface irregularity at a weld restart.
mauvaise reprise. Irrégularité locale de surface à l'endroit d'une reprise.
reanudación defectuosa

base metal. Metal to be joined, or surfaced, by welding, or brazing.
métal de base. Métal destiné à être assemblé ou rechargé par soudage, ou brasage.
metal de base

bead on plate. A single run of weld metal on a surface.
cordon déposé; chenille. Ligne de métal fondu déposée sur une surface.
cordón de metal depositado

bend test. Test for ductility performed by bending or folding, usually by steadily applied forces but in some instances by blows, a specimen having a cross section substantially uniform over a length several times as great as the largest dimension of the cross section.
essai de pliage. Essai consistant à soumettre une éprouvette à une déformation plastique par pliage, sans inversion du sens de la flexion au cours de l'essai. Le pliage est poussé jusqu'à ce que l'une des branches de l'éprouvette fasse avec le prolongement de l'autre un angle déterminé.
ensayo de doblado

blistering. A bulge on the surface caused by the expansion of a gas entrapped within a member and which increases its volume due to heating.
formation de cloques. Soulèvement de la peau d'une pièce provoqué par la dilatation, sous l'effet d'un chauffage, d'un gaz piégé.
formación de ampollas

braze welding. Brazing in which a joint of the open type is obtained, step by step, using a technique similar to fusion welding, with a filler material, the melting point of which is lower than that of the parent metal but higher than 450°C, but neither using capillary action as in brazing nor intentionally melting the parent metal.
soudobrasage. Brasage fort sur préparation ouverte dans lequel le joint brasé est obtenu de proche en proche, par une technique opératoire analogue à celle du soudage par fusion, mais sans action capillaire comme dans le brasage, ni fusion intentionnelle du métal de base. La température de fusion du produit d'apport est inférieure à celle du métal de base, mais supérieure à 450°C.
cobresoldeo

brazing. Processes of joining metals in which, during and after heating, molten metal is drawn by capillary action into the space between closely adjacent surfaces of the parts to be joined. In general, the melting point of the filler metal is above 450 °C, but always below the melting temperature of the parent metal. Flux may or may not be used.
brasage fort. Procédés consistant à assembler des pièces métalliques à l'aide d'un produit d'apport, à l'état liquide, ayant une température de fusion inférieure à celle des pièces à réunir, mais supérieure à 450 °C et mouillant le (les) métal (métaux) de base qui ne participe(nt) pas par fusion à la constitution de la brasure. L'opération se fait avec ou sans flux.
soldeo fuerte

brittle fracture. The sudden rupture without significant absorption of energy. Brittle fracture is principally a cleavage phenomenon and is not accompanied by a significant plastic deformation.
rupture fragile. Rupture brutale intervenant sans absorption importante d'énergie. La rupture fragile se produit principalement par clivage et n'est accompagnée d'aucune déformation plastique importante.
rotura frágil

building-up by welding; building-up. The deposition of metal on a surface until required dimensions are obtained.
rechargement; rechargement par soudage; rechargement curatif. Dépôt de métal sur une surface jusqu'à obtention des dimensions voulues.
soldadura de recargue

burn-through. A collapse of the weld pool resulting in a hole in the weld or at the side of the weld.
trou. Effondrement du bain de fusion entraînant la perforation de la soudure.
hueco

burnt weld. Weld metal which has been exposed to too high a temperature for too long a time and has therefore become excessively oxidized.
soudure brûlée. Métal fondu exposé pendant un temps excessif à une température trop élevée et qui a par conséquent été excessivement oxydé.
soldadura quemada

butt joint. A connection between two members aligned approximately in the same plane in the region of the joint.
assemblage soudé bout à bout; joint soudé bout à bout. Assemblage soudé constitué par deux éléments situés approximativement dans le prolongement l'un de l'autre dans un même plan.
soldadura a tope; unión soldada

butt weld with excess weld metal
soudure bout à bout avec surépaisseur
soldadura a tope con sobreespesor

butt weld. A weld connecting the ends or edges of two parts making an angle of between 0° and 45° inclusive in the region of the joint.
soudure bout à bout; soudure en bout. Soudure unissant deux éléments situés sensiblement dans le prolongement l'un de l'autre dans un même plan.
soldadura a tope; unión a tope

buttering by welding; buttering. A surfacing operation in which one or more layers of weld metal are deposited on the groove face of one or both members to be joined. Buttering provides a metallurgically compatible weld metal for subsequent completion of the weld.
beurrage; beurrage par soudage. Opération de rechargement dans laquelle une ou plusieurs couches de métal fondu sont déposées sur la face à souder de l'une ou des deux pièces à assembler. Le beurrage permet d'obtenir un métal fondu compatible du point de vue métallurgique en vue de l'achèvement ultérieur de la soudure.
untado; acolchado; recarque blando

C

carbon arc welding. Arc welding using a carbon electrode or electrodes.
soudage à l'arc avec électrode de carbone. Soudage à l'arc utilisant une ou plusieurs électrodes en carbone.
soldeo por arco con electrodo de carbón

carburisation. Introduction of carbon into a solid ferrous alloy by holding above Ac1 in contact with a suitable carbonaceous material, which may be solid, liquid or a gas. The carburised alloy is usually quench hardened.
cémentation. Opération consistant à enrichir superficiellement en carbone un produit ferreux maintenu à une température supérieure à Ac 1 dans un milieu carburant qui peut être solide, liquide ou gazeux. Les pièces sont ensuite généralement durcies par trempe.
cementación

cavity; void. The space between overlapping runs or between the underside of runs and the parent metal where insufficient penetration has been achieved.
cavité. Espace situé soit entre passes de soudage se chevauchant soit entre la face inférieure de passes et le métal de base, là où la pénétration est insuffisante.
cavidad

chipping hammer. A hand hammer designed for the removal of slag from weld deposits.
marteau à piquer. Marteau servant à détacher le laitier recouvrant les soudures.
martillo para picar

cladding by welding; surfacing. The deposition of metal on a surface to provide a layer giving properties different from those of the parent metal.
placage par soudage. Dépôt de métal sur une surface en vue d'obtenir un revêtement aux propriétés différentes de celles du substrat.
plaqueado por soldeo; chapeado

cleavage. The fracture of a crystal on a crystallographic plane of low index.
clivage. Rupture des grains suivant des plans cristallographiques simples.
rotura por clivaje

cold crack. A spontaneous fracture resulting from stresses occurring within the metal at or near ambient temperature.
fissure à froid. Fissure spontanée provoquée par des contraintes apparaissant dans le métal à température ambiante.
fisura en frío

cold cracking; delayed cracking. The result of the conjunction of three factors: hardening of the welded steel, hydrogen and stresses associated with the welding operation. Sometimes the phenomenon does not immediately become evident after welding, it is therefore also called "delayed cracking". Note: see also "cold crack".
fissuration à froid; fissuration différée. Résultat de la conjonction de trois facteurs: la trempe de l'acier soudé, l'hydrogène et les contraintes associées à l'opération de soudage. Le phénomène se manifeste parfois tardivement après soudage: c'est pourquoi on l'appelle aussi "fissuration différée". Note: Voir aussi le terme "fissure à froid".
fisuración en frío

consumable insert. A preplaced filler material which is fused to aid the formation of a weld made from one side only.
support à l'envers fusible; insert. Produit d'apport inséré avant soudage à la racine et dont la fusion facilite la formation de la soudure exécutée d'un seul côté.
placa soporte fusíble

corrosion due to welding; weld decay. Corrosive attack on a weldment due to either a modification to the structure or residual welding stresses.
corrosion due au soudage. Corrosion affectant un assemblage soudé et due soit à une modification de la structure, soit aux contraintes résiduelles de soudage.
corrosión debida al soldeo

coupon plate. A test piece made by adding plates to the end of a joint to give an extension of the weld for test purposes.
appendice témoin; coupon d'essai. Pièce d'essai obtenue en ajoutant aux extrémités d'un assemblage des appendices dans lesquels sont ensuite prélevées des éprouvettes.
chapa apéndice de ensayo

crack. A discontinuity produced by a local rupture which may arise from the effect of cooling or stresses.
fissure. Discontinuité à deux dimensions qui peut se produire en cours de refroidissement ou sous l'effet de contraintes.
grieta; fisura

crack arrest. In a piece subjected to static loading the arrest of the unstable propagation of the crack, beginning for example from a fault caused, either by the decrease of the effect of the loading as the extension of the crack continues, or when the crack meets a zone of increased toughness.
arrêt de fissure. Dans une pièce soumise à un chargement statique, arrêt de la propagation instable d'une fissure, amorcée par exemple à partir d'un défaut résultant soit de la diminution des actions résultant du chargement au fur et à mesure de l'extension de la rupture, soit par la rencontre d'une zone à ténacité accrue.
detención de la grieta

crack growth rate. The ratio of the increase of the length of a crack related to the corresponding number of loading cycles in the case of fatigue, or to the corresponding time interval in the case of stress corrosion.
vitesse de propagation de la fissure. Rapport de l'augmentation de longueur d'une fissure au nombre de cycles de sollicitation correspondant dans le cas de fatigue, ou à l'intervalle de temps correspondant dans le cas de corrosion sous tension.
velocidad de propagación de la grieta

crack tip opening displacement (CTOD). The crack displacement due to elastic and plastic deformation near the crack tip, generally to mode I.
ouverture à fond de fissure. Généralement en mode I (par ouverture), déplacement relatif des faces d'une fissure, résultant des déformations élastique et plastique au voisinage du front de fissure.
desplazamiento del fondo de la grieta (CTOD)

crater. In fusion welding, a depression at the termination of a weld bead.
cratère. En soudage par fusion, dépression à la fin d'un cordon de soudure.
cráter

crater crack. A crack in the end crater of a weld which may be longitudinal, transverse or star cracking.
fissure de cratère. Fissure se produisant dans un cratère et qui, suivant le cas, peut être longitudinale, transversale ou en étoile.
fisura de cráter

crater pipe; crater hole. The depression due to shrinkage at the end of a weld run and not eliminated before or during the deposition of subsequent weld passes.

retassure de cratère. Cavité (ou dépression) dans une reprise non éliminée avant ou pendant l'exécution de la passe suivante.

rechupe de cráter

creep. Time dependent strain occurring under stress.

fluage. Allongement progressif dans le temps, se produisant sous l'effet d'une charge.

fluencia

D

deposited metal. Filler metal after it becomes part of a weld or joint.

métal déposé. Métal d'apport après qu'il soit devenu partie intégrante de la soudure ou du joint soudé.

metal depositado

destructive testing. Testing to detect internal or external defects, or to assess mechanical or metallurgical properties by mechanical means, which generally result in the destruction of the material.

essais destructifs; contrôle destructif. Essais réalisés en vue soit de détecter des défauts internes ou externes, soit d'évaluer des caractéristiques mécaniques ou des propriétés métallurgiques, en mettant en oeuvre des moyens mécaniques provoquant généralement la destruction du matériau.

ensayos destructivos

dilution. The alteration of the composition of the metal deposited from a filler wire or electrode due to mixing with the melted parent material. It is usually expressed as the percentage of melted parent metal in the weld metal.

dilution. Modification de la composition du métal déposé par un fil d'apport ou une électrode, provoquée par mélange avec le métal de base fondu. La dilution s'exprime généralement en pourcentage de métal de base fondu dans le métal fondu.

dilución

double -V- groove weld. A butt weld in the preparation for which the edges of both components are double beveled, so that in cross section the fusion faces form two opposing V's.

soudure en X; soudure en V double. Soudure bout à bout avec préparation à double ouverture dans laquelle les bords des éléments sont chanfreinés de telle sorte que le profil du joint forme deux V opposés.

soldadura en X

E

eddy current testing. A non-destructive testing method in which eddy current flow is induced in the test object. Changes in the flow caused by variation in the object are reflected into a nearby coil(s) for subsequent analysis by suitable instrumentation and techniques.
contrôle par courants de Foucault. Méthode de contrôle non destructif consistant à induire des courants de Foucault dans la pièce à contrôler. La présence de défauts crée dans la (les) bobine(s) d'induction des variations d'impédance qui sont ensuite analysées.
inspección por corrientes de Foucault

effective length of weld. The length of continuous weld of specified dimensions.
longueur utile d'un cordon; longueur efficace d'un cordon. Longueur d'un cordon de soudure de dimensions spécifiées.
longitud útil de un cordón

effective throat. The minimum dimension of throat thickness used for purposes of design.
épaisseur nominale de la soudure. Valeur minimale de l'épaisseur de la soudure utilisée en calcul.
espesor nominal de la soldadura

elasto-plastic fracture mechanics. Analysis where one introduces the consolidation of the material to take account of its plastification.
mécanique élastoplastique de la rupture. Analyse dans laquelle on fait intervenir la consolidation du matériau pour tenir compte de sa plastification.
mecánica de fractura elastoplástica

electrode drying oven. A heated receptacle in which the electrodes can be dried in a specified manner, to ensure that the surface coating is free from moisture.
étuve. Four dans lequel les électrodes sont séchées dans des conditions prescrites de façon à éliminer en totalité l'humidité de leur enrobage.
estufa

electron beam welding; EB welding. Radiation welding using a focused beam of electrons.
soudage par faisceau d'électrons; soudage F.E.. Soudage par énergie de rayonnement utilisant un faisceau focalisé d'électrons.
soldeo por haz de electrones

electro-slag welding. Fusion welding using the combined effects of current and electrical resistance in a consumable electrode or electrodes, and a conducting bath of molten slag through which the electrode passes into the molten pool; both the pool and the slag bath being retained in the joint by cooled shoes which move progressively upwards. After the initial arcing period, the end of the electrode is covered by the rising slag; melting then continues until the joint is completed. Electrodes may be bare or cored wire(s) or strip(s) or plate(s).

soudage vertical sous laitier; soudage sous laitier; soudage électroslag; soudage sous laitier électroconducteur. Soudage par fusion utilisant les effets combinés du courant et de la résistance électrique dans une (ou plusieurs) électrode fusible, ainsi qu'un bain de laitier électroconducteur fondu à travers lequel l'électrode pénètre dans le bain de fusion; ce dernier, ainsi que le laitier sont retenus dans le joint par des patins refroidis se déplaçant progressivement de bas en haut. Après la période initiale d'amorçage de l'arc, l'extrémité de l'électrode est recouverte par le laitier montant et la fusion continue jusqu'à l'achèvement de la soudure. Les électrodes peuvent être des fils nus ou fourrés, des feuillards ou des plaques.

soldeo por electroescoria

embrittlement. Reduction in the normal ductility of a metal due to a physical or chemical change.

fragilisation.   Abaissement de la ductilité d'un métal sous l'effet d'un changement physique ou chimique.

fragilización

excess weld metal in a butt weld; reinforcement.   Weld metal lying outside the plane joining the toes.

surépaisseur (d'une soudure bout à bout). Métal fondu situé au-delà du plan joignant les lignes de raccordement.

sobreespesor (en una unión a tope)

excess weld metal in a fillet weld; reinforcement.   Weld metal lying outside the plane joining the toes.

surépaisseur (d'une soudure d'angle); convexité. Métal fondu situé au-delà du plan joignant les lignes de raccordement.

sobreespesor (en una unión de ángulo); convexidad

excessive reinforcement; excess weld metal. An excess of weld metal at the face(s) of the weld.

surépaisseur excessive. Excès d'épaisseur du métal déposé.

sobreespesor excesivo

excessive root reinforcement; excess penetration bead. Excess weld metal protruding through the root of a fusion weld made from one side only.
surépaisseur excessive à la racine; excès de pénétration. Excès de métal à la racine d'une soudure exécutée d'un seul côté.
exceso de penetración; sobreespesor excesivo en la raíz

explosion welding. Pressure welding to make lap joints or cladding in which the overlapping workpieces are welded when impacted together by the detonation of an explosive charge.
soudage par explosion. Soudage par pression utilisé pour réaliser des assemblages à recouvrement ou des placages et dans lequel les pièces sont plaquées l'une contre l'autre par suite de la mise à feu d'une charge explosive.
soldeo por explosión

eye protection
protection de l'oeil
protección ocular

F

face bend test. A bend test in which the specified side of the weld specimen is in tension, namely: a) the side opposite that containing the root is nearer; b) either weld face when the root is central; c) the outer side of a pipe in welds made with pressure.
essai de pliage à l'endroit; pliage endroit. Essai de pliage dans lequel la face correspondant à l'ouverture du chanfrein est mise en extension. Sur les assemblages symétriques (en X par exemple) les essais doivent avoir lieu en mettant en extension chacune des faces de l'assemblage. Sur les tubes soudés avec pression, la paroi extérieure du tube est mise en extension.
ensayo de doblado normal por el anverso

faying surface. The surface of one component which is intended to be in contact with the surface of another component to form a joint assembly.
surface de contact. Surface d'un élément destinée à être en contact avec la surface d'un autre élément en vue de constituer un assemblage.
superficie de contacto

field weld; site weld. A weld made at the location where the assembly is to be installed.
soudure sur chantier. Soudure exécutée sur les lieux où l'assemblage doit être installé.
soldadura en obra; soldadura en campo

filler material. Consumables to be added during the welding operation to form the weld, i.e. welding wires and stick electrodes, etc.
produit d'apport. Produit consommable ajouté au cours d'une opération de soudage et participant à la constitution du joint. Il se présente généralement sous la forme de fil ou de baguette.
material de aportación

filler metal.   Metal added during welding, braze welding, brazing or surfacing.
métal d'apport.   Métal ajouté pendant le soudage, le soudobrasage, le brasage ou le rechargement.
metal de aportación

fillet weld. A fusion weld, other than a butt, edge or fusion spot weld which is approximately triangular in transverse cross section.
soudure d'angle. Soudure par fusion autre qu'une soudure bout à bout, sur chant ou par fusion par points et dont la coupe transversale est sensiblement triangulaire.
soldadura de ángulo

fish eye. A small bright area of cleavage fracture which is only visible on the fractured surface of weld metal.
oeil de poisson (pluriel: oeils de poisson); point blanc. Petite zone brillante de rupture par clivage, visible uniquement sur la cassure du métal fondu.
ojo de pez

fixture. A device to hold the parts to be joined in proper relation to each other.
montage; dispositif de serrage; gabarit de soudage; mannequin. Dispositif permettant la mise en place exacte des divers éléments d'un ensemble en vue de leur assemblage par soudage.
plantilla de soldeo

flame hardening. Quench hardening in which the heat is applied directly by a flame.
trempe à la flamme; trempe au chalumeau. Trempe dans laquelle la chaleur est fournie directement par une flamme.
temple a la llama

flash; arc eye. Irritation of the eye caused by exposure to radiation from an electric arc. Note: not to be confused with arc strike ou stray flash.
coup d'arc (irritation   de l'oeil). Irritation de l'oeil provoquée par l'exposition aux radiations d'un arc électrique. Ne pas confondre avec coup d'arc (sur la pièce).
irritación ocular (debida al arco)

flat position butt weld.
soudure bout à bout à plat; soudure en bout à plat; soudure bout à bout horizontale; soudure en bout horizontale.
<u>*soldadura horizontal a tope*</u>

flux cored wire metal arc welding with active gas shield. MAG welding using a flux cored wire electrode.
soudage MAG avec fil fourré; soudage sous gaz actif avec fil fourré. Soudage MAG dans lequel le fil-électrode est un fil fourré.
<u>*soldeo MAG con alambre tubular*</u>

flux cored wire welding. Metal arc welding using a cored wire electrode without additional gas shielding.
soudage à l'arc avec fil fourré; soudage avec fil fourré. Soudage à l'arc avec fil-électrode fusible fourré sans aucune protection gazeuse extérieure.
<u>*soldeo por arco con alambre tubular*</u>

flux inclusion.   Flux entrapped in weld metal. According to circumstances such inclusions may be linear, isolated, or others.
inclusion de flux. Résidu de flux emprisonné dans la soudure. Suivant le cas, il peut s'agir d'inclusions alignées (ou en chapelet), isolées ou autres.
<u>*inclusión de fundente*</u>

foot protection; protective boots; safety boots.
chaussures de sécurité.
<u>*zapatos de seguridad*</u>

forge welding. Pressure welding whereby the workpieces are heated in a forge and the weld is made by applying blows or some other impulsive force sufficient to cause permanent deformation at the interfaces. The force may be applied manually (blacksmith welding) or mechanically (hammer welding).
soudage à la forge. Soudage par pression dans lequel les pièces sont chauffées à l'air dans une forge et réunies par martelage ou par tout autre type d'effort assez important pour provoquer une déformation permanente au niveau de l'interface. L'effort peut être appliqué manuellement ou mécaniquement.
<u>*soldeo por forja*</u>

fracture mechanics.   All the concepts and law of behavior permitting the description of the behavior of a body containing sharp notches (cracks or surface faults) and subjected to various kinds of loading (static, fatigue, corrosion, etc.) applied singly or together.
mécanique de la rupture. Ensemble des concepts et lois de comportement permettant de décrire le comportement d'un corps comportant des entailles aiguës (fissures ou défauts plans) et soumis à divers types de chargement (statique, fatigue, corrosion,etc.) appliqués isolément ou simultanément.
<u>*mecánica de fractura*</u>

friction welding. Pressure welding in which the interfaces are heated by friction, normally by rotating one or both workpieces in contact with each other or by means of a separate rotating friction element; the weld is completed by an upset force after relative rotation. In the case of inertia friction welding, the rotational energy is stored in a flywheel to which the chuck holding the part to be rotated, is connected. The rotational speed decreases continuously.

soudage par friction.   Soudage par pression dans lequel les faces à assembler sont chauffées par friction, soit par rotation de l'une des pièces serrée contre l'autre (ou par rotation des deux pièces), soit par rotation d'un élément intermédiaire; la soudure est achevée par un effort de refoulement appliqué pendant la rotation ou une fois celle-ci arrêtée. Dans le cas du soudage par friction avec volant d'inertie, l'énergie cinétique est emmagasinée dans un volant auquel est relié le mandrin maintenant la pièce de rotation. La vitesse de rotation diminue continuellement.

soldeo por fricción

full penetration weld; complete penetraction weld.
soudure à pleine pénétration; soudure pénétrée.
soldadura con pentración completa

fusion face. The portion of the surface or of an edge, which is to be fused in making a fusion weld.
face à souder par fusion.   Partie de la face à souder qui doit être fondue lors de l'exécution d'une soudure par fusion.
cara fusible

fusion welding. Welding processes involving localized melting without the application of force and with or without the addition of filler material.
soudage par fusion.   Procédés de soudage comportant une fusion localisée, exécutée avec ou sans addition de produit d'apport, sans intervention de pression.
soldeo por fusión

fusion zone. The part of the parent metal which is melted into the weld metal.
zone de dilution.   Partie du métal de base, fondue et mélangée avec le métal fondu.
zona de difusión

G

gap; opening. Gap in the space between two faces or edges.
écartement des bords; écartement.   Espace entre les faces à souder.
separación de bordes

gas metal arc welding. Metal arc welding in which the arc and molten pool are shielded from the atmosphere by a shroud of gas supplied from an external source.

soudage sous protection gazeuse avec fil fusible; soudage sous protection gazeuse avec fil-électrode fusible; soudage à l'arc sous protection gazeuse avec fil-électrode fusible. Soudage à l'arc avec fil-électrode fusible dans lequel l'arc et le bain de fusion sont isolés de l'atmosphère ambiante par une enveloppe de gaz provenant d'une source extérieure.

soldeo por arco con electrodo fusible bajo protección gaseosa

gas metal arc welding; MIG welding. Metal arc welding using a wire electrode and where the shielding is provided by an inert gas.

soudage MIG. Soudage à l'arc avec fil-électrode fusible, dans lequel la protection est assurée par un gaz inerte.

soldeo MIG; soldeo por arco bajo protección de gas inerte con alambre fusible

gas shielded arc welding. Arc welding processes in which the arc and molten pool are shielded from the atmosphere by an envelope of gas supplied from an external source.

soudage sous protection gazeuse; soudage à l'arc sous protection gazeuse. Procédés de soudage à l'arc dans lesquels l'arc et le bain de fusion sont isolés de l'atmosphère ambiante par une enveloppe de gaz provenant d'une source extérieure.

soldeo por arco bajo protección gaseosa

gas tungsten arc welding; GTA welding; TIG welding. Arc welding using a non-consumable pure or activated tungsten electrode where shielding is provided by inert gas.

soudage TIG. Soudage à l'arc avec électrode non fusible en tungstène pur ou activé dans lequel la protection est assurée par un gaz inerte. On peut utiliser un produit d'apport.

soldeo TIG; soldeo por arco en atmósfera inerte con electrodo de volframio

gas welding. Fusion welding processes in which the heat for welding is produced by the combustion of a fuel gas, or a mixture or fuel gases, with an admixture of oxygen or air.

soudage aux gaz. Procédés de soudage par fusion dans lesquels la chaleur nécessaire au soudage est produite par la combustion d'un gaz ou d'un mélange de gaz combustibles avec de l'oxygène ou de l'air.

soldeo con gas

groove. An opening or channel in the surface of a part or between two components, which provides space to contain a weld.
joint. Espace entre deux éléments ou sillon à la surface d'un élément, destiné à être rempli de métal déposé.
abertura de la unión

groove face. That surface of a member included in the groove.
face à souder; bord à souder. Surface délimitant le joint.
cara a soldar

groove weld; butt weld in a butt joint. A weld between two parts making an angle to one another of 135° to 180° inclusive in the region of the weld such that a line parallel to the surface of one part, perpendicular to the line of the joint, and passing through the center of the fusion face of that part, passes through the fusion face of the other part.
soudure en bout sur bords chanfreinés; soudure bout à bout sur bords chanfreinés. Soudure unissant deux éléments faisant entre eux un angle de 135° à 180° au voisinage de la soudure et telle qu'une ligne parallèle à la surface de l'un des éléments, perpendiculaire à l'axe de la soudure et passant par le milieu de la face à souder de ce même élément, coupe la face à souder de l'autre élément.
soldadura a tope

guided bend test. A bend test made by bending the specimen around a specified form.
essai de guidage plié; essai de pliage sur mandrin. Essai de pliage dans lequel l'éprouvette est fléchie en son milieu à l'aide d'un mandrin ou dégorgeoir de diamètre déterminé.
ensayo de doblado sobre mandril

H

hard zone crack. A crack arising from the effect of modifications in the crystalline structure after welding, causing volumetric changes, increased hardness as well as stresses in the metal.
fissure due au durcissement. Fissure résultant du changement de structure après le soudage qui a pour effet des modifications volumiques, une dureté accrue ainsi que des contraintes dans le métal.
grieta debida al endurecimiento

harden and temper; quench and temper; QT. Treatment of hardening followed by tempering at elevated temperature with the object to obtain the required combination of mechanical properties, in particular of good ductility.
trempe et revenu. Traitement de durcissement par trempe suivi d'un revenu à température élevée ayant pour objet d'obtenir la combinaison recherchée entre les propriétés mécaniques et, en particulier, une bonne ductilité.
temple y revenido

hardening. Increasing the hardness by suitable treatment, usually involving heating and cooling.
durcissement. Augmentation de dureté obtenue par un traitement comportant généralement un chauffage et un refroidissement.
endurecimiento

hardfacing by welding; hardfacing; hard-facing. The application of a hard and/or wear resisting material to a surface of a component.
rechargement dur; rechargement dur par soudage. Dépôt, sur la surface d'une pièce, d'un matériau dur et/ou résistant à l'usure.
recargue duro

hardness test. An indentation test using calibrated machines to force an indenter, under specified conditions, into the surface of the material under test, and to measure the resulting impression after the removal of the load.
essai de dureté. Essai consistant à enfoncer dans une pièce un pénétrateur (bille ou diamant conique ou pyramidal), sous une charge donnée, parfaitement tarée, puis à mesurer l'empreinte laissée sur la surface après enlèvement de la charge.
ensayo de dureza

heat affected zone; HAZ. The part of parent metal which is metallurgically affected by the heat of welding or thermal cutting, but not melted.
zone thermiquement affectée. Partie du métal de base affectée métallurgiquement par la chaleur de soudage ou de coupage thermique, sans toutefois être fondue.
zona térmicamente afectada; ZTA

horizontal butt weld; horizontal-vertical butt weld.
soudure en corniche.
soldadura en cornisa

hot crack. A discontinuity produced by tearing of the metal while at an elevated temperature.
fissure à chaud. Discontinuité provoquée par un déchirement du métal alors qu'il se trouve encore à une température élevée.
fisura en caliente

hydrogen embrittlement. A condition of low ductility in metals resulting from the absorption of hydrogen. This will lead to a modification of the mechanism of the fatigue propagation and thereby causing a considerable acceleration in the propagation rate of a crack.
fragilisation par l'hydrogène. Abaissement de la ductilité d'un métal causée par l'absorption d'hydrogène. Ceci provoque une modification du mécanisme de progression par fatigue d'une fissure, entraînant ainsi une augmentation notable de la vitesse de propagation.
fragilización por hidrógeno

I

image quality indicator; IQI. A graduated device which is employed to permit the quality of a radiographic image to be judged.
indicateur de qualité d'image; IQI. Dispositif employé pour permettre de juger de la qualité de l'image en contrôle radiographique.
indicador de calidad de imagen I.C.I.

inadequate joint penetration; lack of penetration; incomplete penetration. Lack of fusion between the faying surfaces due to failure of weld metal to extend into the root of the joint.
manque de pénétration. Absence partielle de fusion des bords à souder, laissant subsister un interstice entre ces bords parce que le métal fondu n'a pas atteint la racine du joint.
falta de penetración

inclusion; solid inclusion. Slag or other foreign matter entrapped during welding. The defect is usually irregular in shape.
inclusion; inclusion solide. Corps solide étranger emprisonné dans la masse du métal fondu. Le défaut est généralement de forme irrégulière.
inclusión

incomplete fusion; lack of fusion. Lack of union between weld metal and parent metal or weld metal and weld metal. It will be one of the following: lack of side wall fusion, lack of inter-run fusion or lack of fusion at the root of the weld. Note: In certain countries one uses the terms "collage noir" and "collage blanc" depending on the presence or absence of oxide inclusions together with the lack of fusion.
manque de fusion; collage. Manque de liaison entre le métal déposé et le métal de base ou entre deux couches contiguës de métal déposé. Il y a lieu de distinguer le manque de fusion latérale, qui intéresse les bords à souder, le manque de fusion(1) entre passes et le manque de fusion à la racine de la soudure. Note:(1) Dans certains pays, on utilise les termes "collage noir" lorsqu'il y a interposition d'une couche d'oxyde non fondu entre le métal déposé et le métal de base, et "collage blanc" lorsqu'un lien entre le métal d'apport et le métal de base est assuré par une pellicule d'oxyde fondu.
falta de fusión; pegadura

incorrect weld profile. Too large an angle (a) between the plane of the parent metal surface and a plane tangential to the weld bead surface at the toe.
défaut de raccordement. Angle (a) trop important du dièdre formé par le plan tangent au métal de base et le plan tangent au métal d'apport et passant par la ligne de raccordement.
perfil de soldadura incorrecto; ángulo del sobreespesor incorrecto

induction hardening. Quench hardening in which the heat is generated by electrical induction.
trempe par induction. Trempe dans laquelle le chauffage est produit par des courants induits.
temple por inducción

interdendritic shrinkage; solidification void. An elongated shrinkage cavity formed between dendrites during cooling, which may contain entrapped gas. Such a defect is generally to be found perpendicular to the weld face.
retassure interdendritique. Cavité de forme allongée qui se produit entre les dendrites au cours du refroidissement et dans laquelle se trouve emprisonné du gaz. Un tel défaut est généralement perpendiculaire aux faces extérieures de la soudure.
rechupe interdendrítico

intergranular crack; intercrystalline crack. A crack which runs along the boundary between crystals.
fissure intergranulaire; fissure intercristalline. Fissure au joint des grains.
fisura intercristalina

J

J-integral. A mathematical expression (a line or surface integral) used to characterize the local stress-strain field around the crack front.
intégrale "J". Grandeur mathématique (intégrale de surface ou intégrale de contour) utilisée pour définir le champ de contraintes et de déformations au voisinage de l'extrémité d'une fissure.
integral "J"

joint penetration. The depth of penetration plus the root penetration, but excluding the excess metal.
profondeur de pénétration. Pénétration rapportée à la totalité de la soudure (hauteur de la préparation plus pénétration à la racine) sans la surépaisseur.
profundidad de penetración

K

knee protection
genouillère
rodillera

L

lamellar tear. A defect which appears in the base metal, directly under fillet welds in lap or T joints, due principally to the restraint caused by the shrinkage after welding which the weld seam exerts in a perpendicular direction upon the surface of the plate that carries it. The defect is characterized by a system of bands of tears running parallel to the surface, aligned in the direction of the rolling of the plate, and connected by sharp steps of small height.

arrachement lamellaire (défaut). Défaut apparaissant dans le métal de base, à l'aplomb des soudures d'angle ou des soudures des assemblages en T ou par recouvrement, sous l'effet principalement des contraintes que, du fait du retrait après soudage, le cordon de soudure exerce perpendiculairement à la surface de la tôle qui le porte. Le défaut est caractérisé par un système de plages de décohésion, parallèles à cette surface, allongées dans la direction du laminage de la tôle, et raccordées par des ressauts de faible hauteur relative.

desgarre laminar (defecto)

land. The straight portion of a fusion face between the root face and the curved part of a J preparation.

lèvre. Partie rectiligne d'une préparation en J ou en U, comprise entre le méplat et la partie curviligne.

labio

lap joint. A connection between two overlapping parts making an angle to one another of 0° to 5° inclusive in the region of the weld.

assemblage soudé à recouvrement; joint soudé à recouvrement; assemblage soudé à clin; joint soudé à clin. Assemblage soudé constitué par deux éléments se recouvrant et formant entre eux un angle compris entre 0 et 5° au voisinage du joint.

unión soldada a solape; soldadura con solape

laser beam welding. Radiation welding using a coherent beam of monochromatic radiation from a laser.

soudage par faisceau laser; soudage laser; soudage au laser. Soudage par énergie de rayonnement utilisant un faisceau cohérent de lumière monochromatique émis par un laser.

soldeo por rayo láser; soldeo láser

leg length; leg. The distance from the actual or projected intersection of the fusion faces and the toe of a fillet weld, measured across the fusion face.

côté. Distance mesurée à travers la surface de la soudure et comprise entre l'intersection des faces à souder (vu de leur prolongement) et la ligne de raccordement d'une soudure d'angle.

largo de patas

leg protection; leggings; gaiters
jambière-guêtre; guêtre; guêtron
polaina

linear misalignment. Misalignment between two welded pieces such that, although their surfaces are parallel, they are not in the required same plane.
défaut d'alignement. Non respect du niveau prévu entre deux pièces soudées. Ce défaut est exprimé généralement par la mesure d'une dénivellation.
defecto de alineación

liquid penetrant inspection. A method of inspection employing a colored or fluorescent penetrating fluid which allows the detection of faults issuing at the surface of the piece under examination.
contrôle par ressuage; ressuage. Méthode de contrôle permettant à l'aide d'un liquide d'imprégnation coloré ou fluorescent de localiser les défauts débouchant à la surface de la pièce examinée.
inspección por líquidos penetrantes

load bearing weld; structural weld. A weld made to transmit a force or forces from one member of a joint to another or between several adjoining members of a weldment.
soudure de résistance; soudure travaillante. Soudure destinée à transmettre un ou des efforts d'un élément de l'assemblage à un autre ou entre plusieurs éléments adjacents de l'assemblage.
soldadura estructural

local undercut. An intermittent groove at the toe(s) of the weld.
morsure. Manque local de métal, situé sur les bords du cordon de soudure.
mordedura local

longitudinal crack. A crack substantially parallel to the axis of the weld. It may be situated in the weld metal, at the weld junction, in the heat affected zone or in the parent metal.
fissure longitudinale. Fissure dont la direction principale est voisine de celle de l'axe de la soudure. Elle peut se situer, suivant le cas, dans le métal fondu, dans la zone de liaison, dans la zone thermiquement affectée ou dans le métal de base.
grieta longitudinal

longitudinal test specimen. A test specimen which is longitudinally bisected by the portion of the weld included in it.
éprouvette longitudinale. Éprouvette prélevée longitudinalement au joint soudé.
probeta longitudinal

low cycle fatigue. Rate of stress reversals characterized by the presence of macroscopic cyclic plastic deformation producing a fracture normally between 100 and 50 000 cycles.

fatigue oligocyclique. Régime de sollicitation caractérisé par la présence de déformations plastiques macroscopiques cycliques, correspondant en une rupture en un nombre de cycles normalement compris entre 100 et 50 000.

fatiga oligocíclica

M

magnetic particle inspection. A method of inspection which facilitates the detection of superficial defects on a piece of ferrous metal subjected to a magnetic field. Where they occur, these defects cause magnetic disturbances which are revealed by the indicating media put into contact with the surface of the test piece.

magnétoscopie; contrôle magnétoscopique. Méthode de contrôle permettant de localiser les défauts superficiels d'une pièce ferromagnétique soumise à un champ magnétique. Ces défauts provoquent à leur endroit des fuites magnétiques qui sont mises en évidence par les produits indicateurs mis en contact avec la surface de la pièce.

inspección por partículas magnéticas; magnetoscopía

manual welding; hand welding. Welding operation in which the welding variables are controlled by the operator and the means of making the weld are held in the hand.

soudage manuel. Opération de soudage dans laquelle les paramètres de soudage sont commandés par le soudeur et l'appareillage est guidé manuellement.

soldeo manual

measured throat thickness of a fillet weld. This is measured without penetration beyond the root of the preparation.

épaisseur de la soudure d'angle (de la surface du cordon au sommet de l'angle). Valeur mesurée sans tenir compte de la pénétration dans l'angle.

espesor de una soldadura de ángulo

mechanized welding. Welding operation in which all welding and operating variables are controlled by mechanical and/or electronic means.

soudage mécanisé. Opération de soudage dans laquelle tous les paramètres de soudage et les paramètres opératoires sont commandés par des moyens mécaniques et/ou électroniques.

soldeo mecanizado

metal arc welding. Arc welding using a consumable electrode.

soudage à l'arc avec électrode fusible; soudage à l'arc avec électrode consommable. (Définition inutile).

soldeo por arco con electrodo fusible

metallic inclusion. A particle of foreign metal trapped in the weld metal. It may be of tungsten, of copper, or of other metal.
inclusion métallique. Particule de métal étranger emprisonnée dans la masse du métal fondu. On peut trouver du tungstène, du cuivre ou un autre métal.
inclusión metálica

microfissure; microcrack. A crack which can only be detected by microscope giving a six-fold magnification.
microfissure. Fissure ne pouvant être mise en évidence qu'avec un grossissement d'au moins X 6.
microgrieta; microfisura

mode (I, II, III). One of the three classes (I, II, III) of displacement of the surfaces of the crack in relation to each other and in relation to the crack tip.
mode (I, II, III). L'une des trois manières (I, II, III) dont les surfaces de la fissure se déplacent, à la fois l'une par rapport à l'autre et par rapport au front de la fissure.
modo (I, II, III)

N

nick break test. A fracture test in which a specimen is broken from a notch cut at a predetermined position where the interior of the weld is to be examined.
essai de texture; essai de compacité. Essai consistant à rompre une éprouvette dans la soudure en y pratiquant une amorce de rupture par trait de scie. On examine les défauts apparents sur la section rompue.
ensayo de textura

nondestructive testing. Testing to detect internal, surface and concealed defects or flaws in materials using techniques that do not damage or destroy the items being tested.
essais non destructifs; END; contrôle non destructif; CND. Essai en vue de détecter dans les matériaux des défauts internes, superficiels ou cachés, en mettant en oeuvre des techniques ne provoquant ni endommagement ni destruction des biens soumis à l'essai.
ensayos no destructivos; control no destructivo CND

normalizing. Heat treatment consisting of austenitisation followed by cooling in still air.
traitement de normalisation. Traitement thermique comportant une austénitisation suivie d'un refroidissement à l'air calme.
normalización

notch sensitivity. A measure of the reduction in strength of a metal caused by the presence of stress concentration. Values can be obtained from static, impact or fatigue tests.
sensibilité à l'entaille. Une mesure de la réduction de la résistance d'un métal causée par la présence de la concentration de contrainte. Les valeurs peuvent être obtenues par des essais statiques, des essais de choc ou des essais de fatigue.
sensibilidad a la entalla

O

verhead butt weld
soudure bout à bout au plafond; soudure en bout au plafond
soldadura a tope sobre techo

overhead position welding
soudage au plafond
soldeo sobre techo

overheated weld. If a weld is exposed to excessive heat for too long a time during the operation there is danger of the formation of overlarge crystals.
soudure surchauffée. Si, au cours du soudage, une soudure est exposée pendant trop longtemps à une température trop élevée, il risque de s'y former des grains excessivement gros.
soldadura sobrecalentada

overlap. Excess of weld metal of the toe of the weld covering the parent metal surface but not fused to it.
débordement. Excès de métal déposé qui se répand sur la surface du métal de base, sans liaison intime avec celui-ci.
desbordamiento

oxide inclusion. Metallic oxide trapped in the weld metal during solidification.
inclusion d'oxyde. Oxyde métallique emprisonné dans le métal au cours de la solidification.
inclusión de óxido

oxy-acetylene welding. Gas welding where the fuel gas is acetylene.
soudage oxyacétylénique. Soudage au gaz dans lequel on utilise de l'acétylène comme gaz combustible.
soldeo oxiacetilénico

P

partly mechanized welding. Welding operation in which some of the welding variables are mechanically controlled, but manual guidance is necessary.
soudage semi-mécanisé. Opération de soudage dans laquelle certains des paramètres de soudage sont commandés mécaniquement mais un guidage manuel est nécessaire.
soldeo parcialmente mecanizado

penetration. 1) In fusion welding, the depth to which the parent metal has been fused. **2)** In spot, seam or projection welding, the distance from the interface to the edge of the weld nugget, measured in each case on the cross section through the center of the weld and normal to the surface.
pénétration. 1) En soudage par fusion, profondeur jusqu'à laquelle le métal de base est fondu. ***2)*** En soudage par résistance par points, à la molette ou par bossage, distance de l'interface au bord du noyau, mesurée sur la section transversale passant par le centre de la soudure et perpendiculaire à la surface.
penetración

penetration bead. Weld metal protruding through the root of a fusion weld made from one side only.
cordon de pénétration. Cordon formant saillie à la racine d'une soudure par fusion exécutée d'un seul côté.
cordón de penetración

permanent backing. A piece of metal placed at the root and penetrated by the weld metal. It may remain as part of the joint or be removed by machining or other means.
support à l'envers subsistant. Pièce de métal disposée à la racine de la soudure et pénétrée par le métal fondu. Après soudage elle peut soit rester, formant partie intégrante du joint soudé, soit être éliminée par usinage ou autre méthode.
placa soporte permanente

plasma arc welding. Arc welding using the plasma of a constricted arc. Shielding may be supplemented by an auxiliary gas. Filler material may or may not be supplied.
soudage plasma; soudage au plasma; soudage au jet de plasma. Soudage à l'arc utilisant le plasma d'un arc étranglé. La protection peut être complétée par un gaz auxiliaire. On peut utiliser un produit d'apport.
soldeo por plasma

plug weld (circular). A weld made by filling a hole in one component of a workpiece with filler metal so as to join it to the surface of an overlapping component exposed through the hole.

soudure en bouchon. Soudure obtenue en remplissant de métal fondu un trou situé dans l'un des éléments de la pièce de manière à le réunir à la surface de l'élément sous-jacent visible par le trou, qui peut être circulaire ou allongé.

soldadura en tapón

porosity. A group of gas pores.

groupe de soufflures. Groupe de soufflures sphéroïdales.

porosidad

post weld heat treatment; PWHT. Any heat treatment subsequent to welding.

traitement thermique après soudage. Tout traitement thermique postérieur au soudage.

tratamiento térmico postsoldeo

precipitation hardening. Hardening caused by precipitation of a constituent from a supersaturated solid solution.

durcissement par précipitation. Durcissement provoqué par la précipitation d'un composé dans une solution solide sursaturée.

endurecimiento por precipitación

preheating. The application of heat to the base metal immediately before welding, brazing, soldering, thermal spraying or cutting.

préchauffage. Action de chauffer le métal de base immédiatement avant l'opération de soudage, brasage, projection à chaud ou coupage thermique.

precalentamiento

pressure fusion welding. Welding with fusion which employs static or dynamic pressure to complete the union.

soudage par fusion avec pression. Soudage comportant l'intervention combinée d'une pression statique ou dynamique, exercée sur les pièces, et d'une fusion pour assurer la formation du joint.

soldeo por fusión con presión

protective screen
écran de protection; écran protecteur
pantalla de protección

Q

quench hardening. Hardening a ferrous alloy by austenitising, then cooling it rapidly so that some or all of the austenite is transformed into martensite.
durcissement par trempe. Durcissement d'un produit ferreux obtenu par austénitisation suivie d'un refroidissement rapide de manière à transformer tout ou partie de l'austénite en martensite.
temple con autorrevenido

R

radiating cracks. Cracks radiating from a common point. They may be found in the weld metal, in the heat affected zone or in the parent metal. Note: Small cracks of this type are known as star cracks.
fissures rayonnantes. Groupe de fissures issues d'un même point et situées, suivant le cas, dans le métal fondu, dans la zone thermiquement affectée ou dans le métal de base. Note: En anglais, lorsque ces fissures rayonnantes sont de petites dimensions, elles prennent la dénomination "star cracks" (fissures en étoile).
fisuras radiales

radiographic inspection; radiography. A method of inspection which employs X or gamma (g) rays or neutrons, which are able to penetrate a piece of metal to produce an image of a fault within this piece of metal upon a sensitive screen or a radiograph.
contrôle radiographique; radiographie. Méthode d'essai utilisant la transparence d'une pièce métallique aux rayons ionisants (rayons X, rayons gamma, neutrons, etc.) et permettant d'obtenir l'ombre des défauts sur un écran ou un radiogramme.
inspección radiográfica; control radiográfico

recovery annealing. Annealing with heating below Ac1 in order to restore, at least partially, the physical or mechanical properties, without visible modification to the crystalline structure: reduction of hardness, of electrical resistance, work hardening, etc.
traitement de restauration. Recuit effectué au-dessous de Ac1 en vue de restaurer au moins partiellement les propriétés physiques ou mécaniques sans modification apparente de la structure: diminution de la dureté, de la résistivité, de l'écrouissage, etc.
recocido de restauración

recrystallisation annealing. Annealing cold worked metal to produce a new grain structure without phase change.
traitement de recristallisation; recuit de recristallisation. Recuit ayant pour objet de provoquer, sans changement de phase, le développement d'une nouvelle structure dans un métal écroui.
recocido de recristalización

residual welding stress. Stress remaining in a metal part or structure as a result of welding.

contrainte résiduelle de soudage; tension résiduelle de soudage. Contrainte subsistant dans une pièce ou une construction métallique et résultant de l'opération de soudage.

tensión residual de soldeo

resistance seam welding. Resistance welding in which force is applied continuously and current continuously or intermittently to produce a linear weld, the workpiece being between two electrode wheels or an electrode wheel and an electrode bar. The wheels apply the force and current and rotate either continuously to produce a continuous seam weld or on a start and stop program to produce a discontinuous seam weld.

soudage à la molette; soudage au galet. Soudage par résistance dans lequel on applique un effort continu et le courant par intermittence ou en continu pour obtenir une soudure linéaire, les pièces étant placées entre deux molettes ou entre une molette et une barre-électrode. L'effort et le courant sont transmis par les molettes, animées d'un mouvement de rotation soit continu pour obtenir une soudure continue soit intermittent avec temporisation programmée pour obtenir une soudure discontinue.

soldeo por roldana

resistance spot welding. Resistance welding in which the weld is produced at a spot in the workpiece between spot welding electrodes, the weld being of approximately the same area as the electrode tips. Force is applied continuously to the spot by the electrodes during the process.

soudage par résistance par points. Soudage par résistance dans lequel la soudure est exécutée en un point de la pièce situé entre les électrodes de soudage par points, la surface du point de soudure étant approximativement la même que celle des pointes d'électrodes. Un effort continu est appliqué sur le point par les électrodes pendant l'opération de soudage.

soldeo por resistencia por puntos

resistance welding. Welding processes in which the weld is produced by heat obtained from the resistance of the workpieces to electric current in a circuit of which they are part (Joule's heat) and by the application of pressure.

soudage par résistance. Procédés de soudage dans lesquels la soudure est exécutée avec pression en utilisant pour le chauffage la résistance des pièces au passage d'un courant électrique traversant l'assemblage (effet Joule).

soldeo por resistencia

rewelding; repairing the weld; weld repair
soudure de réparation; reprise de la soudure (pour réparer un défaut) retouche de la soudure
soldadura de reparación (por un defecto); retoque

robotic welding.   Welding operation executed by a reprogrammable multifunctional manipulator, i.e. a robot.
soudage robotisé.   Opération de soudage exécutée par un manipulateur reprogrammable polyvalent, c'est-à-dire un robot.
soldeo robotizado

root bend test.   A reverse bend test in which the root of the weld is on the tension side.
essai de pliage mettant la racine de la soudure en extension.   Essai de pliage dans lequel la face correspondant à la racine de la soudure est mise en extension.
ensayo de doblado por la raíz

root concavity.   A shallow groove due to shrinkage of a butt weld at the root.
retassure à la racine.   Manque d'épaisseur à la racine dû au retrait du métal fondu.
rechupe en la raíz

root crack.   A crack in the weld or heat-affected zone occurring at the root of the weld.
fissure à la racine.   Fissure située dans le métal fondu ou dans la zone thermiquement affectée, à la racine de la soudure.
fisura en la raíz

root face.   The portion of a fusion face which is not beveled or grooved.
méplat; talon.   Portion d'une face à souder n'ayant pas été chanfreinée.
talón

root of preparation; root of joint; joint root.   The portion of the weld preparation where the members approach closest to each other.   In cross section, the root of the preparation may be either a point or a line.
racine du joint.   Région du joint où les pièces à assembler sont les plus proches l'une de l'autre.   En coupe transversale, la racine du joint peut être un point ou une ligne.
raíz de la unión

root of the weld.   Zone on the side of the first run farthest from the welder.
racine de la soudure.   Zone la plus éloignée du soudeur, située du côté de la première passe.
raíz de la soldadura

root opening.   The distance between the members to be joined, at the root of the preparation.
écartement à la racine.   Distance entre les faces à souder, mesurée à la racine.
separación en la raíz

root pass. The first run deposited in the root of a multirun weld.
passe de fond.   Première passe déposée à la racine d'une soudure multipasse.
pasada de raíz

root radius.   The radius of the curved portion of the fusion face in a component prepared for a single J, single U, double J, or double U weld.
rayon à fond de chanfrein. Rayon à la racine du profil curviligne d'un joint en J, en U, en double J ou double U.
radio inferior del chaflán

S

safety valve
soupape de sécurité
válvula de seguridad

seal weld. A weld, not being a strength weld, used to make a fluid-tight joint.
soudure d'étanchéité. Soudure ne transmettant pas d'effort et destinée à former un joint étanche.
soldadura de estanqueidad

shielded metal-arc welding.   Arc welding using a consumable covered electrode.
soudage à l'arc avec électrode enrobée.   Soudage à l'arc avec une électrode fusible enrobée.
soldeo por arco con electrodo revestido

shop weld. A weld made within the premises of the manufacturer of the welded assembly.
soudure d'atelier; soudure en atelier. Soudure exécutée chez le constructeur de l'assemblage soudé.
soldadura de taller

side bend test. A bend test in which the face of the transverse section of the weld is in tension.
essai de pliage latéral; essai de pliage de côté. Essai de pliage dans lequel l'une des faces de l'éprouvette correspondant à une coupe perpendiculaire à la surface de l'assemblage est mise en extension.
ensayo de doblado lateral

single U groove weld. A butt weld in the preparation for which the edges of both parts are machined so that in cross section the fusion faces form a U.
soudure en U. Soudure bout à bout avec préparation dans laquelle les bords des éléments sont usinés de telle sorte que profil du joint forme un U.
soldadura en U

single V groove weld. A butt weld in the preparation for which the edges of both parts are beveled so that in cross section the fusion faces form a V.
soudure en V. Soudure bout à bout avec préparation dans laquelle les deux bords sont chanfreinés de telle sorte que le profil du joint forme un V.
soldadura en V

slag inclusion.   Slag entrapped in weld metal. According to the circumstances of their formation such inclusions may be linear, isolated, or other.
inclusion de laitier. Résidu de laitier emprisonné dans la masse du métal fondu. On peut rencontrer, suivant leur répartition, des inclusions de laitier alignées (ou en chapelet), isolées ou autres.
inclusión de escoria

soldering. Processes in which metallic parts are joined by means of a filler material having a melting temperature lower than that of the parts to be joined and in general lower than 450 °C and wetting the parent metal(s). The parent metal(s) does (do) not participate by fusion in making the joint. Soldering involves the use of a flux.
brasage tendre. Procédés consistant à assembler des pièces métalliques à l'aide d'un produit d'apport, à l'état liquide, ayant une température de fusion inférieure à celle des pièces à réunir, en général inférieure à 450 °C et mouillant le (les) métal (métaux) de base qui ne participe(nt) pas par fusion à la constitution de la brasure. L'opération se fait en utilisant un flux.
soldeo blando

solidification crack. A crack which forms during the solidification of the weld pool.
fissure de solidification. Fissure se produisant lors de la solidification du bain de fusion.
grieta de solidificación

solution annealing. Heating an alloy to a suitable temperature, holding at that temperature long enough to allow one or more constituents to enter into solid solution, and then cooling rapidly enough to hold the constituents in solution.
traitement de mise en solution. Traitement consistant à chauffer un alliage à une température appropriée, à l'y maintenir suffisamment longtemps pour permettre la mise en solution solide d'un ou plusieurs constituants et à ensuite refroidir à une vitesse suffisamment rapide pour permettre le maintien en solution du ou des constituants.
recocido de solubilización

source of welding current; welding power source. An apparatus for supplying current and sufficient voltage, and having the required output suitable for the welding process employed.
source de courant de soudage. Appareil fournissant le courant et la tension et ayant des caractéristiques externes appropriées au procédé de soudage auquel il est destiné.
generador de corriente de soldeo

spatter. Globules of metal expelled during welding on to the surface of parent metal or of a weld.
projections (particules). Particules de métal projetées à la surface du métal de base ou de la soudure pendant l'opération de soudage.
proyecciones (partículas)

spheroidising. Heating and cooling to produce a spheroidal or globular form of carbide in steel.
recuit de globulisation; recuit de sphéroïdisation. Traitement comportant un chauffage suivi d'un refroidissement et ayant pour objet de produire des carbures de forme sphérique ou globulaire.
recocido de globulización

spot weld. Denotes one single weld spot.
point de soudure. Soudure par points isolés.
punto de soldadura

stabilization annealing; stabilizing. Heat treatment with the purpose of preventing later variations in the structure or of the dimensions of a ferrous part.
traitement de stabilisation. Traitement thermique ayant pour objet d'éviter des variations ultérieures des dimensions ou de la structure d'un produit ferreux.
recocido de estabilización

stitch weld
soudure en ligne continue par points
soldadura alineada continua por puntos

stress concentration factor. The local increase of stress in the vicinity of a considerable variation, overall or local, of the geometry of the work piece or at the intersection of two pieces, and corresponding to an overall or local disturbance in the transmission of loads.
coefficient de concentration de contrainte. Élévation locale de contrainte au voisinage d'une variation notable, globale ou locale, de la géométrie d'une pièce ou à l'intersection de deux pièces et correspondant à une perturbation globale ou locale dans la transmission des efforts.
factor de concentración de tensiones

stress corrosion cracking. Failure of metals by cracking under combined action of corrosion and stress, residual or applied.

rupture par corrosion sous tension; corrosion sous tension. Rupture d'une pièce métallique soumise aux actions simultanées d'un environnement corrosif et de contraintes de traction, résiduelles ou induites.

rotura por corrosión bajo tensiones

stress intensity factor. Characteristic value (K, K1, K2, K3) of a stress-strain field existing in the vicinity of a crack tip, for a given type of loading in a homogeneous body and under condition of linear elasticity.

facteur d'intensité de contrainte. Grandeur (K, K1, K2, K3) caractéristique du champ de contraintes et de déformations existant au voisinage de l'extrémité d'une fissure, pour un mode de sollicitation donné, dans un solide homogène et sous condition linéaire élastique.

factor de intensidad de tensión

stress relieving. Mechanical or thermal treatment used for reducing internal stresses in metals that have been induced by welding, casting, quenching, cold working, etc.

relaxation des contraintes. Traitement mécanique ou thermique destiné à abaisser dans une pièce métallique les contraintes résiduelles de soudage, de moulage, de trempe, d'écrouissage, etc.

relajación de tensiones

stringer bead. A type of weld bead made without appreciable weaving motion.

passe tirée; passe étroite; cordon tiré. Type de cordon de soudure exécuté sans mouvement de balancement notable.

pasada recta; cordón estirado

stud welding. A general term for the joining of a metal stud or similar part to a workpiece. Welding may be accomplished by arc, resistance, friction, or other suitable welding process, using pressure, with or without external gas shielding. The weld is made over the whole end area of the stud or attachment.

soudage des goujons. Terme générique pour le soudage d'un goujon métallique ou d'un élément similaire sur une pièce. Le soudage peut être effectué à l'arc, par résistance, par friction ou par un autre procédé de soudage approprié mettant en oeuvre une pression avec ou sans atmosphère de protection gazeuse extérieure. La soudure est exécutée sur toute la surface de l'extrémité du goujon ou de la pièce de fixation.

soldeo de espárragos

submerged arc welding. Metal arc welding in which a bare or cored wire or strip electrode, or electrodes, are used, with or without metal powder addition; the arc or arcs are enveloped in a granular flux, some of which fuses to form a removable covering of slag on the weld.

soudage à l'arc sous flux en poudre; soudage à l'arc sous flux; soudage à l'arc submergé.   Soudage à l'arc avec électrode fusible constituée par un (ou plusieurs) fil nu ou fourré ou électrode en bande, avec ou sans addition de métal en poudre; l'arc (ou arcs) est entouré d'un flux en poudre, dont une partie fond en formant sur la soudure une couche de laitier détachable.

soldeo por arco sumergido

surface hardening; case hardening. Hardening a ferrous alloy so that the outer portion, or case, is made substantially harder than the inner portion or core.

trempe superficielle; durcissement superficiel.   Opération ayant pour objet d'accroître en surface la dureté tout en conservant au coeur les caractéristiques initiales.

temple superficial

surfacing weld. A type of weld composed of one or more stringer or weave beads deposited on an unbroken surface to obtain desired properties or dimensions.

rechargement; rechargement par soudage.   Type de soudure en une ou plusieurs passes, tirées ou larges, déposée sur une surface continue, en vue d'obtenir des caractéristiques ou des dimensions désirées.

recargue por soldeo

T

T joint.   A connection where the end or edge of one member abuts the surface of the other member at approximately a right angle in the region of the joint.

assemblage soudé en T; joint soudé en T.   Assemblage soudé dans lequel l'extrémité ou le chant de l'un des éléments est approximativement perpendiculaire à la surface de l'autre au voisinage du joint.

unión soldada a tope en T; soldadura de rincón

tack weld.   A weld made to hold the parts to be welded in proper alignment until the final welds are made.

soudure de pointage; point d'épinglage.   Soudure destinée à maintenir les pièces à assembler dans la position voulue jusqu'à l'exécution de la soudure définitive.

soldadura de punteo

tempering. Reheating a quench hardened or normalized ferrous alloy to a temperature below the transformation range and then cooling at any rate desired.

revenu. Traitement thermique auquel est soumis un produit ferreux après durcissement par trempe ou normalisation. Il consiste à réchauffer le produit jusqu'à une température inférieure au domaine de transformation puis à le refroidir à la vitesse désirée.

revenido

temporary backing. A piece of metal or other material placed at the root and used to control the penetration of a weld, but not intended to become part of the weld.

support à l'envers non subsistant. Pièce de métal ou autre matériau, disposée à la racine de la soudure en vue de contrôler la pénétration mais non destinée à faire partie intégrante du joint soudé.

placa soporte no permanente

thermal spraying. A group of welding or allied processes in which finely divided metallic or nonmetallic materials are deposited in a molten or semi-molten condition to form a coating. The coating material may be in the form of powder, ceramic rod, wire, or molten materials.

projection à chaud. Groupe de techniques connexes au soudage consistant à déposer de fines particules, métalliques ou non, à l'état liquide ou semiliquide, en vue d'obtenir un revêtement. Le matériau à déposer peut se présenter sous plusieurs formes: poudre, baguette de céramique, fil métallique ou matériau fondu.

proyección en caliente

toe crack. A crack in the base metal occurring at the toe of the weld.

fissure au raccordement. Fissure située dans le métal de base et débouchant en surface au raccordement du cordon de soudure et du métal de base.

fisura en el acuerdo de la unión

toe of weld. The boundary between a weld face and the parent metal, or between runs.

raccordement; raccordement du cordon de soudure; ligne de raccordement. Limite, soit entre la surface d'une soudure et le métal de base, soit entre passes.

acuerdo; acuerdo del cordón de soldadura

transgranular or transcrystalline crack. A crack extending through a grain.

fissure transgranulaire; fissure intragranulaire. Fissure traversant les grains.

fisura transcristalina

transverse crack. A crack substantially transverse to the axis of the weld. It may be situated in the weld metal, in the heat affected zone or in the parent metal.

fissure transversale.   Fissure dont la direction est sensiblement perpendiculaire à l'axe de la soudure. Elle peut se situer, suivant le cas, dans le métal fondu, dans la zone thermiquement affectée ou dans le métal de base.

grieta transversal

transverse test specimen. A test specimen which is transversely bisected by the portion of the weld included in it.

éprouvette transversale.   Éprouvette prélevée transversalement au joint soudé.

probeta transversal

travel speed. The length of single or multi-run weld completed in a unit of time.

vitesse de soudage.   Longueur de soudure en une seule passe ou multipasse exécutée par unité de temps.

velocidad de soldeo

U

ultrasonic inspection. A method of inspection to discover interior faults by projecting ultrasonic waves into the test piece and the observation of their reflected paths within it; ultrasonic inspection permits to determine the importance of faults as well as to locate their position.

contrôle par ultrasons; contrôle ultrasonore.   Mise en évidence de défauts par introduction dans la pièce à contrôler d'ondes ultrasonores et par l'observation de leurs parcours. Le contrôle par ultrasons permet également de déterminer l'importance des défauts et de les localiser en profondeur.

inspección por ultrasonidos

underbead crack. A crack in the heat-affected zone generally not extending to the surface of the base metal.

fissure sous cordon. Fissure située dans la zone thermiquement affectée et n'atteignant généralement pas la surface du métal de base.

fisura bajo el cordón

undercut. A groove at the toe(s) of a weld run due to welding. Undercut may be continuous or intermittent but, in English, such distinction is not normally made.

caniveau. Manque de métal en forme de sillon s'étendant sur une certaine longueur des bords de la soudure.

mordedura

underfill. A longitudinal continuous or intermittent channel in the surface of the weld due to insufficient deposition of weld metal. The channel may be along one or both edges of the weld.
manque d'épaisseur. Insuffisance locale ou continue de métal déposé conduisant à un profil de cordon en retrait par rapport au profil correct.
falta de espesor

upset welding. Resistance welding in which the components are butted together under pressure before heating is started. Pressure is maintained and current is allowed to flow until the welding temperature is reached at which upset metal is produced. Current and force are transmitted through clamps.
soudage en bout par résistance pure. Soudage par résistance dans lequel les pièces sont aboutées sous pression avant de commencer le chauffage. La pression est maintenue et les pièces sont traversées par un courant jusqu'à ce que la température de soudage soit atteinte et que se forme un bourrelet. Le courant et l'effort sont transmis par des mâchoires.
soldeo a tope por resistencia

V

vertical downward welding
soudage vertical descendant; soudage en descendant
soldeo vertical descendente

vertical position welding
soudage vertical
soldeo vertical

vertical upward welding
soudage vertical montant; soudage en montant
soldeo vertical ascendente

visual examination; visual inspection. Visual judgment of the condition of the surface, shape and position of weld.
contrôle visuel; examen visuel. Estimation visuelle de l'état de surface, de la forme et de la position de la soudure.
inspección visual

W

weave bead. A type of bead made with transverse oscillation
passe large; cordon balancé. Type de cordon de soudure exécuté avec oscillation transversale.
pasada ancha; cordón con balanceo

weld. The result of the welding operation.
soudure. Résultat de l'opération de soudage.
soldadura

weld bead. A weld deposit resulting from a pass.
cordon de soudure. Métal déposé lors de l'exécution d'une passe de soudage.
cordón de soldadura

weld defects.   Weld defects are understood to be irregularities, incompleteness and deviation from the specified geometric shape, of the weld and of the welded joint.
défauts des soudures. Irrégularités, discontinuités et écarts par rapport à la configuration géométrique théorique de la soudure et du joint soudé.
defectos de las soldaduras

weld face; weld surface. The surface of a fusion weld exposed on the side from which the weld has been made.
surface de la soudure. Surface d'une soudure par fusion, située du côté par lequel la soudure a été exécutée.
superfície de la soldadura

weld junction; weld line; fusion line. The boundary between the fusion zone and the heat affected zone.
zone de liaison. Limite entre la zone fondue et la zone thermiquement affectée.
zona de unión

weld length
longueur du cordon
longitud del cordón

weld metal. All metal melted during the making of a weld and retained in the weld.
métal fondu. Ensemble du métal ayant été fondu au cours de l'opération de soudage et contenu dans la soudure.
metal fundido

weld metal area. The area of the weld metal as measured on the cross section of a weld.
zone fondue; zone de fusion. Surface du métal fondu mesurée sur la coupe transversale de la soudure.
zona de fusión

weld pass. The metal melted or deposited during one passage of an electrode, torch or blowpipe, electron or laser beam, etc.
passe. Métal fondu ou déposé lors du passage d'une électrode, d'une torche, d'un chalumeau, d'un faisceau d'électrons, d'un faisceau laser, etc.
pasada

weld pool. The pool of liquid metal formed during fusion welding.
bain de fusion. Bain de métal liquide formé au cours du soudage par fusion.
baño de fusión

weld spatter. Globules of weld metal or filler expelled during welding and adhering to the surface of parent metal or solidified weld metal.
projection (de soudage); perle. Éclaboussure de métal en fusion projeté pendant l'opération de soudage et qui adhère sur le métal de base ou le métal fondu déjà solidifié.
proyección; salpicadura

weld width of a butt weld. The shortest distance between the outer toes of the weld face.
largeur de la soudure. Plus courte distance entre les lignes de raccordement d'une soudure.
anchura de la soldadura

weld with backing ring.
soudure avec bague support; soudure avec anneau support.
soldadura con anillo soporte soldado

weld zone. The zone containing the weld metal and the heat affected zone.
zone de soudure. Ensemble de la zone fondue et de la zone thermiquement affectée.
zona de soldadura

weldability. A metallic substance is considered to be weldable to a stated degree by a given process and for a given purpose, when metallic continuity can be obtained by welding using a suitable procedure, so that the joints comply with the requirements specified in regard to both their local properties and their influence on the construction of which they form part.
soudabilité. On considère qu'un matériau métallique est soudable à un degré donné par un procédé et pour un type d'application donnés, lorsqu'il se prête, moyennant les précautions correspondant à ce degré, à la réalisation d'une construction entre les éléments de laquelle il est possible d'assurer la continuité métallique par la constitution de joints soudés qui, par leurs caractéristiques locales et les conséquences globales de leur présence, satisfont aux propriétés requises et choisies comme base de jugement.
soldabilidad

welder.   The operator who performs the manual or partly mechanized welding operation.
soudeur. Personne exécutant une opération de soudage manuel ou semi-automatique.
soldador

welder certification; welder certificate. Certification in writing that a welder has produced welds meeting prescribed standards.
certificat de qualification de soudeur; licence de soudeur; B carte de soudeur. Attestation écrite qu'un soudeur a réalisé, dans des conditions déterminées, des assemblages donnés répondant aux critères de qualité requis.
certificado de cualificación del soldador

welder performance qualification. The demonstration of a welder's ability to produce welds meeting the prescribed standards.
qualification d'un soudeur. Capacité reconnue à un soudeur pour réaliser, dans des conditions déterminées, un type d'assemblage donné répondant aux critères de qualité requis.
cualificación del soldador

welder's apron
tablier de soudeur
delantal de soldador

welder's face mask. 1. A protective device worn in front of the face to shield it from injury during welding or cutting. It is fitted with welding glass and plain glass. 2. A protective device supplied with fresh air and worn over the nose and mouth.
masque de soudeur. 1. En francais, ce terme désigne uniquement un dispositif de protection de la face et de la gorge du soudeur. Il est équipé d'un filtre et d'un verre de garde. 2. En anglais, ce terme désigne également un appareil de protection respiratoire recouvrant le bouche et le nez, c'est-a-dire un demi-masque respiratoire.
máscara de soldador

welder's helmet. A protective device supported on the head and arranged to shield the face and throat from injury during welding. It is fitted with a window consisting of welding glass and plain glass, and if necessary, a heat filter.
casque de soudeur. Ensemble constitué d'un casque de protection et d'un masque de soudeur et destiné à protéger la face et la gorge pendant le soudage. Il comporte un châssis dans lequel sont logés le filtre et le verre de garde et, si nécessaire, un verre athermane.
casco de soldador

welding. The union of two or more parts by heat or pressure or a combination of both such that the materials form a continuum. A filler material with a melting point similar to that of the parent material may be used. Materials can be metals, plastics, composites, ceramics, glass, etc. Welding may be employed also for surfacing.

soudage. Opération consistant à assembler deux pièces ou plus par chauffage, par pression ou par combinaison des deux, de manière à assurer la continuité des matériaux. Un produit d'apport dont le point de fusion est du même ordre de grandeur que celui du matériau de base peut être utilisé. Les matériaux peuvent être des métaux, des matières plastiques, des matériaux composites, des céramiques, du verre, etc. Le soudage peut aussi être utilisé pour exécuter des rechargements.

soldeo

welding bench
table de soudage
mesa de soldeo

welding booth
cabine de soudage
cabina de soldeo

welding consumables. All materials used-up during the making of a weld, such as filler material, gases and fluxes.

produits consommables. Ensemble des matériaux consommés au cours de l'exécution d'une soudure, tels que produit d'apport, gaz et flux.

consumibles

welding equipment. Individual items of welding plant, power source, auxilliaries and accessories used in making a weld.

matériel de soudage. Chaque élément d'un appareillage, tels que source de courant, équipements et accessoires utilisés pour exécuter une soudure.

equipo de soldeo

welding gloves. Gloves to protect the hands, or gauntlets to protect the hands and forearms, from heat and metal splashes due to welding or cutting.

gants de soudeur. Gants, avec ou sans crispin, destinés à protéger la main et l'avant-bras ou la main seule contre la chaleur et les projections de métal provoquées par le soudage ou le coupage.

guantes de soldador

welding goggles
lunettes de soudeur
gafas de soldador

welding operator.   One who operates machine or automatic welding equipment.
opérateur. Personne exécutant une opération de soudage au moyen d'une machine ou d'un matériel de soudage automatique.
operario de soldeo

welding position. The orientation of a weld expressed in terms of the limits of weld slope and weld rotation.
position de soudage. Orientation d'une soudure exprimée en fonction des valeurs limites des angles d'inclinaison et de rotation.
posición de soldeo

welding procedure. A specified course of action to be followed in making a weld, including reference to materials, preparation, pre-heating, method of welding and post heat treatment, and where necessary, of tools to be used.
mode opératoire de soudage.   Ensemble des opérations (préparation, préchauffage, soudage, postchauffage et traitement thermique) appliquées à une ou plusieurs pièces métalliques de type, de produit, de nuance, de forme et de dimensions connus, en vue d'obtenir un assemblage soudé répondant à des critères de qualité définis.
procedimiento operatorio de soldeo

welding procedure qualification. Establishment of the correct selection of welding processes, equipment, the employment of materials and personnel in one particular welded structure under consideration, in accordance with the latest technical knowledge.
qualification des soudeurs et du mode opératoire. Essais à effectuer et conditions à satisfaire d'une part par les soudeurs et/ou les opérateurs et, d'autre part, par les métaux, produits et matériels utilisés ainsi que le mode opératoire, en vue de contrôler leur aptitude à réaliser des assemblages soudés préalablement définis et répondant aux critères de qualité requis.
cualificación de soldadores y del procedimiento operatorio

welding procedure specification.   A document providing in detail the required parameters for a specific application to assure repeatability by properly trained welders and welding operators.
programme de soudage.   Énumération chronologique des séquences caractérisant la fabrication d'un ensemble soudé.
especificación del procedimiento de soldeo

welding procedure test.   The making and testing of a welded joint, representative of that to be used in production in order to prove the feasibility of the welding procedure.
essai de pré-qualification (du mode opératoire). Exécution et essai d'un assemblage soudé représentatif de celui destiné à être produit en fabrication, en vue de démontrer la faisabilité du mode opératoire.
ensayo de precualificación del procedimiento

welding screen
écran de soudage
pantalla de soldeo

welding spark
étincelle de soudage
chispa de soldeo

welding techniques. The details of welding procedure which are controlled by the welder or welding operator.
techniques opératoires.   Éléments du mode opératoire de soudage commandés par le soudeur ou l'opérateur.
tecnicas operatorias

welding with weave beads; welding with weaving motion
soudage en passes larges; soudage en passes balancées
soldeo en pasadas con balanceo; soldeo en pasadas con oscilación

welding without using filler material
soudage sans métal d'apport
soldeo sin metal de aportación

weldment. An assembly whose component parts are joined by welding.
ensemble soudé; construction soudée.   Assemblage dont les parties constitutives sont réunies par soudage.
conjunto soldado; construcción soldada

worm-hole. A tubular cavity in weld metal caused by the release of gas. The shape and position of worm-holes is determined by the mode of solidification and the sources of the gas and they may be distributed in a herringbone fashion.
soufflure vermiculaire. Soufflure en forme de galerie de ver dans le métal fondu résultant du cheminement des gaz. La forme et la position de ces soufflures sont déterminées par le mode de solidification et l'origine des gaz; elles sont parfois disposées en arêtes de poisson.
sopladura vermicular

SPANISH	ENGLISH	FRENCH
abertura de la unión	groove	joint
acuerdo; acuerdo del cordón de soldadura	toe of weld	raccordement; raccordement du cordon de soudure; ligne de raccordement
anchura de la soldadura	weld width of a butt weld	largeur de la soudure
anillo soporte	backing ring	bague support; anneau support
baño de fusión	weld pool	bain de fusion
bruto de soldeo	as welded	brut de soudage
cabina de soldeo	welding booth	cabine de soudage
cara a soldar; borde	groove face	face à souder; bord à souder
cara fusible	fusion face.	face à souder par fusion
casco de soldador	welder's helmet	casque de soudeur
cavidad	cavity; void	cavité
cementación	carburisation	cémentation
certificado de cualificación del soldador	welder certification; welder certificate	certificat de qualification de soudeur; licence de soudeur; carte de soudeur
chapa apéndice de ensayo	coupon plate	appendice témoin; coupon d'essai
chispa de soldeo	welding spark	étincelle de soudage
cobresoldeo	braze welding	soudobrasage
conjunto soldado; construcción soldada	weldment	ensemble soudé; construction soudée
consumibles	welding consumables	produits consommables
cordón de metal depositado	bead on plate	cordon déposé; chenille
cordón de penetración	penetration bead	cordon de pénétration
cordón de soldadura	weld bead	cordon de soudure
cordón soporte de raíz	backing weld; support run; backing run	cordon support envers
corrosión debida al soldeo	corrosion due to welding; weld decay	corrosion due au soudage
corte de arco (sobre la pieza)	arc strike; stray flash	coup d'arc (sur la pièce)
cráter	crater	cratère
cualificación de soldadores y del procedimiento operatorio	welding procedure qualification	qualification des soudeurs et du mode opératoire

SPANISH	ENGLISH	FRENCH
cualificación del soldador	welder performance qualification	qualification d'un soudeur
defecto de alineación	linear misalignment	défaut d'alignement
defectos de las soldaduras	weld defects	défauts des soudures
deformación angular	angular misalignment	déformation angulaire
delantal de soldador	welder's apron	tablier de soudeur
desbordamiento	overlap	débordement
desgarre laminar (defecto)	lamellar tear	arrachement lamellaire (défaut)
desplazamiento del fondo de la grieta (CTOD)	crack tip opening displacement (CTOD)	ouverture à fond de fissure
detención de la grieta	crack arrest	arrêt de fissure
dilución	dilution	dilution
endurecimiento	hardening	durcissement
endurecimiento por envejecimiento	age hardening	endurcissement par vieillissement
endurecimiento por precipitación	precipitation hardening	durcissement par précipitation
ensayo de doblado	bend test	essai de pliage
ensayo de doblado lateral	side bend test	essai de pliage latéral; essai de pliage de côté
ensayo de doblado normal por el anverso	face bend test	essai de pliage à l'endroit; pliage endroit
ensayo de doblado por la raíz	root bend test	essai de pliage mettant la racine de la soudure en extension
ensayo de doblado sobre mandril	guided bend test	essai de guidage plié; essai de pliage sur mandrin
ensayo de dureza	hardness test	essai de dureté
ensayo de precualificación del procedimiento	welding procedure test	essai de pré-qualification (du mode opératoire)
ensayo de textura	nick break test	essai de texture; essai de compacité
ensayos destructivos	destructive testing	essais destructifs; contrôle destructif
ensayos no destructivos; control no destructivo CND	nondestructive testing	essais non destructifs; END; contrôle non destructif; CND
envejecimiento	ageing	vieillissement
equipo de soldeo	welding equipment	matériel de soudage

SPANISH	ENGLISH	FRENCH
especificación del procedimiento de soldeo	welding procedure specification	programme de soudage
espesor de una soldadura de ángulo	measured throat thickness of a fillet weld	épaisseur de la soudure d'angle (de la surface du cordon au sommet de l'angle)
espesor nominal de la soldadura	effective throat	épaisseur nominale de la soudure
espesor total de la soldadura (en una unión a tope)	actual throat thickness in a butt weld	épaisseur totale de la soudure (dans une soudure bout à bout)
espesor total de la soldadura (en una unión de ángulo)	actual throat thickness of a fillet weld	épaisseur totale de la soudure (dans une soudure d'angle)
estufa	electrode drying oven	étuve
exceso de penetración; sobreespesor excesivo en la raíz	excessive root reinforcement; excess penetration bead	surépaisseur excessive à la racine; excès de pénétration
factor de concentración de tensiones	stress concentration factor	coefficient de concentration de contrainte
factor de intensidad de tensión	stress intensity factor	facteur d'intensité de contrainte
falta de espesor	underfill	manque d'épaisseur
falta de fusión; pegadura	incomplete fusion; lack of fusion	manque de fusion; collage
falta de penetración	inadequate joint penetration; lack of penetration; Incomplete penetration.	manque de pénétration
fatiga oligocíclica	low cycle fatigue	fatigue oligocyclique
fisura bajo el cordón	underbead crack	fissure sous cordon
fisura de cráter	crater crack	fissure de cratère
fisura en caliente	hot crack	fissure à chaud
fisura en el acuerdo de la unión	toe crack	fissure au raccordement
fisura en frío	cold crack	fissure à froid
fisura en la raíz	root crack	fissure à la racine
fisura intercristalina	intergranular crack; intercrystalline crack	fissure intergranulaire; fissure intercristalline
fisura transcristalina	transgranular crack; transcrystalline crack	fissure transgranulaire; fissure intragranulaire
fisuración en frío	cold cracking; delayed cracking	fissuration à froid; fissuration différée

SPANISH	ENGLISH	FRENCH
fisuras radiales	radiating cracks	fissures rayonnantes
fluencia	creep	fluage
formación de ampollas	blistering	formation de cloques
fragilización	embrittlement	fragilisation
fragilización por hidrógeno	hydrogen embrittlement	fragilisation par l'hydrogène
gafas de soldador	welding goggles	lunettes de soudeur
generador de corriente de soldeo	source of welding current; welding power source	source de courant de soudage
grieta de solidificación	solidification crack	fissure de solidification
grieta debida al endurecimiento	hard zone crack	fissure due au durcissement
grieta debida al envejecimiento	age hardening crack	fissure due au vieillissement
grieta longitudinal	longitudinal crack	fissure longitudinale
grieta transversal	transverse crack	fissure transversale
grieta; fisura	crack	fissure
guantes de soldador	welding gloves	gants de soudeur
hueco	burn-through	trou
inclusión	inclusion; solid inclusion	inclusion; inclusion solide
inclusión de escoria	slag inclusion	inclusion de laitier
inclusión de fundente	flux inclusion	inclusion de flux
inclusión de óxido	oxide inclusion	inclusion d'oxyde
inclusión metálica	metallic inclusion	inclusion métallique
indicador de calidad de imagen I.C.I.	image quality indicator; IQI	indicateur de qualité d'image; IQI
inspección por corrientes de Foucault	eddy current testing	contrôle par courants de Foucault
inspección por líquidos penetrantes	liquid penetrant inspection	contrôle par ressuage; ressuage
inspección por partículas magnéticas; magnetoscopía	magnetic particle inspection	magnétoscopie; contrôle magnétoscopique
inspección por ultrasonidos	ultrasonic inspection	contrôle par ultrasons; contrôle ultrasonore
inspección radiográfica; control radiográfico	radiographic inspection; radiography	contrôle radiographique; radiographie
inspección visual	visual examination; visual inspection	contrôle visuel; examen visuel
integral "J"	J-integral	intégrale "J"
irritación ocular (debida al arco)	flash; arc eye	coup d'arc (irritation de l'oeil)

SPANISH	ENGLISH	FRENCH
labio	land	lèvre
largo de patas	leg length; leg	côté
longitud del cordón	weld length	longueur du cordon
longitud útil de un cordón	effective length of weld	longueur utile d'un cordon; longueur efficace d'un cordon
manguito de soldador	arm protection; sleeve; welding sleeve	manchette; manchette de soudeur
martillo para picar	chipping hammer	marteau à piquer
máscara de soldador	welder's face mask	masque de soudeur
material de aportación	filler material	produit d'apport
mecánica de fractura	fracture mechanics	mécanique de la rupture
mecánica de fractura elastoplástica	elasto-plastic fracture mechanics	mécanique élastoplastique de la rupture
mesa de soldeo	welding bench	table de soudage
metal de aportación	filler metal	métal d'apport
metal de base	base metal	métal de base
metal depositado	deposited metal	métal déposé
metal fundido	weld metal	métal fondu
microgrieta; microfisura	microfissure; microcrack	microfissure
modo (I,II,III)	mode (I, II, III)	mode (I, II, III)
mordedura	undercut	caniveau
mordedura local	local undercut	morsure
normalización	normalizing	traitement de normalisation
ojo de pez	fish eye	oeil de poisson (pluriel: oeils de poisson); point blanc
operario de soldeo	welding operator	opérateur
pantalla de protección	protective screen	écran de protection; écran protecteur
pantalla de soldeo	welding screen	écran de soudage
pasada	weld pass	passe
pasada ancha; cordón con balanceo	weave bead	passe large; cordon balancé
pasada de raíz	root pass	passe de fond
pasada recta; cordón estirado	stringer bead	passe tirée; passe étroite; cordon tiré
penetración	penetration	pénétration
perfil de soldadura incorrecto; ángulo del sobreespesor incorrecto	incorrect weld profile	défaut de raccordement

SPANISH	ENGLISH	FRENCH
placa soporte fusíble	consumable insert	support à l'envers fusible; insert
placa soporte no permanente	temporary backing	support à l'envers non subsistant
placa soporte permanente	permanent backing	support à l'envers subsistant
plantilla de soldeo	fixture	montage ; dispositif de serrage; gabarit de soudage; mannequin
plaqueado por soldeo	cladding by welding; surfacing	placage par soudage
polaina	leg protection; leggings; gaiters	jambière-guêtre; guêtre; guêtron
porosidad	porosity	groupe de soufflures
posición de soldeo	welding position	position de soudage
precalentamiento	preheating	préchauffage
probeta longitudinal	longitudinal test specimen	éprouvette longitudinale
probeta transversal	transverse test specimen	éprouvette transversale
procedimiento operatorio de soldeo	welding procedure	mode opératoire de soudage
profundidad de penetración	joint penetration	profondeur de pénétration
protección ocular	eye protection	protection de l'oeil
proyección en caliente	thermal spraying	projection à chaud
proyección; salpicadura	weld spatter	projection (de soudage); perle
proyecciones (partículas)	spatter	projections (particules)
punto de soldadura	spot weld	point de soudure
radio inferior del chaflán	root radius	rayon à fond de chanfrein
raíz de la soldadura	root of the weld	racine de la soudure
raíz de la unión	root of preparation; root of joint; joint root	racine du joint
reanudación defectuosa	bad restart	mauvaise reprise
recargue duro	hardfacing by welding; hardfacing; hard-facing	rechargement dur; rechargement dur par soudage
recargue por soldeo	surfacing weld	rechargement; rechargement par soudage
rechupe de cráter	crater pipe; crater hole	retassure de cratère

SPANISH	ENGLISH	FRENCH
rechupe en la raíz	root concavity	retassure à la racine
rechupe interdendrítico	interdendritic shrinkage; solidification void	retassure interdendritique
recocido	annealing	recuit
recocido de estabilización	stabilization annealing; stabilizing	traitement de stabilisation
recocido de globulización	spheroidising	recuit de globulisation; recuit de sphéroïdisation
recocido de recristalización	recrystallisation annealing	traitement de recristallisation; recuit de recristallisation
recocido de restauración	recovery annealing	traitement de restauration
recocido de solubilización	solution annealing	traitement de mise en solution.
relajación de tensiones	stress relieving	relaxation des contraintes
revenido	tempering	revenu
rodillera	knee protection	genouillère
rotura frágil	brittle fracture	rupture fragile
rotura por clivaje	cleavage	clivage
rotura por corrosión bajo tensiones	stress corrosion cracking	rupture par corrosion sous tension; corrosion sous tension
sensibilidad a la entalla	notch toughness	ténacité sous entaille
separación de bordes	gap, opening	écartement des bords; écartement
separación en la raíz	root opening	écartement à la racine
sobreespesor (en una unión a tope)	excess weld metal in a butt weld; reinforcement	surépaisseur (d'une soudure bout à bout)
sobreespesor (en una unión de ángulo); convexidad	excess weld metal in a fillet weld; reinforcement	surépaisseur (d'une soudure d'angle); convexité
sobreespesor excesivo	excessive reinforcement; excess weld metal	surépaisseur excessive
soldabilidad	weldability	soudabilité
soldador	welder	soudeur
soldadura	weld	soudure
soldadura a tope	groove weld; butt weld in a butt joint	soudure en bout sur bords chanfreinés; soudure bout à bout sur bords chanfreinés

SPANISH	ENGLISH	FRENCH
soldadura a tope con sobreespesor	butt weld with excess weld metal	soudure bout à bout avec surépaisseur
soldadura a tope sobre techo	overhead butt weld	soudure bout à bout au plafond; soudure en bout au plafond
soldadura a tope; unión soldada	butt joint	assemblage soudé bout à bout; joint soudé bout à bout
soldadura acanalada en la raíz	back-gouged weld	soudure gougée à la racine
soldadura alineada continua por puntos	stitch weld	soudure en ligne continue par points
soldadura con anillo soporte soldado	weld with backing ring	soudure avec bague support; soudure avec anneau support
soldadura con pentración completa	full penetration weld; complete penetraction weld	soudure à pleine pénétration; soudure pénétrée
soldadura de ángulo	fillet weld	soudure d'angle
soldadura de estanqueidad	seal weld	soudure d'étanchéité
soldadura de punteo	tack weld	soudure de pointage; point d'épinglage
soldadura de recargue	building-up by welding; building-up	rechargement; rechargement par soudage; rechargement curatif
soldadura de reparación (por un defecto); retoque	rewelding; repairing the weld; weld repair	soudure de réparation; reprise de la soudure (pour réparer un défaut) retouche de la soudure
soldadura de taller	shop weld	soudure d'atelier; soudure en atelier
soldadura en cornisa	horizontal butt weld; horizontal-vertical butt weld	soudure en corniche
soldadura en obra; soldadura en campo	field weld; site weld	soudure sur chantier
soldadura en tapón	plug weld (circular)	soudure en bouchon
soldadura en U	single U groove weld	soudure en U
soldadura en V	single V groove weld	soudure en V
soldadura en X	double -V- groove weld	soudure en X; soudure en V double
soldadura estructural	load bearing weld; structural weld	soudure de résistance; soudure travaillante

SPANISH	ENGLISH	FRENCH
soldadura horizontal a tope	flat position butt weld	soudure bout à bout à plat; soudure en bout à plat; soudure bout à bout horizontale; soudure en bout horizontale
soldadura quemada	burnt weld	soudure brûlée
soldadura sobrecalentada	overheated weld	soudure surchauffée
soldeo	welding	soudage
soldeo a paseo de peregrino	back-step welding	soudage à pas de pèlerin
soldeo a tope por resistencia	upset welding	soudage en bout par résistance pure
soldeo automático	automatic welding	soudage automatique
soldeo blando	soldering	brasage tendre
soldeo con gas	gas welding	soudage aux gaz
soldeo de espárragos	stud welding	soudage des goujons
soldeo en pasadas con balanceo; soldeo en pasadas con oscilación	welding with weave beads; welding with weaving motion	soudage en passes larges; soudage en passes balancées
soldeo fuerte	brazing	brasage fort
soldeo MAG con alambre tubular	flux cored wire metal arc welding with active gas shield	soudage MAG avec fil fourré; soudage sous gaz actif avec fil fourré
soldeo manual	manual welding; hand welding	soudage manuel.
soldeo mecanizado	mechanized welding	soudage mécanisé
soldeo MIG; soldeo por arco bajo protección de gas inerte con alambre fusible	gas metal arc welding; MIG welding	soudage MIG
soldeo oxiacetilénico	oxy-acetylene welding	soudage oxyacétylénique
soldeo parcialmente mecanizado	partly mechanized welding	soudage semi-mécanisé
soldeo por arco	arc welding	soudage à l'arc
soldeo por arco bajo protección gaseosa	gas shielded arc welding	soudage sous protection gazeuse; soudage à l'arc sous protection gazeuse
soldeo por arco con alambre tubular	flux cored wire welding	soudage à l'arc avec fil fourré; soudage avec fil fourré
soldeo por arco con electrodo de carbón	carbon arc welding	soudage à l'arc avec électrode de carbone

SPANISH	ENGLISH	FRENCH
soldeo por arco con electrodo fusible	metal arc welding	soudage à l'arc avec électrode fusible; soudage à l'arc avec électrode consommable
soldeo por arco con electrodo fusible bajo protección gaseosa	gas metal arc welding	soudage sous protection gazeuse avec fil fusible; soudage sous protection gazeuse avec fil-électrode fusible; soudage à l'arc sous protection gazeuse avec fil-électrode fusible
soldeo por arco con electrodo revestido	shielded metal-arc welding	soudage à l'arc avec électrode enrobée
soldeo por arco con percusión	arc pressure welding	soudage par pression à l'arc; soudage à l'arc par pression
soldeo por arco sumergido	submerged arc welding	soudage à l'arc sous flux en poudre; soudage à l'arc sous flux; soudage à l'arc submergé
soldeo por electroescoria	electro-slag welding	soudage vertical sous laitier; soudage sous laitier; soudage électroslag; soudage sous laitier électroconducteur
soldeo por explosión	explosion welding	soudage par explosion
soldeo por forja	forge welding	soudage à la forge
soldeo por fricción	friction welding	soudage par friction
soldeo por fusión	fusion welding	soudage par fusion
soldeo por fusión con presión	pressure fusion welding	soudage par fusion avec pression
soldeo por haz de electrones	electron beam welding; EB welding	soudage par faisceau d'électrons; soudage F.E.
soldeo por plasma	plasma arc welding	soudage plasma; soudage au plasma; soudage au jet de plasma
soldeo por rayo láser	laser beam welding	soudage par faisceau laser; soudage laser; soudage au laser
soldeo por resistencia	resistance welding	soudage par résistance

SPANISH	ENGLISH	FRENCH
soldeo por resistencia por puntos	resistance spot welding	soudage par résistance par points
soldeo por roldana	resistance seam welding	soudage à la molette; soudage au galett
soldeo robotizado	robotic welding	soudage robotisé
soldeo sin metal de aportación	welding without using filler material	soudage sans métal d'apport
soldeo sobre techo	overhead position welding	soudage au plafond
soldeo TIG; soldeo por arco en atmósfera inerte con electrodo de volframio	gas tungsten arc welding; GTA welding; TIG welding	soudage TIG
soldeo vertical	vertical position welding	soudage vertical
soldeo vertical ascendente	vertical upward welding	soudage vertical montant; soudage en montant
soldeo vertical descendente	vertical downward welding	soudage vertical descendant; soudage en descendant
sopladura vermicular	worm-hole	soufflure vermiculaire
superficie de contacto	faying surface	surface de contact
superfície de la soldadura	weld face; weld surface	surface de la soudure
talón	root face	méplat; talon.
tecnicas operatorias	welding techniques	techniques opératoires
temple a la llama	flame hardening	trempe à la flamme; trempe au chalumeau
temple al aire	air hardening	trempe à l'air
temple con autorrevenido	quench hardening	durcissement par trempe
temple por inducción	induction hardening	trempe par induction
temple superficial	surface hardening; case hardening	trempe superficielle; durcissement superficiel
temple y revenido	harden and temper; quench and temper; QT	trempe et revenu
tensión residual de soldeo	residual welding stress	contrainte résiduelle de soudage; tension résiduelle de soudage
tratamiento térmico postsoldeo	post weld heat treatment; PWHT	traitement thermique après soudage

SPANISH	ENGLISH	FRENCH
unión soldada a solape; soldadura con solape	lap joint	assemblage soudé à recouvrement; joint soudé à recouvrement; assemblage soudé à clin; joint soudé à clin
unión soldada a tope en T; soldadura de rincón	T joint	assemblage soudé en T; joint soudé en T
untado; acolchado; recarque blando	buttering by welding; buttering	beurrage; beurrage par soudage
válvula de seguridad	safety valve	soupape de sécurité
velocidad de propagación de la grieta	crack growth rate	vitesse de propagation de la fissure
velocidad de soldeo	travel speed	vitesse de soudage
zapatos de seguridad	foot protection; protective boots; safety boots	chaussures de sécurité
zona de difusión	fusion zone	zone de dilution
zona de fusión	weld metal area	zone fondue; zone de fusion
zona de soldadura	weld zone	zone de soudure
zona de unión	weld junction; weld line; fusion line	zone de liaison
zona térmicamente afectada; ZTA	heat affected zone; HAZ	zone thermiquement affectée

FRENCH	ENGLISH	SPANISH
appendice témoin; coupon d'essai	coupon plate	chapa apéndice de ensayo
arrachement lamellaire (défaut)	lamellar tear	desgarre laminar (defecto)
arrêt de fissure	crack arrest	detención de la grieta
assemblage soudé à recouvrement; joint soudé à recouvrement; assemblage soudé à clin; joint soudé à clin	lap joint	unión soldada a solape; soldadura con solape
assemblage soudé bout à bout; joint soudé bout à bout	butt joint	soldadura a tope; unión soldada
assemblage soudé en T; joint soudé en T	T joint	unión soldada a tope en T; soldadura de rincón
bague support; anneau support	backing ring	anillo soporte
bain de fusion	weld pool	baño de fusión
beurrage; beurrage par soudage	buttering by welding; buttering	untado; acolchado; recarque blando
brasage fort	brazing	soldeo fuerte
brasage tendre	soldering	soldeo blando
brut de soudage	as welded	bruto de soldeo
cabine de soudage	welding booth	cabina de soldeo
caniveau	undercut	mordedura
casque de soudeur	welder's helmet	casco de soldador
cavité	cavity; void	cavidad
cémentation	carburisation	cementación
certificat de qualification de soudeur; licence de soudeur; carte de soudeur	welder certification; welder certificate	certificado de cualificación del soldador
chaussures de sécurité	foot protection; protective boots; safety boots	zapatos de seguridad
clivage	cleavage	rotura por clivaje
coefficient de concentration de contrainte	stress concentration factor	factor de concentración de tensiones
contrainte résiduelle de soudage; tension résiduelle de soudage	residual welding stress	tensión residual de soldeo
contrôle par courants de Foucault	eddy current testing	inspección por corrientes de Foucault

FRENCH	ENGLISH	SPANISH
contrôle par ressuage; ressuage	liquid penetrant inspection	inspección por líquidos penetrantes
contrôle par ultrasons; contrôle ultrasonore	ultrasonic inspection	inspección por ultrasonidos
contrôle radiographique; radiographie	radiographic inspection; radiography	inspección radiográfica; control radiográfico
contrôle visuel; examen visuel	visual examination; visual inspection	inspección visual
cordon de pénétration	penetration bead	cordón de penetración
cordon de soudure	weld bead	cordón de soldadura
cordon déposé; chenille	bead on plate	cordón de metal depositado
cordon support envers	backing weld; support run; backing run	cordón soporte de raíz
corrosion due au soudage	corrosion due to welding; weld decay	corrosión debida al soldeo
côté	leg length; leg	largo de patas
coup d'arc (irritation de l'oeil)	flash; arc eye	irritación ocular (debida al arco)
coup d'arc (sur la pièce)	arc strike; stray flash	corte de arco (sobre la pieza)
cratère	crater	cráter
débordement	overlap	desbordamiento
défaut d'alignement	linear misalignment	defecto de alineación
défaut de raccordement	incorrect weld profile	perfil de soldadura incorrecto; ángulo del sobreespesor incorrecto
défauts des soudures	weld defects	defectos de las soldaduras
déformation angulaire	angular misalignment	deformación angular
dilution	dilution	dilución
durcissement	hardening	endurecimiento
durcissement par précipitation	precipitation hardening	endurecimiento por precipitación
durcissement par trempe	quench hardening	temple con autorrevenido
écartement à la racine	root opening	separación en la raíz
écartement des bords; écartement	gap, opening	separación de bordes
écran de protection; écran protecteur	protective screen	pantalla de protección
écran de soudage	welding screen	pantalla de soldeo
endurcissement par vieillissement	age hardening	endurecimiento por envejecimiento

FRENCH	ENGLISH	SPANISH
ensemble soudé; construction soudée	weldment	conjunto soldado; construcción soldada
épaisseur de la soudure d'angle (de la surface du cordon au sommet de l'angle)	measured throat thickness of a fillet weld	espesor de una soldadura de ángulo
épaisseur nominale de la soudure	effective throat	espesor nominal de la soldadura
épaisseur totale de la soudure (dans une soudure bout à bout)	actual throat thickness in a butt weld	espesor total de la soldadura (en una unión a tope)
épaisseur totale de la soudure (dans une soudure d'angle)	actual throat thickness of a fillet weld	espesor total de la soldadura (en una unión de ángulo)
éprouvette longitudinale	longitudinal test specimen	probeta longitudinal
éprouvette transversale	transverse test specimen	probeta transversal
essai de dureté	hardness test	ensayo de dureza
essai de guidage plié; essai de pliage sur mandrin	guided bend test	ensayo de doblado sobre mandril
essai de pliage	bend test	ensayo de doblado
essai de pliage à l'endroit; pliage endroit	face bend test	ensayo de doblado normal por el anverso
essai de pliage latéral; essai de pliage de côté	side bend test	ensayo de doblado lateral
essai de pliage mettant la racine de la soudure en extension	root bend test	ensayo de doblado por la raíz
essai de pré-qualification (du mode opératoire)	welding procedure test	ensayo de precualificación del procedimiento
essai de texture; essai de compacité	nick break test	ensayo de textura
essais destructifs; contrôle destructif	destructive testing	ensayos destructivos
essais non destructifs; END; contrôle non destructif; CND	nondestructive testing	ensayos no destructivos; control no destructivo CND
étincelle de soudage	welding spark	chispa de soldeo
étuve	electrode drying oven	estufa
face à souder par fusion	fusion face.	cara fusible
face à souder; bord à souder	groove face	cara a soldar; borde

FRENCH	ENGLISH	SPANISH
facteur d'intensité de contrainte	stress intensity factor	factor de intensidad de tensión
fatigue oligocyclique	low cycle fatigue	fatiga oligocíclica
fissuration à froid; fissuration différée	cold cracking; delayed cracking	fisuración en frío
fissure	crack	grieta; fisura
fissure à chaud	hot crack	fisura en caliente
fissure à froid	cold crack	fisura en frío
fissure à la racine	root crack	fisura en la raíz
fissure au raccordement	toe crack	fisura en el acuerdo de la unión
fissure de cratère	crater crack	fisura de cráter
fissure de solidification	solidification crack	grieta de solidificación
fissure due au durcissement	hard zone crack	grieta debida al endurecimiento
fissure due au vieillissement	age hardening crack	grieta debida al envejecimiento
fissure intergranulaire; fissure intercristalline	intergranular crack; intercrystalline crack	fisura intercristalina
fissure longitudinale	longitudinal crack	grieta longitudinal
fissure sous cordon	underbead crack	fisura bajo el cordón
fissure transgranulaire; fissure intragranulaire	transgranular crack; transcrystalline crack	fisura transcristalina
fissure transversale	transverse crack	grieta transversal
fissures rayonnantes	radiating cracks	fisuras radiales
fluage	creep	fluencia
formation de cloques	blistering	formación de ampollas
fragilisation	embrittlement	fragilización
fragilisation par l'hydrogène	hydrogen embrittlement	fragilización por hidrógeno
gants de soudeur	welding gloves	guantes de soldador
genouillère	knee protection	rodillera
groupe de soufflures	porosity	porosidad
inclusion d'oxyde	oxide inclusion	inclusión de óxido
inclusion de flux	flux inclusion	inclusión de fundente
inclusion de laitier	slag inclusion	inclusión de escoria
inclusion métallique	metallic inclusion	inclusión metálica
inclusion; inclusion solide	inclusion; solid inclusion	inclusión
indicateur de qualité d'image; IQI	image quality indicator; IQI	indicador de calidad de imagen I.C.I.
intégrale "J"	J-integral	integral "J"
jambière-guêtre; guêtre; guêtron	leg protection; leggings; gaiters	polaina

FRENCH	ENGLISH	SPANISH
joint	groove	abertura de la unión
largeur de la soudure	weld width of a butt weld	anchura de la soldadura
lèvre	land	labio
longueur du cordon	weld length	longitud del cordón
longueur utile d'un cordon; longueur efficace d'un cordon	effective length of weld	longitud útil de un cordón
lunettes de soudeur	welding goggles	gafas de soldador
magnétoscopie; contrôle magnétoscopique	magnetic particle inspection	inspección por partículas magnéticas; magnetoscopía
manchette; manchette de soudeur	arm protection; sleeve; welding sleeve	manguito de soldador
manque d'épaisseur	underfill	falta de espesor
manque de fusion; collage	incomplete fusion; lack of fusion	falta de fusión; pegadura
manque de pénétration	inadequate joint penetration; lack of penetration; Incomplete penetration.	falta de penetración
marteau à piquer	chipping hammer	martillo para picar
masque de soudeur	welder's face mask	máscara de soldador
matériel de soudage	welding equipment	equipo de soldeo
mauvaise reprise	bad restart	reanudación defectuosa
mécanique de la rupture	fracture mechanics	mecánica de fractura
mécanique élastoplastique de la rupture	elasto-plastic fracture mechanics	mecánica de fractura elastoplástica
méplat; talon.	root face	talón
métal d'apport	filler metal	metal de aportación
métal de base	base metal	metal de base
métal déposé	deposited metal	metal depositado
métal fondu	weld metal	metal fundido
microfissure	microfissure; microcrack	microgrieta; microfisura
mode (I, II, III)	mode (I, II, III)	modo (I,II,III)
mode opératoire de soudage	welding procedure	procedimiento operatorio de soldeo
montage ; dispositif de serrage; gabarit de soudage; mannequin	fixture	plantilla de soldeo
morsure	local undercut	mordedura local
oeil de poisson (pluriel: oeils de poisson); point blanc	fish eye	ojo de pez

FRENCH	ENGLISH	SPANISH
opérateur	welding operator	operario de soldeo
ouverture à fond de fissure	crack tip opening displacement (CTOD)	desplazamiento del fondo de la grieta (CTOD)
passe	weld pass	pasada
passe de fond	root pass	pasada de raíz
passe large; cordon balancé	weave bead	pasada ancha; cordón con balanceo
passe tirée; passe étroite; cordon tiré	stringer bead	pasada recta; cordón estirado
pénétration	penetration	penetración
placage par soudage	cladding by welding; surfacing	plaqueado por soldeo
point de soudure	spot weld	punto de soldadura
position de soudage	welding position	posición de soldeo
préchauffage	preheating	precalentamiento
produit d'apport	filler material	material de aportación
produits consommables	welding consumables	consumibles
profondeur de pénétration	joint penetration	profundidad de penetración
programme de soudage	welding procedure specification	especificación del procedimiento de soldeo
projection (de soudage); perle	weld spatter	proyección; salpicadura
projection à chaud	thermal spraying	proyección en caliente
projections (particules)	spatter	proyecciones (partículas)
protection de l'oeil	eye protection	protección ocular
qualification d'un soudeur	welder performance qualification	cualificación del soldador
qualification des soudeurs et du mode opératoire	welding procedure qualification	cualificación de soldadores y del procedimiento operatorio
raccordement; raccordement du cordon de soudure; ligne de raccordement	toe of weld	acuerdo; acuerdo del cordón de soldadura
racine de la soudure	root of the weld	raíz de la soldadura
racine du joint	root of preparation; root of joint; joint root	raíz de la unión
rayon à fond de chanfrein	root radius	radio inferior del chaflán

FRENCH	ENGLISH	SPANISH
rechargement dur; rechargement dur par soudage	hardfacing by welding; hardfacing; hard-facing	recargue duro
rechargement; rechargement par soudage	surfacing weld	recargue por soldeo
rechargement; rechargement par soudage; rechargement curatif	building-up by welding; building-up	soldadura de recargue
recuit	annealing	recocido
recuit de globulisation; recuit de sphéroïdisation	spheroidising	recocido de globulización
relaxation des contraintes	stress relieving	relajación de tensiones
retassure à la racine	root concavity	rechupe en la raíz
retassure de cratère	crater pipe; crater hole	rechupe de cráter
retassure interdendritique	interdendritic shrinkage; solidification void	rechupe interdendrítico
revenu	tempering	revenido
rupture fragile	brittle fracture	rotura frágil
rupture par corrosion sous tension; corrosion sous tension	stress corrosion cracking	rotura por corrosión bajo tensiones
soudabilité	weldability	soldabilidad
soudage	welding	soldeo
soudage à l'arc	arc welding	soldeo por arco
soudage à l'arc avec électrode de carbone	carbon arc welding	soldeo por arco con electrodo de carbón
soudage à l'arc avec électrode enrobée	shielded metal-arc welding	soldeo por arco con electrodo revestido
soudage à l'arc avec électrode fusible; soudage à l'arc avec électrode consommable	metal arc welding	soldeo por arco con electrodo fusible
soudage à l'arc avec fil fourré; soudage avec fil fourré	flux cored wire welding	soldeo por arco con alambre tubular
soudage à l'arc sous flux en poudre; soudage à l'arc sous flux; soudage à l'arc submergé	submerged arc welding	soldeo por arco sumergido
soudage à la forge	forge welding	soldeo por forja

FRENCH	ENGLISH	SPANISH
soudage à la molette; soudage au galett	resistance seam welding	soldeo por roldana
soudage à pas de pèlerin	back-step welding	soldeo a paseo de peregrino
soudage au plafond	overhead position welding	soldeo sobre techo
soudage automatique	automatic welding	soldeo automático
soudage aux gaz	gas welding	soldeo con gas
soudage des goujons	stud welding	soldeo de espárragos
soudage en bout par résistance pure	upset welding	soldeo a tope por resistencia
soudage en passes larges; soudage en passes balancées	welding with weave beads; welding with weaving motion	soldeo en pasadas con balanceo; soldeo en pasadas con oscilación
soudage MAG avec fil fourré; soudage sous gaz actif avec fil fourré	flux cored wire metal arc welding with active gas shield	soldeo MAG con alambre tubular
soudage manuel.	manual welding; hand welding	soldeo manual
soudage mécanisé	mechanized welding	soldeo mecanizado
soudage MIG	gas metal arc welding; MIG welding	soldeo MIG; soldeo por arco bajo protección de gas inerte con alambre fusible
soudage oxyacétylénique	oxy-acetylene welding	soldeo oxiacetilénico
soudage par explosion	explosion welding	soldeo por explosión
soudage par faisceau d'électrons; soudage F.E.	electron beam welding; EB welding	soldeo por haz de electrones
soudage par faisceau laser; soudage laser; soudage au laser	laser beam welding	soldeo por rayo láser
soudage par friction	friction welding	soldeo por fricción
soudage par fusion	fusion welding	soldeo por fusión
soudage par fusion avec pression	pressure fusion welding	soldeo por fusión con presión
soudage par pression à l'arc; soudage à l'arc par pression	arc pressure welding	soldeo por arco con percusión
soudage par résistance	resistance welding	soldeo por resistencia
soudage par résistance par points	resistance spot welding	soldeo por resistencia por puntos

FRENCH	ENGLISH	SPANISH
soudage plasma; soudage au plasma; soudage au jet de plasma	plasma arc welding	soldeo por plasma
soudage robotisé	robotic welding	soldeo robotizado
soudage sans métal d'apport	welding without using filler material	soldeo sin metal de aportación
soudage semi-mécanisé	partly mechanized welding	soldeo parcialmente mecanizado
soudage sous protection gazeuse avec fil fusible; soudage sous protection gazeuse avec fil-électrode fusible; soudage à l'arc sous protection gazeuse avec fil-électrode fusible	gas metal arc welding	soldeo por arco con electrodo fusible bajo protección gaseosa
soudage sous protection gazeuse; soudage à l'arc sous protection gazeuse	gas shielded arc welding	soldeo por arco bajo protección gaseosa
soudage TIG	gas tungsten arc welding; GTA welding; TIG welding	soldeo TIG; soldeo por arco en atmósfera inerte con electrodo de volframio
soudage vertical	vertical position welding	soldeo vertical
soudage vertical descendant; soudage en descendant	vertical downward welding	soldeo vertical descendente
soudage vertical montant; soudage en montant	vertical upward welding	soldeo vertical ascendente
soudage vertical sous laitier; soudage sous laitier; soudage électroslag; soudage sous laitier électroconducteur	electro-slag welding	soldeo por electroescoria
soudeur	welder	soldador
soudobrasage	braze welding	cobresoldeo
soudure	weld	soldadura
soudure à pleine pénétration; soudure pénétrée	full penetration weld; complete penetraction weld	soldadura con pentración completa

FRENCH	ENGLISH	SPANISH
soudure avec bague support; soudure avec anneau support	weld with backing ring	soldadura con anillo soporte soldado
soudure bout à bout à plat; soudure en bout à plat; soudure bout à bout horizontale; soudure en bout horizontale	flat position butt weld	soldadura horizontal a tope
soudure bout à bout au plafond; soudure en bout au plafond	overhead butt weld	soldadura a tope sobre techo
soudure bout à bout avec surépaisseur	butt weld with excess weld metal	soldadura a tope con sobreespesor
soudure brûlée	burnt weld	soldadura quemada
soudure d'atelier; soudure en atelier	shop weld	soldadura de taller
soudure d'angle	fillet weld	soldadura de ángulo
soudure d'étanchéité	seal weld	soldadura de estanqueidad
soudure de pointage; point d'épinglage	tack weld	soldadura de punteo
soudure de réparation; reprise de la soudure (pour réparer un défaut) retouche de la soudure	rewelding; repairing the weld; weld repair	soldadura de reparación (por un defecto); retoque
soudure de résistance; soudure travaillante	load bearing weld; structural weld	soldadura estructural
soudure en bouchon	plug weld (circular)	soldadura en tapón
soudure en bout sur bords chanfreinés; soudure bout à bout sur bords chanfreinés	groove weld; butt weld in a butt joint	soldadura a tope
soudure en corniche	horizontal butt weld; horizontal-vertical butt weld	soldadura en cornisa
soudure en ligne continue par points	stitch weld	soldadura alineada continua por puntos
soudure en U	single U groove weld	soldadura en U
soudure en V	single V groove weld	soldadura en V
soudure en X; soudure en V double	double -V- groove weld	soldadura en X
soudure gougée à la racine	back-gouged weld	soldadura acanalada en la raíz

FRENCH	ENGLISH	SPANISH
soudure sur chantier	field weld; site weld	soldadura en obra; soldadura en campo
soudure surchauffée	overheated weld	soldadura sobrecalentada
soufflure vermiculaire	worm-hole	sopladura vermicular
soupape de sécurité	safety valve	válvula de seguridad
source de courant de soudage	source of welding current; welding power source	generador de corriente de soldeo
support à l'envers fusible; insert	consumable insert	placa soporte fusíble
support à l'envers non subsistant	temporary backing	placa soporte no permanente
support à l'envers subsistant	permanent backing	placa soporte permanente
surépaisseur (d'une soudure bout à bout)	excess weld metal in a butt weld; reinforcement	sobreespesor (en una unión a tope)
surépaisseur (d'une soudure d'angle); convexité	excess weld metal in a fillet weld; reinforcement	sobreespesor (en una unión de ángulo); convexidad
surépaisseur excessive	excessive reinforcement; excess weld metal	sobreespesor excesivo
surépaisseur excessive à la racine; excès de pénétration	excessive root reinforcement; excess penetration bead	exceso de penetración; sobreespesor excesivo en la raíz
surface de contact	faying surface	superficie de contacto
surface de la soudure	weld face; weld surface	superfície de la soldadura
table de soudage	welding bench	mesa de soldeo
tablier de soudeur	welder's apron	delantal de soldador
techniques opératoires	welding techniques	tecnicas operatorias
ténacité sous entaille	notch toughness	sensibilidad a la entalla
traitement de mise en solution.	solution annealing	recocido de solubilización
traitement de normalisation	normalizing	normalización
traitement de recristallisation; recuit de recristallisation	recrystallisation annealing	recocido de recristalización
traitement de restauration	recovery annealing	recocido de restauración
traitement de stabilisation	stabilization annealing; stabilizing	recocido de estabilización

FRENCH	ENGLISH	SPANISH
traitement thermique après soudage	post weld heat treatment; PWHT	tratamiento térmico postsoldeo
trempe à l'air	air hardening	temple al aire
trempe à la flamme; trempe au chalumeau	flame hardening	temple a la llama
trempe et revenu	harden and temper; quench and temper; QT	temple y revenido
trempe par induction	induction hardening	temple por inducción
trempe superficielle; durcissement superficiel	surface hardening; case hardening	temple superficial
trou	burn-through	hueco
vieillissement	ageing	envejecimiento
vitesse de propagation de la fissure	crack growth rate	velocidad de propagación de la grieta
vitesse de soudage	travel speed	velocidad de soldeo
zone de dilution	fusion zone	zona de difusión
zone de liaison	weld junction; weld line; fusion line	zona de unión
zone de soudure	weld zone	zona de soldadura
zone fondue; zone de fusion	weld metal area	zona de fusión
zone thermiquement affectée	heat affected zone; HAZ	zona térmicamente afectada; ZTA

HARDNESS CONVERSION NUMBERS FOR STEELS

APPROXIMATE HARDNESS CONVERSION NUMBERS FOR NONAUSTENTIC STEELS[a, b]

Rockwell C 150 kgf Diamond HRC	Vickers HV	Brinell 3000 kgf 10mm ball HB	Knoop 500 gf HK	Rockwell A 60 kgf Diamond HRA	Rockwell Superficial Hardness				Approximate Tensile Strength ksi (MPa)
					15 kgf Diamond HR15N	30 kgf Diamond HR30N	45 kgf Diamond HR45N		
68	940	---	920	85.6	93.2	84.4	75.4	---	
67	900	---	895	85.0	92.9	83.6	74.2	---	
66	865	---	870	84.5	92.5	82.8	73.3	---	
65	832	739d	846	83.9	92.2	81.9	72.0	---	
64	800	722d	822	83.4	91.8	81.1	71.0	---	
63	772	706d	799	82.8	91.4	80.1	69.9	---	
62	746	688d	776	82.3	91.1	79.3	68.8	---	
61	720	670d	754	81.8	90.7	78.4	67.7	---	
60	697	654d	732	81.2	90.2	77.5	66.6	---	
59	674	634d	710	80.7	89.8	76.6	65.5	351 (2420)	
58	653	615	690	80.1	89.3	75.7	64.3	338 (2330)	
57	633	595	670	79.6	88.9	74.8	63.2	325 (2240)	
56	613	577	650	79.0	88.3	73.9	62.0	313 (2160)	
55	595	560	630	78.5	87.9	73.0	60.9	301 (2070)	
54	577	543	612	78.0	87.4	72.0	59.8	292 (2010)	
53	560	525	594	77.4	86.9	71.2	58.6	283 (1950)	
52	544	512	576	76.8	86.4	70.2	57.4	273 (1880)	
51	528	496	558	76.3	85.9	69.4	56.1	264 (1820)	
50	513	482	542	75.9	85.5	68.5	55.0	255 (1760)	
49	498	468	526	75.2	85.0	67.6	53.8	246 (1700)	
48	484	455	510	74.7	84.5	66.7	52.5	238 (1640)	
47	471	442	495	74.1	83.9	65.8	51.4	229 (1580)	
46	458	432	480	73.6	83.5	64.8	50.3	221 (1520)	
45	446	421	466	73.1	83.0	64.0	49.0	215 (1480)	

APPROXIMATE HARDNESS CONVERSION NUMBERS FOR NONAUSTENITIC STEELS[a,b] (Continued)									
Rockwell C 150 kgf Diamond HRC	Vickers HV	Brinell 3000 kgf 10mm ball HB	Knoop 500 gf HK	Rockwell A 60 kgf Diamond HRA	Rockwell Superficial Hardness				Approximate Tensile Strength ksi (MPa)
					15 kgf Diamond HR15N	30 kgf Diamond HR30N	45 kgf Diamond HR45N	
44	434	409	452	72.5	82.5	63.1	47.8	208 (1430)
43	423	400	438	72.0	82.0	62.2	46.7	201 (1390)
42	412	390	426	71.5	81.5	61.3	45.5	194 (1340)
41	402	381	414	70.9	80.9	60.4	44.3	188 (1300)
40	392	371	402	70.4	80.4	59.5	43.1	182 (1250)
39	382	362	391	69.9	79.9	58.6	41.9	177 (1220)
38	372	353	380	69.4	79.4	57.7	40.8	171 (1180)
37	363	344	370	68.9	78.8	56.8	39.6	166 (1140)
36	354	336	360	68.4	78.3	55.9	38.4	161 (1110)
35	345	327	351	67.9	77.7	55.0	37.2	156 (1080)
34	336	319	342	67.4	77.2	54.2	36.1	152 (1050)
33	327	311	334	66.8	76.6	53.3	34.9	149 (1030)
32	318	301	326	66.3	76.1	52.1	33.7	146 (1010)
31	310	294	318	65.8	75.6	51.3	32.5	141 (970)
30	302	286	311	65.3	75.0	50.4	31.3	138 (950)
29	294	279	304	64.6	74.5	49.5	30.1	135 (930)
28	286	271	297	64.3	73.9	48.6	28.9	131 (900)
27	279	264	290	63.8	73.3	47.7	27.8	128 (880)
26	272	258	284	63.3	72.8	46.8	26.7	125 (860)
25	266	253	278	62.8	72.2	45.9	25.5	123 (850)
24	260	247	272	62.4	71.6	45.0	24.3	119 (820)
23	254	243	266	62.0	71.0	44.0	23.1	117 (810)
22	248	237	261	61.5	70.5	43.2	22.0	115 (790)
21	243	231	256	61.0	69.9	42.3	20.7	112 (770)
20	238	226	251	60.5	69.4	41.5	19.6	110 (760)

The Metals Blue Book

APPROXIMATE HARDNESS CONVERSION NUMBERS FOR NONAUSTENITIC STEELS[a, b] (Continued)

a. This table gives the approximate interrelationships of hardness values and approximate tensile strength of steels. It is possible that steels of various compositions and processing histories will deviate in hardness-tensile strength relationship from the data presented in this table. The data in this table should not be used for austenitic stainless steels, but have been shown to be applicable for ferritic and martensitic stainless steels. Where more precise conversions are required, they should be developed specially for each steel composition, heat treatment, and part.

b. All relative hardness values in this table are averages of tests on various metals whose different properties prevent establishment of exact mathematical conversions. These values are consistent with ASTM A 370-91 for nonaustenitic steels. It is recommended that ASTM standards A 370, E 140, E 10, E 18, E 92, E 110 and E 384, involving hardness tests on metals, be reviewed prior to interpreting hardness conversion values.

c. Carbide ball, 10mm.

d. This Brinell hardness value is outside the recommended range for hardness testing in accordance with ASTM E 10.

APPROXIMATE HARDNESS CONVERSION NUMBERS FOR NONAUSTENITIC STEELS [a, b]

Rockwell B 100 kgf 1/16" ball HRB	Vickers HV	Brinell 3000 kgf 10 mm HB	Knoop 500 gf HK	Rockwell A 60 kgf Diamond HRA	Rockwell Superficial Hardness			Approximate Tensile Strength ksi (MPa)
					15 kgf 1/16" ball HR15T	30 kgf 1/16" ball HR30T	45 kgf 1/16" ball HR45T	
100	240	240	251	61.5	93.1	83.1	72.9	116 (800)
99	234	234	246	60.9	92.8	82.5	71.9	114 (785)
98	228	228	241	60.2	92.5	81.8	70.9	109 (750)
97	222	222	236	59.5	92.1	81.1	69.9	104 (715)
96	216	216	231	58.9	91.8	80.4	68.9	102 (705)
95	210	210	226	58.3	91.5	79.8	67.9	100 (690)
94	205	205	221	57.6	91.2	79.1	66.9	98 (675)
93	200	200	216	57.0	90.8	78.4	65.9	94 (650)
92	195	195	211	56.4	90.5	77.8	64.8	92 (635)
91	190	190	206	55.8	90.2	77.1	63.8	90 (620)
90	185	185	201	55.2	89.9	76.4	62.8	89 (615)
89	180	180	196	54.6	89.5	75.8	61.8	88 (605)
88	176	176	192	54.0	89.2	75.1	60.8	86 (590)
87	172	172	188	53.4	88.9	74.4	59.8	84 (580)
86	169	169	184	52.8	88.6	73.8	58.8	83 (570)
85	165	165	180	52.3	88.2	73.1	57.8	82 (565)
84	162	162	176	51.7	87.9	72.4	56.8	81 (560)
83	159	159	173	51.1	87.6	71.8	55.8	80 (550)
82	156	156	170	50.6	87.3	71.1	54.8	77 (530)
81	153	153	167	50.0	86.9	70.4	53.8	73 (505)
80	150	150	164	49.5	86.6	69.7	52.8	72 (495)
79	147	147	161	48.9	86.3	69.1	51.8	70 (485)
78	144	144	158	48.4	86.0	68.4	50.8	69 (475)
77	141	141	155	47.9	85.6	67.7	49.8	68 (470)
76	139	139	152	47.3	85.3	67.1	48.8	67 (460)

APPROXIMATE HARDNESS CONVERSION NUMBERS FOR NONAUSTENITIC STEELS [a,b] (Continued)

Rockwell B 100 kgf 1/16" ball HRB	Vickers HV	Brinell 3000 kgf 10 mm HB	Knoop 500 gf HK	Rockwell A 60 kgf Diamond HRA	Rockwell Superficial Hardness			Approximate Tensile Strength ksi (MPa)
					15 kgf 1/16" ball HR15T	30 kgf 1/16" ball HR30T	45 kgf 1/16" ball HR45T	
75	137	137	150	46.8	85.0	66.4	47.8	66 (455)
74	135	135	147	46.3	84.7	65.7	46.8	65 (450)
73	132	132	145	45.8	84.3	65.1	45.8	64 (440)
72	130	130	143	45.3	84.0	64.4	44.8	63 (435)
71	127	127	141	44.8	83.7	63.7	43.8	62 (425)
70	125	125	139	44.3	83.4	63.1	42.8	61 (420)
69	123	123	137	43.8	83.0	62.4	41.8	60 (415)
68	121	121	135	43.3	82.7	61.7	40.8	59 (405)
67	119	119	133	42.8	82.4	61.0	39.8	58 (400)
66	117	117	131	42.3	82.1	60.4	38.7	57 (395)
65	116	116	129	41.8	81.8	59.7	37.7	56 (385)
64	114	114	127	41.4	81.4	59.0	36.7	---
63	112	112	125	40.9	81.1	58.4	35.7	---
62	110	110	124	40.4	80.8	57.7	34.7	---
61	108	108	122	40.0	80.5	57.0	33.7	---
60	107	107	120	39.5	80.1	56.4	32.7	---
59	106	106	118	39.0	79.8	55.7	31.7	---
58	104	104	117	38.6	79.5	55.0	30.7	---
57	103	103	115	38.1	79.2	54.4	29.7	---
56	101	101	114	37.7	78.8	53.7	28.7	---
55	100	100	112	37.2	78.5	53.0	27.7	---
54	---	---	111	36.8	78.2	52.4	26.7	---
53	---	---	110	36.3	77.9	51.7	25.7	---
52	---	---	109	35.9	77.5	51.0	24.7	---
51	---	---	108	35.5	77.2	50.3	23.7	---

The Metals Blue Book

APPROXIMATE HARDNESS CONVERSION NUMBERS FOR NONAUSTENITIC STEELS [a,b] (Continued)

Rockwell B 100 kgf 1/16" ball HRB	Vickers HV	Brinell 3000 kgf 10 mm HB	Knoop 500 gf HK	Rockwell A 60 kgf Diamond HRA	Rockwell Superficial Hardness			Approximate Tensile Strength ksi (MPa)
					15 kgf 1/16" ball HR15T	30 kgf 1/16" ball HR30T	45 kgf 1/16" ball HR45T	
50	---	---	107	35.0	76.9	49.7	22.7	---
49	---	---	106	34.6	76.6	49.0	21.7	---
48	---	---	105	34.1	76.2	48.3	20.7	---
47	---	---	104	33.7	75.9	47.7	19.7	---
46	---	---	103	33.3	75.6	47.0	18.7	---
45	---	---	102	32.9	75.3	46.3	17.7	---
44	---	---	101	32.4	74.9	45.7	16.7	---
43	---	---	100	32.0	74.6	45.0	15.7	---
42	---	---	99	31.6	74.3	44.3	14.7	---
41	---	---	98	31.2	74.0	43.7	13.6	---
40	---	---	97	30.7	73.6	43.0	12.6	---
39	---	---	96	30.3	73.3	42.3	11.6	---
38	---	---	95	29.9	73.0	41.6	10.6	---
37	---	94	29.5	78.0	41.0	9.6		---
36	---	---	93	29.1	72.3	40.3	8.6	---
35	---	---	92	28.7	72.0	39.6	7.6	---
34	---	---	91	28.2	71.7	39.0	6.6	---
33	---	---	90	27.8	71.4	38.3	5.6	---
32	---	---	89	27.4	71.0	37.6	4.6	---
31	---	---	88	27.0	70.7	37.0	3.6	---
30	---	---	87	26.6	70.4	36.3	2.6	---

a. This table gives the approximate interrelationships of hardness values and approximate tensile strength of steels. It is possible that steels of various compositions and processing histories will deviate in hardness-tensile strength relationship from the data presented in this table. The data in this table should not be used for austenitic stainless steels, but have been shown to be applicable for ferritic and martensitic stainless steels. Where more precise conversions are required, they should be developed specially for each steel composition, heat treatment, and part.

The Metals Blue Book

APPROXIMATE HARDNESS CONVERSION NUMBERS FOR NONAUSTENITIC STEELS [a,b] (Continued)

b.   All relative hardness values in this table are averages of tests on various metals whose different properties prevent establishment of exact mathematical conversions. These values are consistent with ASTM A 370-91 for nonaustenitic steels. It is recommended that ASTM standards A 370, E 140, E 10, E 18, E 92, E 110 and E 384, involving hardness tests on metals, be reviewed prior to interpreting hardness conversion values.

APPROXIMATE HARDNESS NUMBERS FOR AUSTENITIC STEELS [a]

Rockwell C 150 kgf, Diamond HRC	Rockwell A 60 kgf, Diamond HRA	Rockwell Superficial Hardness		
		15 kgf, Diamond HR15N	30 kgf, Diamond HR30N	45 kgf, Diamond HR45N
48	74.4	84.1	66.2	52.1
47	73.9	83.6	65.3	50.9
46	73.4	83.1	64.5	49.8
45	72.9	82.6	63.6	48.7
44	72.4	82.1	62.7	47.5
43	71.9	81.6	61.8	46.4
42	71.4	81.0	61.0	45.2
41	70.9	80.5	60.1	44.1
40	70.4	80.0	59.2	43.0
39	69.9	79.5	58.4	41.8
38	69.3	79.0	57.5	40.7
37	68.8	78.5	56.6	39.6
36	68.3	78.0	55.7	38.4
35	67.8	77.5	54.9	37.3
34	67.3	77.0	54.0	36.1
33	66.8	76.5	53.1	35.0
32	66.3	75.9	52.3	33.9
31	65.8	75.4	51.4	32.7
30	65.3	74.9	50.5	31.6
29	64.8	74.4	49.6	30.4

APPROXIMATE HARDNESS NUMBERS FOR AUSTENITIC STEELS[a] (Continued)

Rockwell C 150 kgf, Diamond HRC	Rockwell A 60 kgf, Diamond HRA	Rockwell Superficial Hardness		
		15 kgf, Diamond HR15N	30 kgf, Diamond HR30N	45 kgf, Diamond HR45N
28	64.3	73.9	48.8	29.3
27	63.8	73.4	47.9	28.2
26	63.3	72.9	47.0	27.0
25	62.8	72.4	46.2	25.9
24	62.3	71.9	45.3	24.8
23	61.8	71.3	44.4	23.6
22	61.3	70.8	43.5	22.5
21	60.8	70.3	42.7	21.3
20	60.3	69.8	41.8	20.2

a. All relative hardness values in this table are averages of tests on various metals whose different properties prevent establishment of exact mathematical conversions. These values are consistent with ASTM A 370-91 for austenitic steels. It is recommended that ASTM standards A 370, E 140, E 10, E 18, E 92, E 110 and E 384, involving hardness tests on metals, be reviewed prior to interpreting hardness conversion values.

Appendix

2

UNIT CONVERSIONS

METRIC CONVERSION FACTORS

To Convert From	To	Multiply By
Angle		
degree	rad	1.745 329 E -02
Area		
in.²	mm²	6.451 600 E + 02
in.²	cm²	6.451 600 E + 00
in.²	m²	6.451 600 E - 04
ft²	m²	9.290 304 E - 02
Bending moment or torque		
lbf - in.	N - m	1.129 848 E - 01
lbf - ft	N - m	1.355 818 E + 00
kgf - m	N - m	9.806 650 E + 00
ozf - in.	N-m	7.061 552 E - 03
Bending moment or torque per unit length		
lbf - in./in.	N - m/m	4.448 222 E + 00
lbf - ft/in.	N - m/m	5.337 866 E + 01
Corrosion rate		
mils/yr	mm/yr	2.540 000 E - 02
mils/yr	µ/yr	2.540 000 E + 01
Current density		
A/in.²	A/cm²	1.550 003 E - 01
A/in.²	A/mm²	1.550 003 E - 03
A/ft²	A/m²	1.076 400 E + 01
Electricity and magnetism		
gauss	T	1.000 000 E - 04
maxwell	µWb	1.000 000 E - 02
mho	S	1.000 000 E + 00

To Convert From	To	Multiply By
Mass per unit time		
lb/h	kg/s	1.259 979 E - 04
lb/min	kg/s	7.559 873 E - 03
lb/s	kg/s	4.535 924 E - 01
Mass per unit volume (includes density)		
g/cm³	kg/m³	1.000 000 E + 03
lb/ft³	g/cm³	1.601 846 E - 02
lb/ft³	kg/m³	1.601 846 E + 01
lb/in.³	g/cm³	2.767 990 E + 01
lb/in.³	kg/m³	2.767 990 E + 04
Power		
Btu/s	kW	1.055 056 E + 00
Btu/min	kW	1.758 426 E - 02
Btu/h	W	2.928 751 E - 01
erg/s	W	1.000 000 E - 07
ft - lbf/s	W	1.355 818 E + 00
ft - lbf/min	W	2.259 697 E - 02
ft - lbf/h	W	3.766 161 E - 04
hp (550 ft - lbf/s)	kW	7.456 999 E - 01
hp (electric)	kW	7.460 000 E - 01
Power density		
W/in.²	W/m²	1.550 003 E + 03
Pressure (fluid)		
atm (standard)	Pa	1.013 250 E + 05
bar	Pa	1.000 000 E + 05
in. Hg (32 F)	Pa	3.386 380 E + 03

METRIC CONVERSION FACTORS (Continued)

To Convert From	To	Multiply By
Electricity and magnetism (Continued)		
Oersted	A/m	7.957 700 E + 01
Ω - cm	Ω - m	1.000 000 E - 02
Ω circular - mil/ft	μΩ - m	1.662 426 E - 03
Energy (impact other)		
ft•lbf	J	1.355 818 E + 00
Btu (thermochemical)	J	1.054 350 E + 03
cal (thermochemical)	J	4.184 000 E + 00
kW - h	J	3.600 000 E + 06
W - h	J	3.600 000 E + 03
Flow rate		
ft³/h	L/min	4.719 475 E - 01
ft³/min	L/min	2.831 000 E + 01
gal/h	L/min	6.309 020 E - 02
gal/min	L/min	3.785 412 E + 00
Force		
lbf	N	4.448 222 E + 00
kip (1000 lbf)	N	4.448 222 E + 03
tonf	kN	8.896 443 E + 00
kgf	N	9.806 650 E + 00
Force per unit length		
lbf/ft	N/m	1.459 390 E + 01
lbf/in.	N/m	1.751 268 E + 02
Fracture toughness		
ksi √in.	MPa √m	1.098 800 E + 00
Heat content		
Btu/lb	kJ/kg	2.326 000 E + 00

To Convert From	To	Multiply By
Pressure (fluid) (Continued)		
in. Hg (60 F)	Pa	3.376 850 E + 03
lbf/in.² (psi)	Pa	6.894 757 E + 03
torr (mm Hg, 0 C)	Pa	1.333 220 E + 02
Specific heat		
Btu/lb - F	J/kg - K	4.186 800 E + 03
cal/g - C	J/kg - K	4.186 800 E + 03
Stress (force per unit area)		
tonf/in.² (tsi)	MPa	1.378 951 E + 01
kgf/mm²	MPa	9.806 650 E + 00
ksi	MPa	6.894 757 E + 00
lbf/in.² (psi)	MPa	6.894 757 E - 03
N/mm²	MPa	1.000 000 E + 00
Temperature		
F	C	5/9 (F - 32)
R	K	5/9
Temperature interval		
F	C	5/9
Thermal conductivity		
Btu - in./s - ft² - F	W/m - K	5.192 204 E + 02
Btu/ft - h - F	W/m - K	1.730 735 E + 00
Btu - in./h . ft² - F	W/m - K	1.442 279 E - 01
cal/cm - s - C	W/m - K	4.184 000 E + 02
Thermal expansion		
in./in. - C	m/m - K	1.000 000 E + 00
in./in. - F	m/m - K	1.800 000 E + 00

METRIC CONVERSION FACTORS (Continued)

To Convert From	To	Multiply By	To Convert From	To	Multiply By
Heat content (Continued)			**Velocity**		
cal/g	kJ/kg	4.186 800 E + 00	ft/h	m/s	8.466 667 E - 05
Heat input			ft/min	m/s	5.080 000 E - 03
J/in.	J/m	3.937 008 E + 01	ft/s	m/s	3.048 000 E - 01
kJ/in.	kJ/mm	3.937 008 E - 02	in./s	m/s	2.540 000 E - 02
Length			km/h	m/s	2.777 778 E - 01
A	nm	1.000 000 E - 01	mph	km/h	1.609 344 E + 00
μin.	μm	2.540 000 E - 02	**Velocity of rotation**		
mil	μm	2.540 000 E + 01	rev/min (rpm)	rad/s	1.047 164 E - 01
in.	mm	2.540 000 E + 01	rev/s	rad/s	6.283 185 E + 00
in.	cm	2.540 000 E + 00	**Viscosity**		
ft	m	3.048 000 E - 01	poise	Pa - s	1.000 000 E - 01
yd	m	9.144 000 E - 01	stokes	m^2/s	1.000 000 E - 04
mile	km	1.609 300 E + 00	ft^2/s	m^2/s	9.290 304 E - 02
Mass			$in.^2/s$	mm^2/s	6.451 600 E + 02
oz	kg	2.834 952 E - 02	**Volume**		
lb	kg	4.535 924 E - 01	$in.^3$	m^3	1.638 706 E - 05
ton (short 2000 lb)	kg	9.071 847 E + 02	ft^3	m^3	2.831 685 E - 02
ton (short 2000 lb)	kg x 10^3	9.071 847 E - 01	fluid oz	m^3	2.957 353 E - 05
ton (long 2240 lb)	kg	1.016 047 E + 03	gal (U.S. liquid)	m^3	3.785 412 E - 03
kg x 10^3 = 1 metric ton			**Volume per unit time**		
Mass per unit area			ft^3/min	m^3/s	4.719 474 E - 04
$oz/in.^2$	kg/m^2	4.395 000 E + 01	ft^3/s	m^3/s	2.831 685 E - 02
oz/ft^2	kg/m^2	3.051 517 E - 01	$in.^3/min$	m^3/s	2.731 177 E - 07
oz/yd^2	kg/m^2	3.390 575 E - 02	**Wavelength**		
lb/ft^2	kg/m^2	4.882 428 E + 00	A	nm	1.000 000 E - 01

SI PREFIXES

Prefix	Symbol	Exponential Expression	Multiplication Factor
exa	E	10^{18}	1 000 000 000 000 000 000
peta	P	10^{15}	1 000 000 000 000 000
tera	T	10^{12}	1 000 000 000 000
giga	G	10^{9}	1 000 000 000
mega	M	10^{6}	1 000 000
kilo	k	10^{3}	1 000
hecto	h	10^{2}	100
deka	da	10^{1}	10
Base Unit	---	10^{0}	1
deci	d	10^{-1}	0.1
centi	c	10^{-2}	0.01
milli	m	10^{-3}	0.001
micro	μ	10^{-6}	0.000 001
nano	n	10^{-9}	0.000 000 001
pico	p	10^{-12}	0.000 000 000 001
femto	f	10^{-15}	0.000 000 000 000 001
atto	a	10^{-18}	0.000 000 000 000 000 001

THE GREEK ALPHABET

A, α - Alpha	I, ι - Iota	P, ρ - Rho
B, β - Beta	K, κ - Kappa	Σ, σ - Sigma
Γ, γ - Gamma	Λ, λ - Lambda	T, τ - Tau
Δ, δ - Delta	M, μ - Mu	Y, υ - Upsilon
E, ε - Epsilon	N, ν - Nu	Φ, ϕ - Phi
Z, ζ - Zeta	Ξ, ξ - Xi	X, χ - Chi
H, η - Eta	O, o - Omicron	Ψ, ψ - Psi
Θ, θ - Theta	Π, π - Pi	Ω, ω - Omega

Appendix

3

PIPE DIMENSIONS

DIMENSIONS OF WELDED AND SEAMLESS PIPE[a]

Nominal Pipe Size	Outside Diameter	Nominal Wall Thickness (in)							
		Schedule 5S	Schedule 10S	Schedule 10	Schedule 20	Schedule 30	Schedule Standard	Schedule 40
1/8	0.405	---	0.049	---	---	---	0.068	0.068
1/4	0.540	---	0.065	---	---	---	0.088	0.088
3/8	0.675	---	0.065	---	---	---	0.091	0.091
1/2	0.840	0.065	0.083	---	---	---	0.109	0.109
3/4	1.050	0.065	0.083	---	---	---	0.113	0.113
1	1.315	0.065	0.109	---	---	---	0.133	0.133
1 1/4	1.660	0.065	0.109	---	---	---	0.140	0.140
1 1/2	1.900	0.065	0.109	---	---	---	0.145	0.145
2	2.375	0.065	0.109	---	---	---	0.154	0.154
2 1/2	2.875	0.083	0.120	---	---	---	0.203	0.203
3	3.5	0.083	0.120	---	---	---	0.216	0.216
3 1/2	4.0	0.083	0.120	---	---	---	0.226	0.226
4	4.5	0.083	0.120	---	---	---	0.237	0.237
5	5.563	0.109	0.134	---	---	---	0.258	0.258
6	6.625	0.109	0.134	---	---	---	0.280	0.280
8	8.625	0.109	0.148	---	0.250	0.277	0.322	0.322
10	10.75	0.134	0.165	---	0.250	0.307	0.365	0.365
12	12.75	0.156	0.180	---	0.250	0.330	0.375	0.406
14 O.D.	14.0	0.156	0.188	0.250	0.312	0.375	0.375	0.438
16 O.D.	16.0	0.165	0.188	0.250	0.312	0.375	0.375	0.500
18 O.D.	18.0	0.165	0.188	0.250	0.312	0.438	0.375	0.562
20 O.D.	20.0	0.188	0.218	0.250	0.375	0.500	0.375	0.594
22 O.D.	22.0	0.188	0.218	0.250	0.375	0.500	0.375	---
24 O.D.	24.0	0.218	0.250	0.250	0.375	0.562	0.375	0.688
26 O.D.	26.0	---	---	0.312	0.500	---	0.375	---
28 O.D.	28.0	---	---	0.312	0.500	0.625	0.375	---

DIMENSIONS OF WELDED AND SEAMLESS PIPE (Continued)								
				Nominal Wall Thickness (in)				
Nominal Pipe Size	Outside Diameter	Schedule 5S	Schedule 10S	Schedule 10	Schedule 20	Schedule 30	Schedule Standard	Schedule 40
30 O.D.	30.0	0.250	0.312	0.312	0.500	0.625	0.375	---
32 O.D.	32.0	---	---	0.312	0.500	0.625	0.375	0.688
34 O.D.	34.0	---	---	0.312	0.500	0.625	0.375	0.688
36 O.D.	36.0	---	---	0.312	0.500	0.625	0.375	0.750
42 O.D.	42.0	---	---	---	---	---	0.375	---

a.   See next page for heavier wall thicknesses; all units are inches.

DIMENSIONS OF WELDED AND SEAMLESS PIPE[a]

Nominal Pipe Size	Outside Diameter	Schedule 60	Extra Strong	Schedule 80	Schedule 100	Schedule 120	Schedule 140	Schedule 160	XX Strong
1/8	0.405	---	0.095	0.095	---	---	---	---	---
1/4	0.540	---	0.119	0.119	---	---	---	---	---
3/8	0.675	---	0.126	0.126	---	---	---	---	---
1/2	0.840	---	0.147	0.147	---	---	---	0.188	0.294
3/4	1.050	---	0.154	0.154	---	---	---	0.219	0.308
1	1.315	---	0.179	0.179	---	---	---	0.250	0.358
1 1/4	1.660	---	0.191	0.191	---	---	---	0.250	0.382
1 1/2	1.900	---	0.200	0.200	---	---	---	0.281	0.400
2	2.375	---	0.218	0.218	---	---	---	0.344	0.436
2 1/2	2.875	---	0.276	0.276	---	---	---	0.375	0.552
3	3.5	---	0.300	0.300	---	---	---	0.438	0.600
3 1/2	4.0	---	0.318	0.318	---	---	---	---	---
4	4.5	---	0.337	0.337	---	0.438	---	0.531	0.674
5	5.563	---	0.375	0.375	---	0.500	---	0.625	0.750
6	6.625	---	0.432	0.432	---	0.562	---	0.719	0.864
8	8.625	0.406	0.500	0.500	0.594	0.719	0.812	0.906	0.875
10	10.75	0.500	0.500	0.594	0.719	0.844	1.000	1.125	1.000
12	12.75	0.562	0.500	0.688	0.844	1.000	1.125	1.312	1.000
14 O.D.	14.0	0.594	0.500	0.750	0.938	1.094	1.250	1.406	---
16 O.D.	16.0	0.656	0.500	0.844	1.031	1.219	1.438	1.594	---
18 O.D.	18.0	0.750	0.500	0.938	1.156	1.375	1.562	1.781	---
20 O.D.	20.0	0.812	0.500	1.031	1.281	1.500	1.750	1.969	---
22 O.D.	22.0	0.875	0.500	1.125	1.375	1.625	1.875	2.125	---
24 O.D.	24.0	0.969	0.500	1.218	1.531	1.812	2.062	2.344	---
26 O.D.	26.0	---	0.500	---	---	---	---	---	---
28 O.D.	28.0	---	0.500	---	---	---	---	---	---

Nominal Wall Thickness (in)

DIMENSIONS OF WELDED AND SEAMLESS PIPE (Continued)

Nominal Pipe Size	Outside Diameter	Schedule 60	Extra Strong	Nominal Wall Thickness (in)						
				Schedule 80	Schedule 100	Schedule 120	Schedule 140	Schedule 160	XX Strong	
30 O.D.	30.0	---	0.500	---	---	---	---	---	---	
32 O.D.	32.0	---	0.500	---	---	---	---	---	---	
34 O.D.	34.0	---	0.500	---	---	---	---	---	---	
36 O.D.	36.0	---	0.500	---	---	---	---	---	---	
42 O.D.	42.0	---	0.500	---	---	---	---	---	---	

a.   All units are inches.

Appendix

4

TECHNICAL SOCIETIES & ASSOCIATIONS LIST

AA	The Aluminum Association (202) 862-5100
AEE	The Association of Energy Engineers (404) 447-5083
AFS	American Foundrymen's Society (312) 824-0181
AISI	Association of Iron and Steel Engineers (412) 281-6323
AIChE	American Institute of Chemical Engineers (212) 705-7338
AMEC	Advanced Materials Engineering Centre (902) 425-4500
ANSI	American National Standards Institute (212) 354-3300
ASM	ASM International - The Materials Information Society (800) 336-5152 or (216) 338-5151
ASME	American Society of Mechanical Engineers (212) 705-7722
ASNT	American Society for Nondestructive Testing (614) 274-6003
ASQC	American Society for Quality Control (414) 272-8575
ASTM	American Society for Testing and Materials (215) 299-5400
AWS	American Welding Society (305) 443-9353 or (800) 443-9353
CAIMF	Canadian Advanced Industrial Materials Forum (416) 798-8055
CASI	Canadian Aeronautic & Space Institute (613) 234-0191
CCA	Canadian Construction Association (613) 236-9455
CCPE	Canadian Council of Professional Engineers (613) 232-2474
CCS	Canadian Ceramics Society (416) 491-2886
CDA	Copper Development Association (212) 251-7200
CIE	Canadian Institute of Energy (403) 262-6969
CIM	Canadian Institute for Mining and Metallurgy (514) 939-2710
CMA	Canadian Manufacturing Association (416) 363-7261
CNS	Canadian Nuclear Society (416) 977-6152
CPI	Canadian Plastics Institute (416) 441-3222
CPIC	Canadian Professional Information Centre (905) 624-1058
CSA	Canadian Standards Association (416) 747-4082
CSCE	Canadian Society for Chemical Engineers (613) 526-4652
CSEE	Canadian Society of Electronic Engineers (514) 651-6710
CSME	Canadian Society of Mechanical Engineers (514) 842-8121
CSNDT	Canadian Society for Nondestructive Testing (416) 676-0785
EI	Engineering Information Inc. (212) 705-7600

TECHNICAL SOCIETIES & ASSOCIATIONS LIST (Continued)

FED	Federal & Military Standards (215) 697-2000
IEEE	Institute of Electrical & Electronic Engineers (212) 705-7900
IES	Institute of Environmental Sciences (312) 255-1561
IIE	Institute of Industrial Engineers (404) 449-0460
IMMS	International Material Management Society (705) 525-4667
ISA	Instrument Society of America (919) 549-8411
ISA	Instrument Society of America (919) 549-8411
ISO	International Organization for Standardization 41 22 749 01 11 Geneva, Switzerland
ISS	Iron and Steel Society (412) 776-1535
ITI	International Technology Institute (412) 795-5300
ITRI	International Tin Research Institute (614) 424-6200
MSS	Manufactures Standardization Society of Valves & Fittings Industry (703) 281-6613
MTS	Marine Technology Society (202) 775-5966
NACE	National Association of Corrosion Engineers (713) 492-0535
NAPE	National Association of Power Engineers (212) 298-0600
NAPEGG	Association of Professional Engineers, Geologists and Geophysicists of theNorthwest Territories (403) 920-4055
NiDI	Nickel Development Institute (416) 591-7999
PIA	Plastics Institute of America (201) 420-5553
RIA	Robotic Industries Association (313) 994-6088
SAE	Society of Automotive Engineers (412) 776-4841
SAME	Society of American Military Engineers (703) 549-3800
SAMPE	Society for the Advancement of Materials and Processing Engineering (818) 331-0616
SCC	Standards Council of Canada (800) 267-8220
SCTE	Society of Carbide & Tool Engineers (216) 338 5151
SDCE	Society of Die Casting Engineers (312) 452-0700
SME	Society of Manufacturing Engineers (313) 271-1500
SPE	Society of Petroleum Engineers (214) 669-3377
SSPC	Steel Structures Painting Council (412) 268-3327
STC	Society for Technical Communications (202) 737-0035
STLE	Society of Tribologists and Lubrication Engineers (312) 825-5536
TDA	Titanium Development Association (303) 443-7515
TMS	The Minerals, Metals, and Materials Society (412) 776-9000
WIC	Welding Institute of Canada (905) 257-9881

INDEX